The Stochastic Nature of Environmental Phenomena and Processes

Georgy Golitsyn · Costas Varotsos

The Stochastic Nature of Environmental Phenomena and Processes

Springer

Georgy Golitsyn
A. M. Obukhov Institute of Atmospheric Physics
Moscow, Russia

Costas Varotsos
Environmental Physics and Meteorology
National and Kapodistrian University of Athens
Athens, Greece

ISBN 978-3-031-77014-2 ISBN 978-3-031-77015-9 (eBook)
https://doi.org/10.1007/978-3-031-77015-9

This Springer imprint is published by the registered company Springer Nature Switzerland AG
The registered company address is: Gewerbestrasse 11, 6330 Cham, Switzerland

Preface

During my long scientific career, which began in 1955, I worked on a number of processes in the atmosphere, in the ocean, and in astrophysics, and in geophysics on some of them, I would like to think I left a mark. About 350 articles have been written, of which over 60 were written by me only and six monographs. I taught a lot at the Department of Atmospheric Physics of the Faculty of Physics of Moscow M. B. Lomonosov State University.in 1975—2010 and from 1973 to 2019 at the Department of Ocean Thermohydrodynamics of the Moscow Institute of Physics and Technology. Teaching served as a stimulus for writing monographs.

My main achievements until 2011 were published in the book "Statistics and Dynamics of Natural Processes and Phenomena", M.: Krasand, 2012, 398 pages. During my scientific activity, I devoted a lot of time to scientific, public, and scientific-organizational topics not only in the country but also abroad, participating in various councils, committees, commissions, editorial boards, etc. From 1988 to 2002, I was elected three times as a member of the Presidium of the Academy. From 01.01.1990 to 31.12.2008, I was the director of the Institute of Atmospheric Physics of the Russian Academy of Sciences, i.e., day after day for 19 years. In the years of the Presidium, my employment was there, and in the turbulent 1990s, I led the Institute voluntarily, drawing no money, albeit meager, from the IFA salary fund. During this period, our IFA received various foreign grants. But in these years, I did significant work, for example, explaining the energy spectrum of cosmic rays, Gutenberg—Richter law of frequency—magnitude of earthquakes, I found patterns of hurricane development—necessary, but not sufficient, conditions for their development and others. The basis for the construction of these theories was the theory of similarity and dimension, as in previous work.

However, I remembered the late 1950s—early 1960s, when many scientists said that "the theory of similarity is a similarity of a theory" and even in the late 1990s, after my lecture at the Moscow Lebedev Institute of Physics on the spectrum of cosmic rays, they demanded the physical model, i.e. kinetic equations. Only in 2017, I did suddenly realize that such a model could be the probabilistic laws of Andrey Nikolaevich Kolmogorov and his school, developed by his students A. M. Obukhov, then A. M. Yaglom, A. S. Monin, then G. I. Barenblatt, who developed the general principles of their teacher into practical methods and applied these methods to specific phenomena and processes, primarily to turbulence (see Chap. 2). Now the theory of similarity and dimension is an exact theory (see Barenblatt, 2003: Scaling (Vol. 34). Cambridge University Press.)

Therefore, the time has come to present a description of many processes precisely from the point of view of Andrei Nikolayevich's main two-page work in 1934, which contains the principles of a description of not only turbulence, as I understood by mid-2017, but also sea wind waves, tropical and polar hurricanes, mini-hurricanes on the sea surface in the form of spiral eddies discovered by satellites only at the end of the twentieth century, the statistical structure of inhomogeneities of the gravitational field and the surface topography of celestial bodies, so called the Kaula's rule, etc. The main understanding to apply is that the mean square of velocity is the process energy per unit mass.

These are all the latest discoveries, deeply understood only in the context of the ideas of A. N. Kolmogorov in 1934, brought by his students listed above to practical methods for describing specific phenomena and processes. I would hope, as the scientific grandson of a great mathematician and physicist and as a student and colleague of his students mentioned above, that the examples presented here describing specific processes and phenomena, many of which remained mysteries for decades, will serve as examples for understanding future discoveries. Probabilistic laws and the theory of similarity and dimensions are the ways to understand the world around us. I have two important epigraphs from the classics of science. Initially, I wanted to have a third epigraph. Here it is:

> Upon this gifted age, in its dark hour,
> Rains from the sky a meteoric shower
> Of facts … they lie unquestioned, uncombined
> Wisdom enough to leech us of our ill
> Is daily spun, but there exists no loom
> To weave it into fabric.
> Edna St. Vincent Millay (1892–1950).

Edna is a wonderful American poet, known not only for her magnificent poems but also for her stormy romances. These lines clearly reflect the dramatic situation in the science of the surrounding world of the twentieth century. With too many poets, inexplicable by them the existing science. The wisdom to cure us is the daily bustle.

This book is something like an attempt to develop a unified view of the macroworld. Each of the paragraphs was written in such a way that after Chap. 1 it could be understood independently.

I would like to thank Vera Grigorievna Kochina for her invaluable assistance in the work, for the repeated typing and re-typing of the chapters of the book, which were discussed many times with Evgeniy Borisovich Gledzer and Otto Guramovich Chkhetiani, long-term collaborators.

The following is a list of the most frequently occurring links in the text, which are marked with abbreviations:

AHK34. Kolmogorov A. N. Zufällige Bewegungen // Ann. Math. 1934. V. 35. P. 116–117.
G18. Golitsyn G. S. The laws of random motions by A.N. Kolmogorov. Meteorology and Hydrology 2018. №3, 5–15.
MY71, 75. Monin A. S. Yaglom A. M. Statistical Hydromechanics V.1, V2 MIT Press. 1971. 1975.
GLG10. Gledzer E. B. and Golitsyn G. S. Scaling and finite ensembles of particles in motion with energy influx. Dokl. (2010). 433, (3), 466-470.
BPW. Bridgman P. W. (1932) Dimensional Analysis—Yale Univ. Press. 2nd Ed.
B09. Barenblatt G. I. Scaling—CUP, 2003—171 p.).
G12. Golitsyn G. S. Statistics and Dynamics of processes and phenomena in Nature. Moscow: Krasand Poll, 2012 (in Russian). 198 p.

Prof. Georgy Golitsyn
Academician of the Russian Academy of Sciences
Moscow, Russia

The 2021 Nobel Prize in Physics has been awarded to Klaus Hasselmann, Syukuro Manabe, and Giorgio Parisi for their Study of Climate Change, and Complex Systems. In particular, the Nobel Committee for the first two said: "for the physical modelling of Earth's climate, quantifying variability and reliably predicting global warming" and for the third one "for the discovery of the interplay of disorder and fluctuations in physical systems from atomic to planetary scales".

This award was in recognition of the contribution of Complex Systems science to the Climate Change phenomenon.

This book illustrates not only the current knowledge about the stochasticity of nature but also underlines the most basic unanswered relevant questions of environmental dynamics. The second part of the book is devoted to the climate system considered as a complex and complicated system. It mainly focuses on the major contributions of my research group to the topics mentioned in the first part of the book which is excellently written by Prof. Georgy Golitsyn. I must confess that writing this book with him was an unforgettable venture in which I gained an excellent academic experience.

Prof. Costas Varotsos
Athens, Greece

About this book

The first part of the book introduces the equation of the random motions by A. N. Kolmogorov proposed in 1934. In 1959 A. M. Obukhov found the second moments of probability distribution for it as $\langle u_i^2(t)\rangle = \varepsilon t$, $\langle u_i x_i(t)\rangle = \varepsilon t^2$, $\langle x_i^2(t)\rangle = \varepsilon t^3$, where ε is the diffusion coefficient in the velocity space, the rate of the energy (or any intensity) generation per unit mass. From these moments A. M. Obukhov has obtained the laws of inertial turbulence of 1941 including the Richardson—Obukhov one. From the energy (intensity) moment is producing histograms and differential probability distributions for the events flux such as earthquakes and this makes the Gutenberg—Richter empirical law as the theory of probability law, TPL. A number of other empirical laws become TPL. Pareto's rule in social sciences and the Zipf law can be understood also in this way. The third moment describes the random areas, such as histograms for lithospheric plates size distributions, mass of spiral galaxies, some features of clouds. Cosmic rays spectra, Kaula's rule for relief spectra of celestial bodies etc. also became TPLs. The empirics and those moments may be used for ε.

The book's second part highlights significant findings in the study of nonlinear dynamics within climate system components. It begins by introducing a tool that separates dynamical and chemical variability in ozone, challenging the previous belief that chemical processes were the primary influence on the ozone layer. Then, a key discovery of the second author of this book from 2002 revealed that major stratospheric sudden warmings could occur in the Southern Hemisphere, leading to the splitting of the Antarctic ozone hole. This emphasized the importance of accounting for long-range correlations in predictive models of global geophysical variables. This understanding is crucial for grasping the complexities of geophysical dynamics over extended time scales.

In the following, it examines the linearity assumption in reducing solar and volcanic forcings to radiative equivalents. It finds that variability in models is weak at centennial scales, and that solar and volcanic forcings combine in a nonlinear manner over longer periods, affecting model sensitivity.

Finally, the introduction of the concept of "natural time" is presented as a valuable tool for predicting extreme geophysical events. This approach has led to the development of a nowcasting tool that successfully analyzes the evolution of complex systems, with examples including El Niño, solar radiation, air pollution, cyclones, and heatwaves, all of which have significant societal impacts.

Contents

Chapter 1
Necessary Notions from the Theory of Stochastic Processes

1.1 Correlation and Structure Functions, Energy Spectra

To facilitate the reader's understanding, we will now present the fundamental formulas of the theory of stochastic processes that are essential for the subsequent discussions. While some of these formulas are well-known, others are only found in obscure publications, and a few are being cited for the first time. Although these formulas primarily focus on the initial two moments of probability distributions, they possess a straightforward structure and content, yet their level of familiarity is not widespread. Certain formulas can be derived by considering similarity and dimensionality, thus the material presented here may offer further validation for their accuracy and applicability limits, as well as for the analysis of empirical data. The exposition will specifically address temporal processes, which pertain to processes in a one-dimensional space. The statistical theory of random vector fields was originally developed by A. M. Obukhov in the 1940s, and a comprehensive explanation can be found in volume II of the book authored by A. S. Monin and A. M. Yaglom, referred to as MY75.

Consider a time-stationary stochastic process $a(t)$ for which there is an average value:

$$\langle a \rangle = \lim_{T \to \infty} \frac{1}{T} \int_0^T a(t) dt \tag{1.1}$$

and correlation function

$$B_a(\tau) = \langle a(t+\tau)a(t) \rangle = \sigma_a^2 f(\tau), \quad \sigma_a^2 \equiv \langle a^2 \rangle. \tag{1.2}$$

It corresponds to the Fourier transform, called the spectral energy density of the process in question:

G. Golitsyn and C. Varotsos, *The Stochastic Nature of Environmental Phenomena and Processes*, https://doi.org/10.1007/978-3-031-77015-9_1

$$E_a(\omega) = \frac{2}{\pi}\int_0^\infty B_a(\tau)\cos\omega\tau d\tau, \tag{1.3}$$

and vice versa

$$B_a(\tau) = \int_0^\infty E(\omega)\cos\omega\tau d\omega. \tag{1.4}$$

The last formula shows that:

$$B_a(\tau = 0) = \langle a^2 \rangle = \int_0^\infty E_a(\omega)d\omega \equiv \sigma_a^2, \tag{1.5}$$

from where it is evident that the spectral density, the function $E_a(\omega)$ accurately reflects its purpose by providing the energy distribution based on the frequency $\omega = 2\pi/\tau$ where τ is the time period. The correlation function is normalized to its variance, and therefore in (1.2) the function $f(\tau)$ cannot exceed one.

For the convergence of integrals in (1.3)–(1.5), the corresponding behavior of sub-integral functions at zero and infinity is necessary. Since the function $E_a(\omega)$ describes the distribution across the spectrum of the square (energy) of our random process $a\ (t)$, it must be non-negative everywhere: $E_a(\omega) \geq 0 for\, 0 < \omega < \infty$. This imposes limitations on the kind of correlation function $B_a(\tau)$, according to (1.3) and (1.4). For example, the function $B_a(\tau)$ in (1.3) cannot fall linearly to zero at some τ, and then is zero precisely because, as is easy to see, its Fourier transform then would have negative regions on the frequency axis ω.

Here are three examples MY75 of "correct" correlation functions and their corresponding spectral densities normalized to dispersion σ_α^2:

$$B(\tau) = e^{-a\tau} \tag{1.6}$$

$$E(\omega) = \frac{2a}{\pi\left(a^2 + \omega^2\right)} \tag{1.7}$$

$$B(\tau) = e^{-a\tau^2}, \quad E(\omega) = \frac{e^{-\omega^2/4a}}{(a\pi)^{1/2}}; \tag{1.8}$$

$$B(\tau) = (a\tau)^\nu K_\nu(a\tau), \quad E(\omega) = \frac{\Gamma(\nu+1)a^{2\nu}}{\pi\left(a^2 + \omega^2\right)^{\nu+1/2}}, \tag{1.9}$$

where K_ν is the MacDonald function with an index of ν. The spectrum (1.7) is also called the Cauchy–Lorentz distribution.

Consider now the process $u(t)$, such that:

$$\dot{u}(t) = a(t), \text{ or } u(t) = \int a(t)dt. \tag{1.10}$$

If *the process a* (*t*) was stationary in time, then the integral from it will be a random process with stationary increments. The theory of such processes was developed by A. N. Kolmogorov in 1940, MY75, when he began to create a mathematical tool for studying the theory of turbulence, starting at the invitation of O. Y. Schmidt in 1939 to work in the new Institute of Theoretical Geophysics of the Academy USSR of Sciences. For such processes, it is impossible to determine any average value or correlation function. However, it is possible to construct (and measure) the probability distribution for the average values of the differences in the quantities u (t) taken at two points in time: t_1 and t_2. The corresponding second moment was called by A. N. Kolmogorov a structure function. He showed that for processes with stationary increments, this function depends only on the difference in difference $t_2 - t_1 = \tau$:

$$D_u(\tau) = \left\langle [u(t + \tau) - u(t)]^2 \right\rangle. \tag{1.11}$$

The structure function of this formula can also be determined for stationary stochastic processes. Then the connection between the structure and correlation functions of such processes is obvious:

$$D(\tau) = 2[B(0) - B(\tau)]. \tag{1.12}$$

For the structure function of processes with first-order stationary increments, it is possible to introduce a spectral density with a ratio, also MY 75:

$$D(\tau) = 2\int_0^\infty (1 - \cos\omega\tau)E(\omega)d\omega. \tag{1.13}$$

Most important to us in the future is the example of a power structure function.

$$D(\tau) = A\tau^\gamma, \quad 0 < \gamma < 2, \tag{1.14}$$

the spectrum of which is determined by the ratios:

$$E(\omega) = C/\omega^{\gamma+1}, C = A\pi^{-1}\Gamma(\gamma + 1)\sin(\pi\gamma/2), \tag{1.15}$$

where $\Gamma(\gamma + 1) = \gamma\Gamma(\gamma)$—gamma function.

Power regularities cannot be implemented over the entire infinite interval of times and their inverse frequencies $\omega = 2\pi/t$. There are always internal and external boundaries for the execution of power regularities. The internal scale is usually associated with the dissipative mechanisms inherent in the system under consideration, and the external scale is determined by boundary conditions, for example, the lifetime of the system or its dimensions.

In the development of these concepts, Barenblatt and Zeldovich (1972), see also B09 introduced the concept of intermediate asymptotics, which is performed on the finite interval of the quantity under study. Sometimes, as in the case of the statistical description of the relief (see Chap. 9), one asymptotic may pass into another due to a change like the forces acting in the system, Chap. 8.

In the 1950s, Yaglom developed a general theory of stochastic processes with stationary increments of arbitrary order *n*. For example, for the characteristics of sea waves, oceanographers use the frequency spectrum of elevations of the water surface. This spectrum of vertical displacements is the spectrum of the correlation function of velocities arising from $g \sin\beta$, $g = 9.8\,\text{m/s}^2$, and β—the slope of the surface in the wave, which is a random stationary process. At $n = 2$, the growth of the structure function is limited from above by a power index equal to 4, not 2, as in $n = 1$, see (1.14). This leaves the ratio (1.13) with $n = 2$, binding structure function and spectral density. Using known trigonometric relationships, it can also be rewritten as

$$D(\tau) = 4\int_0^\infty \sin^2\left(\frac{\omega\tau}{2}\right)E(\omega)d\omega. \tag{1.16}$$

Given the limited interval of execution of power regularities both from above and below, the relation (1.16) can be reasonably corrected for its convergence at near-zero frequencies and at infinity.

For random processes with second-order stationary increments, it is possible to determine structure functions and spectra similarly (for particle displacement) (1.13) to (1.15) in the following form (Golitsyn and Fortus 2020):

$$D^{(2)}(\tau) = 2^3\int_0^\infty (1 - \cos\omega t)^2 E(\omega)d\omega \tag{1.17}$$

and with the power form of these functions

$$D^{(2)}(\tau) = A_2\tau^\gamma, \quad 2 < \gamma < 4,$$

connection is maintained $E^{(2)}(\omega) = C_2/\omega^{\gamma+1}$, and the relationship between the constants A_2 and C_2 found in Golitsyn and Fortus (2020):

$$A_2 = \frac{\Gamma(\gamma+1)\sin\left(-\frac{\pi\gamma}{2}\right)C_2}{\pi\left(2^{\gamma+1} - 2^3\right)}. \tag{1.18}$$

Definitions of Fourier-type transformations such as (1.13) or (1.17) exclude divergences at zero. It should be remembered, however, that power dependencies are always intermediate asymptotics (Barenblatt and Zeldovich, 1972).

1.2 Delta-Correlated Stochastic Processes

In nature, most of the processes (for example: turbulence, earthquakes, landslides, etc.) are random in time and space. The primary basis of randomness is the forceful effects on the system. The correlation time of these influences is usually much shorter than the reaction time of the system. For times $\tau_0 << \tau$, an internal time scale, which is often also the time of correlation of random forces acting on the system, in a first approximation their correlation function can be approximated by a delta function (in probability theory, such approximations are called Markovian)

$$B_a(\tau) = \varepsilon\delta(\tau), \quad \varepsilon = \sigma_a^2 \tau_0 \tag{1.19}$$

It should be remembered that the delta function has a dimension inverse of the dimension of its argument. For the process $u(t)$, the integral of $a(t)$, a random process with stationary increments, the structure function (1.11) is (see MY75):

$$D_u(\tau) = 2\varepsilon\tau \tag{1.20}$$

This formula was first published in 1944 in the first edition of the book "Mechanics of Continuous Media" by L. D. Landau and E. M. Lifshitz (the history of many of its rediscoveries is set out in MY75). Turbulence researchers have usually obtained it for reasons of similarity and dimension when u is the increment of velocity, ε is the rate of generation/dissipation of the kinetic energy of turbulence per unit mass of the liquid.

With the delta-shaped correlation function (1.19), according to the formula (1.3), we calculate the spectral density of the stationary process $a(t)$, which turns out to be a frequency-independent constant equal to $2\varepsilon/\pi$, that is, white noise. According to (1.13), the structure function (1.20), proportional to time, corresponds to the spectral density:

$$E_u(\omega) = \varepsilon(2/\pi)\omega^{-2} \approx \varepsilon\omega^{-2}. \tag{1.21}$$

Below we omit the multipliers $2/\pi = 0.637$ everywhere, since in all real situations the theoretical multipliers must be compared with the data of specially staged experiments. For the average square of the process,

$$x(t) = \int u(t)dt$$

equal to

$$\langle [\Delta x(\tau)]^2 \rangle$$

we have a spectrum:

$$E_x(\omega) = \varepsilon\omega^{-4}. \tag{1.22}$$

This expression can be obtained without a numerical coefficient for reasons of dimensionality, or from the formula (1.21) for the spectral density of the process $u(\tau)$ given that the spectrum of the process $x(t)$ for which the equation below exists:

$$\dot{x}(t) = u(t).$$

and associated with the spectrum:

$$\omega^2 E_x(\omega) = E_u(\omega), \tag{1.23}$$

where the multiplier ω^2 arises from the quadratic nature of the structure function (1.11).

For the structure function of displacements $x(t)$, a random process with second-order random increments, $n = 2$, the formula (1.13) gives

$$D_x(\tau) = \varepsilon\tau^3 = \langle(\Delta x)^2\rangle \equiv r^2, \tag{1.24}$$

which is also obtained for reasons of dimensionality.

With formula (1.20) it is instructive to compare Einstein's formula for the average square of displacement of a Brownian particle.

$$\langle x^2\rangle = 2nD\tau, \tag{1.25}$$

where n is the dimension of space, D is the diffusion coefficient. It is this identity of the formulas (1.20) and (1.25) that allowed Obukhov (1959) in the case when $u(\tau)$ is the velocity field, in the Lagrange description *of* T^*, to call the value $\varepsilon/2$, the rate of generation/ dissipation of the kinetic energy of turbulence, the coefficient of diffusion in the space of velocities.

The question arises, what in specific numbers means the condition of the smallness of the times of correlation of the effects τ_0 compared to the reaction time of the system τ. To answer this question, at least in order of magnitude, let's choose the correlation function of the stationary random process $a(t)$ in its simplest form:

$$B_a(\tau) = \frac{\varepsilon}{\tau_0}\exp\left(-\frac{\tau}{\tau_0}\right), \tag{1.26}$$

which at $\tau_0 \to 0$ tends to the δ-function. The correlation function (1.26) corresponds to Langevin's stochastic equation (MY75). This equation differs from the equation $u = a$ by adding linear friction on the right—the term λu, where $\lambda = \tau_0^{-1}$. The frequency spectrum of such a process is described by the formula (1.7) with $\alpha = \lambda = \tau_0^{-1}$.

The structure function of the *process* $u(t)$ will be:

$$D_u(t) = \varepsilon\tau_0\left[\frac{\tau}{\tau_0} - 1 + \exp\left(-\frac{\tau}{\tau_0}\right)\right]. \tag{1.27}$$

With an accuracy of a better than 1% already at $\tau = 5\tau_0$, you can neglect the exponential here and get:

$$D_u(\tau) \cong \varepsilon(\tau - \tau_0) \tag{1.28}$$

and the linear dependence on the constant $\varepsilon\tau_0$ is not difficult to take into account in further formulas.

The formulas (1.19)–(1.28) are correct in the probabilistic-theoretic sense, i.e., for infinite ensembles of events or times of observations over time τ, when the number of events can be taken as the value $N = T/\tau_0$, where T is the total period of observations. However, in practice, ensembles are always finite, often the number of events is only on the order of a few dozen. In the work of GlG10, analytically and by numerical counting, the validity of these formulas is checked, and it is shown that asymptotically they work satisfactorily already at $N \geq 10$. This concludes the consideration of continuous processes. Note, in order not to return to this further, that if the correlation time of the effects of the order or greater than the reaction time of the system, then the Boltzmann equilibrium of GlG10 is established in it, which is also carried out if the influx of energy is balanced by its dissipation.

1.3 Moments of Distribution Functions of A. N. Kolmogorov

All these results are derived from considerations of dimensionality, which, generally speaking, need a more rigorous justification. Such a justification is given by the works of A. N. Kolmogorov of the early 1930s. Their crown is the work of 1934, ANK34, in which, to describe the evolution of the probability density function, PDF, of the system $p(x_i\, u_i,\ t)$ the Fokker–Planck equation is used in the form:

$$\frac{\partial p}{\partial t} + u_i\frac{\partial p}{\partial x_i} = D\frac{\partial^2 p}{\partial u_i^2} \tag{1.29}$$

which in Soviet literature was called the FPK equation, where K corresponds to the surname of Andrei Nikolaevich. Here, x_i, u_i are components of a 6-dimensional vector. It is obtained by the decomposition of the PDF into a Taylor series by a random parameter with an accuracy of a second-order term of smallness and is solved with initial and boundary conditions for the semi-infinite space. At the lower boundary of space, the values of u are set over time, that is, the accelerations distributed according to Markov, or δ-correlated in time and space.

The fundamental solution of the equation (1.29) is of the form (see MY75, and Appendix to Chap. 1):

$$p(u_i, x_i, t) = \left(\frac{\sqrt{3}}{2\pi Dt}\right)^3 \exp\left[-\left(\frac{u_i^2}{Dt} - \frac{3u_i x_i}{Dt^2} + \frac{3x_i^2}{Dt^3}\right)\right]. \tag{1.30}$$

A. M. Obukhov was the first to analyze this equation. He showed that the coefficient $D = \varepsilon/2$ (see Lifshitz and Pitaevski 1979) is proportional to the dissipation rate of the kinetic energy of turbulence ε. Solution (1.30) shows that the desired probability distribution is normal. This solution has three scales (angle brackets mean the average for the distribution over the ensemble):

$$\left\langle u_i^2 \right\rangle = \varepsilon t, \tag{1.31}$$

$$\langle u_i x_i \rangle = \varepsilon t^2 \equiv K, \tag{1.32}$$

$$\left\langle x_i^2 \right\rangle = \varepsilon t^3 \equiv r^2, \tag{1.33}$$

where the scale (1.32) in dimension is equal to the coefficient of turbulentmixing. Expressing the time from (1.33) and substituting it in (1.31) and (1.32), we get (1.34) and (1.35):

$$\left\langle u_i^2 \right\rangle = (\varepsilon r)^{2/3} \tag{1.34}$$

$$K(r) = \varepsilon^{1/3} r^{4/3}. \tag{1.35}$$

i.e., the Kolmogorov–Obukhov law of 1941 for a structure function with zero (small) initial conditions and the Richardson–Obukhov law for turbulent mixing (vortex diffusion). The scales (1.31) and (1.33) are also manifested in the fact that the substitution of the variables:

$$u_i = \tilde{u}_i (Dt)^{1/2}$$
$$x_i = \tilde{x}_i \left(Dt^3\right)^{1/2}$$

(where ~ is the dimensionless symbol), excludes from (1.29) the diffusion coefficient *D*, i.e., the description becomes completely self-similar, GlG10.

Time dependencies (1.31)–(1.33) were checked numerically for ensembles of *N* randomly moving particles in accordance with the formulation of the problem by A. N. Kolmogorov ANK34, GlG10. Figure 1.1 shows that even at $N = 10$ dependencies (1.31) and (1.33) begin to be fulfilled, and at $N = 100$ numerical dependencies are practically non-existent and are different from the theoretical ones. When there is no motion in the system, the Eq. (1.28) takes on a simplified Fokker-Plank form for the

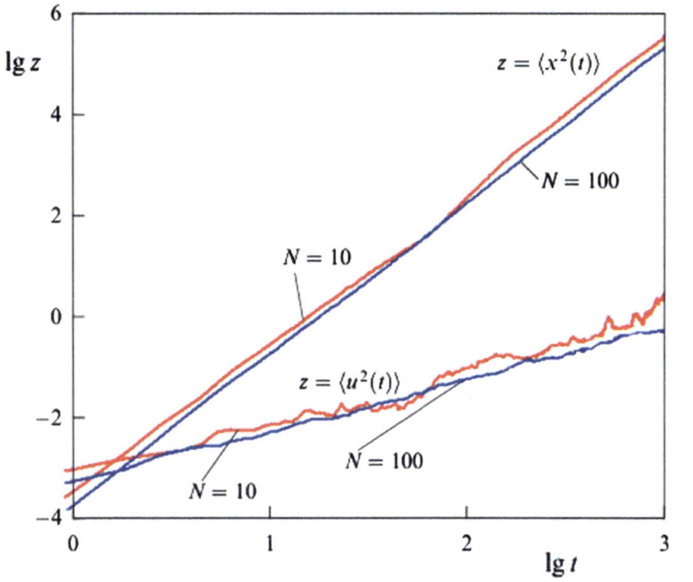

Fig. 1.1 Second moments of the probability density function for velocities $< u^2(t) >$ and distances between particles $< x^2(t) >\sim t^3$, from GIC10

probability distribution $p(t, x_i)$:

$$\frac{\partial p}{\partial t} = D\frac{\partial^2 p}{\partial x_i^2}, \tag{1.36}$$

the solution of which

$$\left\langle x^2 \right\rangle = 2nDt, \tag{1.37}$$

where n is the dimension of space (Phillips 1958), which is the average area of the spot occupied by random movements.

This section with formulas (1.19)–(1.37) in theoretical physics is called diffusion approximation (Lifshitz and Pitaevski 1979), but in fact, is a general approach to the consideration of random influences on the system under consideration correlated in one way or another (MY75).

1.4 Stochastic Events Flow

A statistical description of the flow of events requires a significant part of what has already been said since events require continuous preparation for their implementation. For example, earthquakes occur when the stress reaches a certain critical value in the crust, depending on the properties of the rock, first of all, and the stress arises as a result of random influences. Events must be characterized by a certain "size", intensity, and energy E, which must be a positively defined quadratic value.

Usually, in practice, histograms of the number of events $N(E)$ in the energy range from E to $E + dE$ for some period of observations T_0 are constructed.

Then the value $N(E)dE$, normalized by the full number of events during time T_0 will be an empirical estimate of the probability of an event with energy E. Here, probability $N(E)$ is associated with the frequency of occurrence of an event with energy E. In reality, usually the sample size is not large enough, and the number of events in the range from E to $E + dE$ is small and can fluctuate strongly in adjacent intervals. Therefore, in practice, a cumulative histogram is used:

$$N(\geq E) = \int_{E}^{E_{\max}} N(E)dE, \tag{1.38}$$

i.e., the number of events with an energy greater than or equal to E, a value with the dimension of the reverse time, i.e., frequency. Due to the operation of integration, in practice—summation, the cumulative distribution is much more stable, that is, with a smaller spread than the differential distribution $N(E)$. The function $N(E)$ obviously has a dimension inverse of the dimensions of time and its argument and is an empirical irregular estimate of the probability density for the flow of events characterized by parameter E, and the cumulative distribution has a frequency dimension, i.e., inverse time. Usually, if the integral in (1.29) converges well, then with an accuracy of a multiplier O (1)

$$N(E) = E^{-1}N(\geq E). \tag{1.39}$$

With random influences of the "white noise", or Markovian type, as discussed above, the increment of energy in the system in question occurs linearly with time according to (1.20).

Until we normalize the formula type (1.20)

$$E = \varepsilon\tau \tag{1.40}$$

for the full period of observations T_0, i.e., we do not proceed to estimates of empirical probability:

$$p_e(\geq E) = \tau(\geq E)/T_0 = [N(\geq E)T_0]^{-1}, \tag{1.41}$$

that time $\tau(\geq \mathrm{E})$, according to §13.9 of Feller's book (1957), can be considered the average waiting time of an event with an energy $\geq E$. After normalization by T_0 this value will be an estimate of the dimensionless probability $p(\geq E)$. As a result, in (1.40) time $\tau \equiv \tau(\geq E)$, and the inverse of it will be an estimate of the cumulative frequency

$$N(\geq E) = [\tau(\geq E)]^{-1}.$$

Without an excursion into the concept of the average waiting time of an event, characterized by the parameter $\geq E$, which is rarely mentioned in many standard books on probability theory, we would not be able to write down the ratio (1.41).

Now (1.40) with (1.39) can be rewritten as

$$N(\geq E) = \varepsilon/E, \quad N(E) = \varepsilon/E^2 \tag{1.42}$$

with precision to the multipliers to be found from observations. The value ε in (1.40) can be interpreted as the rate of generation of the quantity E both in dimension and in meaning, considering the right side (1.40) as the first term of the decomposition into the Taylor series by the time of increment of the quadratic parameter E.

An empirical estimate of the value ε is given by the formulas (1.42):

$$\varepsilon = EN(\geq E) = E^2N(E), \tag{1.43}$$

down to a multiplier of the order of one. As we shall see later, many empirical frequency-size distributions have a shape close to those of (1.42). This formula and (1.43) were first published by Golitsyn (2004).

The mean velocity square (1.31) has the dimension of energy per unit mass and we denote it as E. If we deal with the flux of events with frequency of dimension t^{-1}, then from (1.41) we may write (1.42) and for events with random arias S, described by (1.33) denoted by S:

$$N(> S) = \left(\frac{\varepsilon}{S}\right)^{1/3}, \tag{1.44}$$

$$N(S) = \frac{1}{S}\left(\frac{\varepsilon}{S}\right)^{1/3}. \tag{1.45}$$

In practice the distributions of random values $N(\leq A)$ are called cumulative ones or histograms and $N(A)$ differential ones. In Sect. 11.3 we shall find the histogram for lithospheric plates $N(\leq S) \sim S^{-n}$, $n = -0.33$. The similar form has integral distributions for spiral galaxies, Fig. 11.4.

When the text of this book was in printing stage the first author nearing 90 found the direct way of obtaining the CR spectrum from (1.42) taking into account the unit area as $s^{-1} = \left(\frac{w}{E}\right)^{2/3}$. The final form is the product of $\frac{G}{E}$ which is a unit time and $(w/E)^{2/3}$ is a unit area:

$$I(\geq E) = \frac{G}{E}\left(\frac{w}{E}\right)^{2/3} \sim E^{-5/3},$$

$$I(E) = \frac{dI(E)}{dE} \sim E^{-8/3}.$$

The multiyear measurements by PAMELA (Golitsyn 2004) presented $I(E) \sim E^{-n}$, $n = 2.67 \pm 0.02$. Unfortunately they have not presented data for $I(\geq E)$, there uncertainties in n would be at least an order of magnitude smaller.

1.5 Special Spectral Exponents and Their Sense

The difference between the spectrum of excitations and "white noise" should lead to a difference in indicators from one for the structure functions of continuous processes and for histograms, that is, frequency-intensity distributions. Here we will briefly and formally address this issue.

Let the frequency spectrum of the effects be in the form where *the* β is related to the index ν in (1.9)

$$E_a(\omega) \sim \omega^{-\beta} \tag{1.46}$$

where the cases $\beta > 0$ and $\beta < 0$ will be discussed separately. The time correlation function corresponding to this spectrum exists only for $\beta > 0$, i.e., the spectrum growing towards low frequencies. It has the form, MY75:

$$B_a(\tau) = \tau^{-\beta}\Gamma(1+\beta)\sin\frac{\pi(1-\beta)}{2} \tag{1.47}$$

and at $\beta \to 0$ tends to δ -function. If the spectrum of the process $a(t)$ is (1.46), then the spectrum of the process $u(\tau) = \int \alpha(t)dt$, α calculation will be of the form:

$$E_u(\omega) \sim \omega^{-2-\beta}, \tag{1.48}$$

with a corresponding structure function:

$$D_u(\tau) = \left\langle \Delta u^2(t) \right\rangle \sim \tau^{1+\beta}, \tag{1.49}$$

from where acting by analogy with the derivation of formulas (1.42), we obtain:

$$N(\geq E) = [\tau(\geq E)]^{-1} \sim E^{-n}, \quad n = (1+\beta)^{-1} \tag{1.50}$$

Thus, when $\beta > 0$, the number of large events increases compared to the case of "white noise" $\beta = 0$. When $\beta > 0$, the energy of low-frequency fluctuations increases with a decrease in frequency, and the correlation increases, which causes an increase in the number of large events that manage to form more often than with a uniform distribution energy impact across the entire spectrum. By analogy with the theory of electromagnetic radiation in physics of the late XIX century, this situation can be called an "infrared catastrophe". In any case, the spectrum at low frequencies must fall due to natural causes, since infinite energy cannot accumulate there, that is, i.e., $\omega < \omega_0$ and how the wave collapses in wind waves (see Chap. 6).

The case *of* $\beta < 0$, i.e. an increase in the spectrum of influences towards high frequencies, leads to an increase in the indicator n in (1.50) compared to one, i.e. to a decrease in the number of large events compared to the case of "white noise" $\beta = 0$ and to a predominant increase in the number of small events, which is expressed in an increase in the exponent in the distribution (1.50).

From the point of view of similarity and dimensionality, the presence in the spectrum of influences (1.46) of an indicator of the degree of $\beta \neq 0$ means that in the process under consideration, there is a dimensionless parameter depending on the frequency and other dimensional quantities, which does not disappear from consideration, no matter how large or small it may be. This is an example of self-similarity of the second kind according to the terminology of G. I. Barenblatt B02). Of the possible others, for more common reasons, the difference between the indicator n and the unit in the empirical cumulative distributions of the frequency-intensity of events type, it is necessary to mention the insufficient sample size (therefore, confidence intervals for the value of the indicator n should always be estimated), the presence of geometric factors, as for the statistics of earthquakes in thin plates near the mid-ocean ridges in contrast to their statistics in faults, that is, the boundaries of much thicker ones. San Andreas Fault Type Plates in Southern California (see Chap. 3). Another parameter of the difference is the distribution, where the number of events is measured not only per unit of time but also per unit area, which itself may depend on the parameter by which the distribution is sought, as will be seen in the case of cosmic ray statistics (Chap. 4).

Here we have considered various, practically useful questions of statistics of temporal random processes. However, in reality, many tasks require knowledge of the statistics of processes in their spatial manifestation, for example, turbulence, although, as a rule, we have only a time record of the signal at the point. In the latter case, the simplest connection between temporal and spatial characteristics for spatially homogeneous processes is given by the hypothesis of GI. Taylor on so-called "frozen" turbulence (see MY75). Formally, this hypothesis uses the relation:

$$r = U\tau, \tag{1.51}$$

or in wave representation:

$$\omega = kU, k = 2\pi/r, \omega = 2\pi/\tau, \tag{1.52}$$

where U is the average flow velocity, e.g., wind in the atmosphere. It is assumed that the turbulence carried by the wind past the measurement point remains virtually unchanged during the measurement τ.

In general, it is necessary to use a dispersion ratio $\omega = \omega(\tau)$, which does not have to be linear as (1.52). For example, the ratio (1.20) for the inertial turbulence interval can be compared (again, with an accuracy of a numerical coefficient) to the dispersion equation:

$$\omega = \varepsilon^{1/3} k^{2/3} \tag{1.53}$$

Ratios of the dispersion type, as, for example, for waves on the surface of the sea, in the transition from frequency characteristics to spatial ones, can be considered as an operation of replacing variables in the distribution of probability or its moments. Due to the conservation of probability, equality must be fulfilled.

$$P(x)dx = P(y)dy,\ y = y(x). \tag{1.54}$$

Similar relationships can be written for distribution moments, for example, for spectral densities. In the latter case, a physical interpretation of such an equality is also possible: due to (1.5), the integral of the spectrum is equal to the variance of the quantity under study.

Therefore, under the assumption of ergodicity, the magnitude of the variance should not depend on whether we estimate it using temporal or spatial characteristics, that is, whether the amount of variance is estimated using temporal or spatial characteristics,

$$E(\omega)d\omega = E(k)dk. \tag{1.55}$$

When studying the spatial structure of random fields, the question of the diffusion of particles in such fields is practically important. As is known, the motion of the Brownian particle in a random velocity field with a spectrum of "white noise" is carried out with a constant diffusion coefficient - see (1.25). Formally, the diffusion coefficient can be defined as (see Chap. 7 for details):

$$K = \frac{1}{2}\frac{d}{dt}\left\langle (\Delta x)^2 \right\rangle = \frac{1}{2}\frac{dr^2}{dt} = r\frac{dr}{dt} = ru, \tag{1.56}$$

or a time derivative of the area. In this spirit, the old, Taylor (1915), definition works.

Consider the general case of the impact spectrum (1.44): $\mathrm{E}_\alpha(\omega) \sim \omega^{-\beta}$, $\beta > 0$. It corresponds to the spectrum of spatial displacements:

$$E_k(\omega) \sim \omega^{-4-\beta} \tag{1.57}$$

for which the generalized structure function with second-order stationary increments $(0 < \beta < 1)$ will be:

$$D_x(\tau) \sim \tau^{3+\beta} \equiv r^2. \tag{1.58}$$

From here we express $\tau = \tau(r)$ and, substituting in (1.56), we get

$$K \sim r^m, m = \frac{2(2+\beta)}{3+\beta} \tag{1.59}$$

At $\beta = 0$, "white noise", we have the classical Richardson-Obukhov law (see "white noise") (Chap. 6):$K \sim \varepsilon^{1/3} r^{4/3}$. Diffusion in the field of random wind waves, depending on their age, has slightly different asymptotics (see Chap. 6, Sect. 6.2) with indicators from 4/3 to 7/6 depending on the age of the waves. The case of molecular diffusion, i.e. a constant transfer coefficient, corresponds to $\beta = -2$ as can be seen in (1.58) and (1.59). Let us also mention the so-called flicker noise, corresponding to the exponent of the degree in the spectrum of influence $\beta = 1$. Note that according to (1.5) such noise will be "white", that is, constant, but in the space $\ln\omega$. Displacement spectrum, according to (1.57), then it would be ω^{-5}. This form of spectrum was proposed by Philips (1958) for the spectrum of marine surface waves in the high-frequency region for dimensionality reasons, taking frequency ω and acceleration $g$as determining parameters. We will leave its discussion for a special Chap. 6. Index depending on the mixing coefficient on the spot size r will be at this is equal to $m = 3/2$. This issue is discussed in Chap. 7 (Sect. 7.2), where it will be shown that $1 < m < 3/2$.

This concludes the description of the tools we need for the future. A much more complete set of concepts and methods can be found in the book by Sornette (2003) with numerous examples from many areas of physics and other natural sciences.

1.6 Some Consequences of the Results of A. N. Kolmogorov in 1934

Here we will consider some consequences useful for the statistical processing of results, and moments of distribution functions described in paragraph 1.1.3. Scales (1.31) and (1.33) can be thought of as temporary structure functions of processes with zero (or small) initial data. Then the formulas (1.14) and (1.17) make it possible to determine their spectra, which is done in Sect. 1.3.

Another useful consequence is the construction on their basis of cumulative distributions *of* $N(\geq E)$ of magnitude E. The latter has a dimension of inverse time, and then the distributions associated with energy will be:

$$N(\geq E) \sim \varepsilon E^{-1},$$

and with the event area:

$$N(\geq S) \sim (\varepsilon/S)^{1/3}.$$

Examples of the last two distributions are given in Chap. 4, for the distributions of the number of lithospheric plates by area and the galaxies closest to us by mass.

The first of these two formulas is given in the form (1.42), the practical meaning of which became clear only at the last stage of processing the manuscript of this book. If we assume that in a particular case, the value of the rate of generation of the process ε is more or less constant, then $EN(\geq E)$ will be functions inverse of each other. We have this situation in Chap. 10 when considering damage from floods, when:

$N(\geq Y) \sim Y^{-2/3}$, and the damage is proportional to the flooded area associated with the catchment area. For a long time, it remained unclear to the author that the area of mud fungi on the surface of the seas and oceans formed at some distance from the mouths of rivers is proportional to their catchment to the degree S^b, where the degree index for different samples of rivers in California, Morocco, Florida is close to 2/3 with an accuracy of about 10–20% (see Chap. 10). Most of the results of this chapter are contained in G12.

In conclusion, we will describe the process of obtaining a solution to equation (1.30), about which it is written in MY75 that it is known, but no references are given. The solution below was found by A. A. Lushnikov, and I sincerely thank him for his permission to reproduce it here.

A number of results can be also obtained by dimensional analysis and it can be used to check the validity of formulas obtained from probability arguments. First are histograms $N(\geq E)$, their dimension is reverse time or frequency. The differential distributions have dimensions $(\text{TA})^{-1}$, where A is a parameter on which the distribution is sought (seeked?). Of course, these quantities are related to the second probability moments (1.31)–(1.33). The first moment $\langle u_i^2(t)\rangle = \varepsilon t = E$, where E is the energy (or any intensity) per unit mass and can be applied for different processes due to their self-similarity discussed above. So the histogram will be

$$N(\geq E) = \frac{\varepsilon}{E}, \qquad N(E) = \frac{\varepsilon}{E^2}.$$

The distributions (1.42) are known over a century as Pareto law, Zipf law, for population of the cities, avalanches, etc. other empirical laws, but here we describe their probability nature, see [13] where there many empirical distributions with theoretical estimates of their power indices.

As (1.42) and strictly speaking to the right side should be applied a non-dimensional multiplier C_l. If the analyzed process is governed by several factors which can form a non-dimensional similarity number Π, then $C_l = f(\Pi)$ as for earthquakes, and normally $f(\Pi) \sim \Pi$ starts with a linear term.

For the processes with random areas the histograms are

$$N(\geq A) = C_a\left(\frac{\varepsilon}{A}\right)^{1/3}.$$

The illustrations for this are lithospheric plates and spiral galaxies, see 11.3 in Chap. 11 and differential distributions for the horizontal lengths of clouds and blue sky voids between the clouds in Fig. 14.2.

The left side of Eq. (1.31) in general case should be understood as an intensity of the process, and ε as its generation rate. Their product (1.31) and (1.33) being multiplied by a mass of rotating column will evaluate the total energy of the object (see Ch. 9). The right side of (1.33) considered as a 3-rd order structure function for the wind sea waves can immediately produce its frequency spectrum as ω^{-4} (see Sect. 6.3).

The important point in these studies is that all three second probability moments can contain non-dimensional numbers C_1, C_2, C_3 which may depend on non-dimensional similarity members, Π, like the Reynolds number, if the process in consideration may depend on some additional governing parameters. An analysis of the value of Π, experimentally or theoretically can obtain the limits of the found result (see Ch. 3) as intermediate asymptotes [1]. We should recall the very informative paper by Newman [2] where one can find over 20 power dependencies of various sorts and mathematical estimates for the possible limits on the power indices.

1.6.1 Application

If the Eq. (1.29) is non-dimentionalised by the substitutions:

$$x = \tilde{x}\left(\varepsilon t^3\right)^{1/2}, \; u = \tilde{u}(\varepsilon t)^{1/2}$$

where the sign ~ means a dimensionless quantity, then this equation will not contain ε, and the non-dimentionalization itself already gives the scales of A. N. Kolmogorov (see (1.31) and (1.33)).

$$\frac{\partial p}{\partial t} + \tilde{u}_i \frac{\partial p}{\partial \tilde{x}_i} = \frac{\partial^2 p}{\partial \tilde{u}_i^2}. \tag{1.60}$$

We are looking for a self-similar solution in the form of:

$$\ln p = A + ar^2 + b\boldsymbol{r} \cdot \boldsymbol{u} + cu^2, \tag{1.61}$$

and its substitution into the Eq. (1.60) gives:

$$\dot{A} = \dot{a}r^2 + \dot{b}\boldsymbol{r} \cdot \boldsymbol{u} + \dot{c}u^2 = 6c + b^2r^2 + 4bc\boldsymbol{r} \cdot \boldsymbol{u} + 4c^2u^2. \tag{1.62}$$

Equating the coefficients on the left and right, we get a system of ordinary differential equations:

$$\dot{A} = 6c, \quad \dot{a} = b^2, \quad \dot{b} + 2a = 4cb, \quad \dot{c} + b = 4c^2. \tag{1.63}$$

The solution to this system is:

$$a = -3/t^3, \quad b = 3/t^2, \quad c = -1/t, \quad A = -6 \ln t. \tag{1.64}$$

Substituting these solutions into (1.61) and normalizing the probability p per unit gives the solution in the form of (1.30) and note that in ANK34 the average term of the exponent in parentheses was omitted, but MY75 gives it correctly.

It is known that equations of the parabolic type have self-similar solutions, and the method proposed here, previously unknown to the first author, provides a useful way to find them.

References

Barenblatt GI, Zeldovich Ya B (1972) Self-similar solutions as intermediate asymptotics. Ann Rev Fluid Mech 4:285–312

Feller W (1957) An introduction in probability theory and its applications. Wiley, NY

Golitsyn GS (2004) Statistical laws of macroprocesses: random walks in the momentum space. In: Doklady physics, vol 49, no 9

Golitsyn GS, Fortus MI (2020) Random processes with stationary increments and composite spectra. Izv Atmos Ocean Phys 56:364–372

Lifshitz EM, Pitaevski LP (1979) Physical kinetics, chap 2. Fizmatlit

Newman MEJ (2005) Power laws, Pareto distributions and Zipf's law. Contemp Phys 46:323–351

Obukhov AM (1959) Description of turbulence in terms of Lagrangian variables. Adv Geophys 6:113–115

Phillips OM (1958) The equilibrium range in the spectrum of wind-generated waves. J Fluid Mech 4:426–434

Sornette D (2003) Critical phenomena in natural sciences. Chaos, fractals, self-organization and disorder: concepts and tools. Springer, 434 pp

Taylor GI (1915) Eddy motion in the atmosphere. Philos Trans R Soc 215(A):1–26

Яглом АМ (1955) Корреляционная теория процессов со случайными стационарными приращениями n-го порядка // Матем. сб. 1955. Т. 37, № 1. С. 141–196

Yaglom AM (1955) Correlation theory for processes with stochastic stationary increments of the n-th order. Math Collection 37(79), №1, 141–196

Chapter 2
Turbulence

2.1 Kolmogorov–Obukhov Turbulence

The 1941 results on small-scale turbulence were the first test of the potential capabilities of ANK34, undertaken by Obukhov (1959). He presented them in July 1958 at the international symposium in Oxford "Atmospheric Pollution and Turbulent Diffusion." A lengthy discussion of these ideas can be found in §24.4 of the book MY75. A short note [1] shows that there are three velocity scales in turbulence:

$$\left\langle u_i^2 \right\rangle = \varepsilon t \equiv r^2$$

diffusion

$$\langle u_i x_i \rangle = \varepsilon t^2$$

and length (see Chap. 1 paragraph 1.3 here)

$$\left\langle x_i^2 \right\rangle = \varepsilon t^3$$

Expressing from the last scale the time and substituting it into the first one, we get:

$$\left\langle x_i^2 \right\rangle = \varepsilon \left(r^2 / \varepsilon \right)^{1/3} = (\varepsilon r)^{2/3}, \tag{2.1}$$

that there is a structure-function for velocity with zero initial data. To the question of Batchelor, what is new here in comparison with the results of 1941, obtained by similarity and dimension theory, Alexander Mikhailovich replied that there is a lot that is new. However, this was not explained in MY75. This book contains the answers to the questions asked. The magnitude is:

G. Golitsyn and C. Varotsos, *The Stochastic Nature of Environmental Phenomena and Processes*, https://doi.org/10.1007/978-3-031-77015-9_2

$$\varepsilon = \frac{d}{dt}\frac{u_i^2}{2} = \frac{1}{2}\nu\phi_{ik}^2 = \frac{\nu}{2}\left\langle \left(\frac{\partial u_i}{\partial x_k} + \frac{\partial u_k}{\partial x_i} \right)^2 \right\rangle, \tag{2.2}$$

where ϕ_{ik} is the deformation tensor of the velocity field.

There are many temporary random processes and random events in nature, subordinate to the AHK34. Their cumulative energy distributions over time are described by area S proportional to

$S^{-1/3}$, by energy ~ E^{-1} (see Chap. 1).

Results of ANK34, their interpretation by A. M. Obukhov, and generalization here in Chap. 1 serve as a probabilistic substantiation of the results obtained earlier by methods of similarity and dimension theory. In these cases, you should always remember the conditions of local homogeneity, local isotropy, and intermediate asymptotics, which should be tested on real experimental data. From similarity and dimension, one can obtain the shortened dynamic equation of A.N. Kolmogorov, MY75, the so-called "law – 4/5":

$$\langle u_{lll}^3(r) \rangle = -\frac{4}{5}\varepsilon r$$

that is valid for Re >> 1, but not the magnitude and sign before εr, which are obtained with the exact solution of the dynamic problem from the Navier–Stokes equations for the longitudinal n component of speed. The sign of—is determined by the fact that the value of ε, the dissipation rate kinetic energy, generated on a large scale and transferred through the inertial interval of turbulence to dissipation interval:

$$l_\nu = \left(\nu^3/\varepsilon\right)^{1/4}.$$

The dependence of the average longitudinal velocity component (cm/s) on the distance (cm) between observation points is illustrated in Fig. 2.1, while the normalized velocity spectrum is shown in Fig. 2.2.

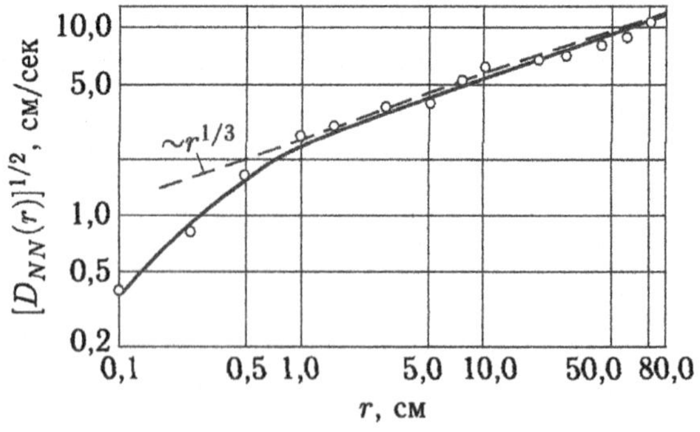

Fig. 2.1. Dependence of the average longitudinal velocity component (cm/s) on the distance (cm) between observation points. (Obukhov 1949, G12).

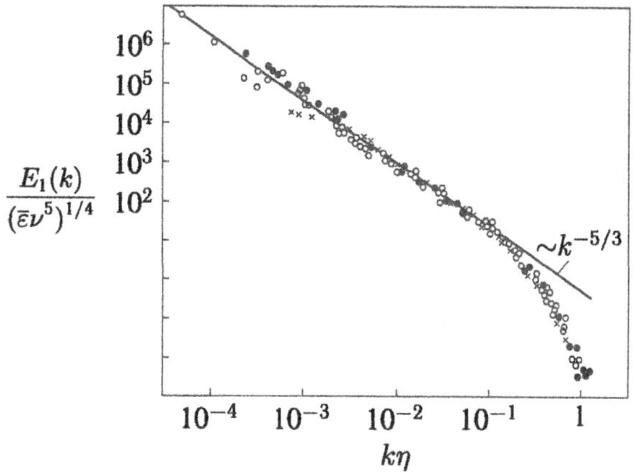

Fig. 2.2. Normalized velocity spectrum (MY75).

2.2 Passive scalar Turbulence

A passive scalar, such as water vapor or temperature, does not affect the velocity field but is transported by velocities. Such an admixture obeys the diffusion equation in as a moving flow, which makes it possible to introduce the concept of generation/dissipation rate of a quadratic measure of heterogeneity impurity field concentration:

$$N = \frac{d}{dt}\frac{\langle \Theta^2 \rangle}{2} = \chi \left\langle \left(\frac{\partial \Theta}{\partial x_i} \right)^2 \right\rangle, \tag{2.3}$$

where Θ is the impurity concentration, χ is the diffusion coefficient. This quantity is a measure of the generation of passive fluctuations of the field Θ. The flow itself in the inertial interval is characterized by the rate of generation/dissipation of kinetic energy ε according to (2.2) and has a time scale:

$$\tau = \left(r^2/\varepsilon\right)^{1/3}.$$

For a passive scalar in a turbulent flow, the same relations as for the velocity field, in instead of ε there is the quantity $N\varepsilon^{-1/3}$, which was established by Obukhov (1949) for the temperature structure function:

$$D_\theta(r) = \left\langle [\Theta(x+r) - \Theta(x)]^2 \right\rangle = C_0 N \varepsilon^{-1/3} r^{2/3}, \tag{2.4}$$

where C_0 is the numerical coefficient O(1) for the atmosphere, depending primarily on atmospheric stratification. Its value was determined many times and in different conditions, since it determines the intensity of fluctuations of wave signals passing through the medium. Corrsin (1951), without knowing [3] (note that the dimension:

$$\left[N\varepsilon^{-1/3}\right] = \Theta^2 L^{-2/3}$$

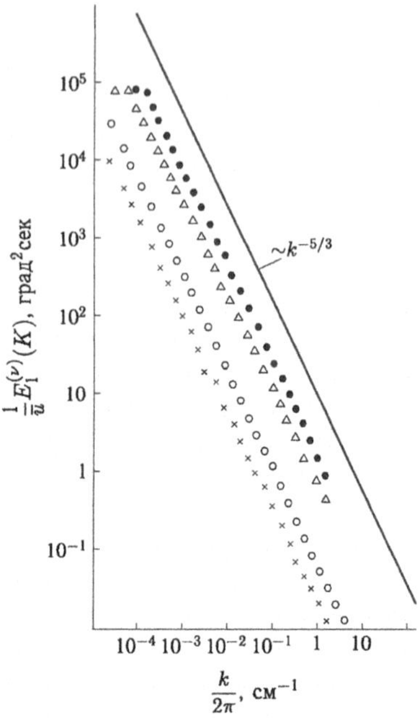

Fig. 2.3. Spectra of temperature fluctuations in the surface layer of the atmosphere (MY75) at day and night.

i.e., the structure-function determines the variance $\sigma_{\Theta}{}^2$), obtained a formula for the spectrum of impurity fluctuations [3], equivalent to (2.4):

$$E_\theta(k) = C_0 N \varepsilon^{-1/3} k^{-5/3}, \tag{2.5}$$

which has been tested a thousand times in various fields of science and technology.

The most impressive results are from measurements of electron concentration fluctuations in our Galaxy. Amstrong et al. (1981) found that their spectrum agrees with (2.5) over 12 orders of magnitude of sizes, from 10^6 to 10^{18} m! Figure 2.3 presents the spectrum of temperature fluctuations in the surface layer of the atmosphere.

2.3 Helicity and Spiral Turbulence

Helicity is a hydrodynamic quantity that is preserved in the inertial range, i.e., at sufficiently large Reynolds numbers. According to the definition by Moffat (1969) this is the scalar product of the velocity vector and its vortex:

$$h = \omega_i u_i = \boldsymbol{u} \cdot \operatorname{rot} \boldsymbol{u}, \quad \omega_i = \varepsilon_{ijk} \frac{\partial u_j}{\partial x_k}. \tag{2.6}$$

The designation h corresponds to the first letter of the word helicity.

From its definition it follows that in purely two-dimensional flows the helicity is identically zero, i.e., it can only be present in three-dimensional flows, for example, in the Ekman boundary layer, where the wind turns with height, approaching its geostrophic value. In the atmospheric boundary layers, it occurs continuously due to the rotation of the Earth and friction with the surface.

It is associated with the phenomena of reverse cascade and large-scale restructuring of currents. Helicity plays a significant role in the processes of magnetic field generation fields in a conducting liquid (Moffat 1978). Therefore, this value is calculated in mesoscale models for hurricanes and typhoons, polar mesocyclones, jet streams, thermal convection, etc. A review of such literature can be found in Vazaeva et al. (2021).

A fairly complete theoretical study of small-scale spiral turbulence was carried out by Chkhetiani (2008), and he also initiated numerical modeling of the role of helicity in atmospheric boundary layers, and then its direct measurements there (Koprov et al. 2005). Although direct consequences for fluctuations in helicity from ANK34 have not yet been found, the ideas that emerged from the study of ordinary turbulence help to understand spiral turbulence. Therefore, let us briefly turn to the most important concepts here.

The dimension of helicity coincides, as can be seen from (2.6), with the dimension of acceleration, i.e., in the SI system it is m/s^2.

Based on modeling and measurements h values are estimated to range from 0.01 m/s^2 (rotating thermals), 0.1 m/s^2 (hurricanes), and 10 m/s^2 ("dust devils" in deserts). But the amount of helicity is pseudoscalar, i.e., it can both be generated and damped Chkhetiani (2008). The note of helicity dissipation, derived from the Navier-Stokes equations, is equal to:

$$\eta = \frac{dh}{dt} = \nu(\boldsymbol{\omega} \cdot \mathrm{rot}\,\boldsymbol{\omega}) \tag{2.7}$$

and has the dimension m/s^3.

Helicity has a tensor character. Chkhetiani used the most general form of tensors of the second and third order, taking into account antisymmetry, and obtained several fairly general and simple visual results, for example, a dynamic equation for helicity, similar to Kolmogorov's "–4/5 law" for the velocity field and Yaglom's "–4/3 law" for passive scalar.

To study the fluctuation fields of helicity and their spectra, the velocity and vortex components were taken at different points separated by a distance r:

$$h(r) = u_i(\mathbf{x})\omega_i(\mathbf{x} + r). \tag{2.8}$$

This quantity is quadratic and can be considered an analog of the structural functions of a passive scalar. For the dimension of helicity, we take the notation $[h] = H$. Then the dimension of the dissipation/generation rate helicity will be $[\eta] = \mathrm{HT}^{-1}$. Dimensional considerations, which involve the rate of dissipation of kinetic energy ε and the distance r, give, under the assumption of helicity passivity:

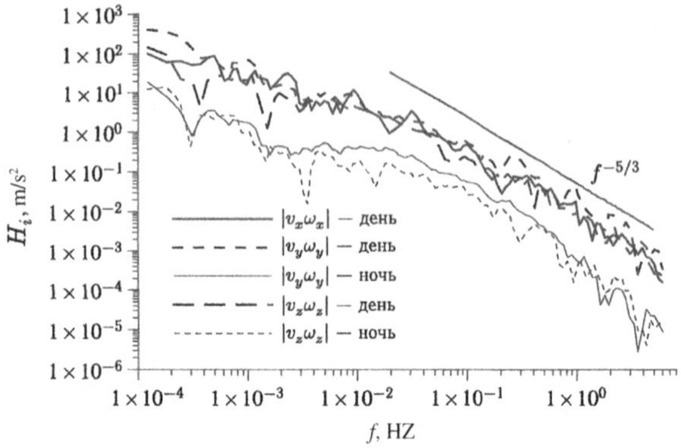

Fig. 2.4. Frequency spectrum of helicity (G12) (Koprov et al. 2005).

$$h(r) = \alpha_h \eta \varepsilon^{-1/3} r^{2/3} = \alpha_h \eta \tau_0, \tag{2.9}$$

where $\tau_0 = \left(r^2/\varepsilon\right)^{1/3}$—time scale in the inertial interval (compare with the rule actions of the fastest reaction in Sect. 11.1).

The quantity (2.9), proportional to $r^{2/3}$, corresponds to the spatial spectrum k $^{-5/3}$. Considering Taylor's "frozen turbulence" hypothesis $f = kU$, it is realistic the measured spectra will be ~ $f^{-5/3}$. Figure 2.3 represents actually measured helicity spectra obtained in the surface layer under different conditions of stratification and for various components $u_i\ \omega_i$. It is directly proportional to $f^{-5/3}$. Since helicity is determined by the field of motion itself, so we may say, its topology, then as a passive scalar it has an energy spectrum of – 5/3 (Fig. 2.4).

2.4 Two-Dimensional Turbulence

Natural processes can differ by 3–4 orders of magnitude on horizontal and vertical scales, exhibiting features that bear little resemblance to conditions for the three-dimensional turbulence. In 1967, Professor Victor Starr's book "Physics of Phenomena with Negative Viscosity" appeared, which described many-large-scale phenomena powered by energy from smaller-scale movements: the Gulf Stream, jet streams in the atmosphere, several astrophysical phenomena on the Sun and in spiral galaxies.

The possibility of such a direction of energy flows was pointed out back in 1949 by Onsager (1949) and in 1953 by Fjortoft (1953). So, it was more than enough to start developing a theory of two-dimensional turbulence Corrsin (1951) and Starr (1968). The main results were obtained by Batchelor (1969), presented by him in January 1964 at a lecture in Moscow, and Kraichnan (1967). Two-dimensional hydrodynamics has several specific features: there is no intensification of vortex filaments due to their stretching, as in the three-dimensional case. The equation for the vortex field $\omega(x, y, t)$ is:

$$\frac{\partial\omega(x_i,t)}{\partial t} + u_i(x_i,t)\nabla\omega(x_i,t) = \nu\Delta\omega(x_i,t), \tag{2.10}$$

where ω—vertical component of the velocity vortex, unique in 2D, i = 1, 2, ν—kinematic viscosity. In the absence of viscosity and incompressibility $d\omega/dt = 0$, i.e., the vortex is retained in the liquid particle. In a statistically homogeneous flow, the sign of vortex is not defined, so the flow is characterized by the mean square vorticity—enstrophy $\Omega = \langle\omega^2\rangle/2$. In the presence of viscosity:

$$\frac{d\Omega}{dt} = -\nu\langle|\nabla\omega|^2\rangle \equiv \varepsilon_\omega, \tag{2.11}$$

and for average energy:

$$\frac{dE}{dt} = -2\nu\,\Omega \equiv -\varepsilon, \tag{2.12}$$

where ε is the average rate of dissipation of kinetic energy. At large Reynolds numbers, energy transfer from small to large wave numbers is impossible. At the same time, for enstrophy and its spectrum such a transfer is possible as most simply and elegantly was shown by Novikov (1978), which is confirmed by numerical models (see Chap. 26 of the second edition of the book MЯ and review (Mirabel and Monin 1979). Lindborg (1999) derived the dynamic equation for the enstrophy transfer interval:

$$D_{lll}(r) = \frac{1}{8}\varepsilon_\omega r^3, \tag{2.13}$$

where ε_ω—is the rate of dissipation (in this case, transfer along the spectrum) of enstrophy and the cube of the distance between observation points. This is an analog of the "law –4/5" for the third moment of velocity in the three-dimensional case. The + sign on the right in (2.13) indicates the presence of a reverse cascade, i.e., energy transfer from small to large scales in purely two-dimensional flows.

In the presence of kinematic viscosity ν and the rate of enstrophy dissipation ε_ω it is possible to form a microscale of length

$$l_\omega = \left(\nu^3/\varepsilon_\omega^{-1}\right)^{1/6}. \tag{2.14}$$

Let us estimate the magnitude scales for the atmosphere. Per unit mass E $\approx$ 300m^2/s^2 (average wind speed according to Oort 1964 17 m/s), $\Omega \approx 10^{-10}\text{s}^{-2}$ with $\nu = 10^{-5}\text{m/s}^2$ and time life of vortices of order 10^5, with, we get $\varepsilon_\omega \approx 10^{-15}$ s. In this case, formula (2.5) gives $l_\omega \approx 1$ m (note that $10^{1/6} \approx 1.5$). It reminds us that this is the scale of the drifting snow eddies that each of us saw.

If there is a scale of enstrophy input into the flow, in the atmosphere, and in the ocean, this is the scale formation of synoptic vortices L_ω, then in the interval $l_\omega \ll l \ll L_\omega$ will act enstrophy transfer. Then, from the dimensional analysis, the value of the spectral density kinetic energy will be:

$$E(k) = C\varepsilon_{\omega}^{2/3}k^{-3}. \tag{2.15}$$

Spectra k^{-3} lead to divergences both at zero and at infinity. Kraichnan (1967), in the fight against them, used the hypothesis that nonlocal interactions between spectral components from a certain interval of wave numbers around one are involved in enstrophy transfer, and proposed a correction to the spectrum (2.6) in the form:

$$E(k) = C\varepsilon^{2/3}k^{-3}[\ln(k/k_0)]^{-1/3}, \tag{2.16}$$

where k_0—is the wave number from the enstrophy transfer interval.

In the 1990s, European countries conducted a large aircraft experiment MOZAIC, where, along with other measurements, registration measurements were taken for overload vector (Lindborg 1999). In them, for the horizontal velocity components there was discovered, again after Gage (1979) similar experiments in the USA, the spectrum $k^{-5/3}$ for scale interval from 2 to about 500 km, later these results were reproduced in a high-resolution atmospheric numerical model. For scales of 500–3000 km, the spectrum becomes steeper and approaches k^{-3}. The European results also show a logarithmic trace. This is discussed in sufficient detail in the article by Golitsyn and Fortus (2020).

It is characteristic that the spectrum k^{-3} in nature was observed not on a medium and small scale, but, on the contrary, on a large scale. But then it is not two-dimensional, but geostrophic turbulence, proposed by Charney (1971). In geostrophic movements, when the Coriolis force term is balanced by the pressure gradient, the potential vortex remains:

$$\Omega_p = \frac{(\boldsymbol{\Omega} + 2\boldsymbol{\omega})\nabla s}{\rho}$$

where Ω_p up to normalization, it is a component of the absolute vortex, i.e., the sum of the planet's own rotation and the hydrodynamic vortex, in the direction of the "thermodynamic vertical," that is, the entropy gradient s. Argumentation similar to Batchelor and Kraichnan allowed (Charney 1971) to obtain the spectrum here as well k^{-3}. A detailed discussion of these and subsequent results can be found in the review by Mirabel and Monin (1979).

Laboratory and numerical experiments developed quite intensively. A review of them up to 2002 can be found in Keller and Goldburg (2002), and for the subsequent 15 years in article by Kellay (2017). However, in real nature, it was not possible to detect clear traces of two-dimensional turbulence. Moreover, the spectra k^{-3}, found for scales from 500 to 3000 km in measurements by Lindborg (1999) and in detailed numerical simulations with high resolution and with energy budget analysis (Koshyk and Hamilton 2001), can be freely interpreted according to Charney's theory of geostrophic turbulence. Observed in aircraft measurements spectra $k^{-5/3}$ on a scale of 2–500 km—this is a new discovery of properties of turbulence in stably stratified media, which can be horizontal turbulence with $k^{-5/3}$. This is well-confirmed by numerical calculations (Koshyk and Hamilton 2001) and corresponding theoretical arguments (Tseskis 2008). Note that in degenerate two-dimensional turbulence (Lindborg 2006; Riley and Lindborg 2008) the spectral intervals of enstrophy and kinetic energy transfer can change places.

References

Armstrong JW, Cordes JM, Rickett BJ (1981) Density power spectrum in the local interstellar medium. Nature 291(5816):561–564

Batchelor GK (1969). Computation of the energy spectrum in homogeneous two-dimensional turbulence. Phys Fluids 12(12):II-233

Charney J (1971) Geostrophic turbulence. J Atmos Sci 29(6):1087–1095

Chkhetiani OG (2008) On the local structure of spiral turbulence // Dokl. RAS. T. 422, No. 5. C. 618–621

Corrsin S (1951) On the spectrum of isotropic temperature fluctuations in an isotropic turbulence. J Appl Phys 22(4):469–473

Fjørtoft R (1953) On the changes in the spectral distribution of kinetic energy for twodimensional, nondivergent flow. Tellus 5(3):225–230

Gage KS (1979) Evidence for $k^{-5/3}$ law inertial range in mesoscale two-dimensional turbulence. J Atmos Sci 36(7):1950–1954

Golitsyn GS, Fortus MI (2020) Random processes with stationary increments and composite spectra. Izvestiya Atmos Oceanic Phys 56:364–372

Kellay H, Goldburg WI (2002) Two-dimensional turbulence: a review of some recent experiments. Rep Prog Phys 65(5):845

Kellay H (2017) Hydrodynamics experiments with soap films and soap bubbles: a short review of recent experiments. Phys Fluids 29(11)

Koprov BM, Koprov VM, Ponomarev VM, Chkhetiani OG (2005) Experimental studies of turbulent helicity and its spectrum in the atmospheric boundary layer. In Doklady Phys 50:419–422 (Nauka/Interperiodica)

Koshyk JN, Hamilton K (2001) The horizontal kinetic energy spectrum and spectral budget simulated by a high-resolution troposphere–stratosphere–mesosphere GCM. J Atmos Sci 58(4):329–348

Kraichnan RH (1967) Inertial ranges in two-dimensional turbulence. Phys Fluids 10(7):1417

Lindborg E (1999) Can the atmospheric kinetic energy spectrum be explained by two-dimensional turbulence? J Fluid Mech 388:259–288

Lindborg E (2006) The energy cascade in a strongly stratified fluid. J Fluid Mech 550:207–242

Mirabel AP, Monin AS (1979) Two-dimensional turbulence // Adv. in Mechanics. T. 2. № 3. 47–95

Moffatt HK (1969) The degree of knottedness of tangled vortex lines. J Fluid Mech 35(1):117–129

Moffatt HK (1978) Field generation in electrically conducting fluids, vol 2. Cambridge University Press, Cambridge, London, New York, Melbourne, p 5-1

Novikov EA (1978) Spectral inequalities for two-dimensional turbulence. Akad Nauk SSSR Fizika Atmosfery i Okeana 14:668–671

Obukhov AM (1949) Structure of the temperature field in a turbulent flow. Izv Akad Nauk SSSR Ser Geogr Geofiz 13(1):58–69

Obukhov AM (1959) Turbulence in Lagrangian terms. Adv Geophys 7:115–117

Onsager L (1949) Statistical hydrodynamics. Il Nuovo Cimento (1943–1954) 6(Suppl 2):279–287

Oort AH (1964) On estimates of the atmospheric energy cycle. In: Renewable energy. Routledge, pp Vol1_214–Vol1_235
Riley JJ, Lindborg E (2008) Stratified turbulence: a possible interpretation of some geophysical turbulence measurements. J Atmos Sci 65(7):2416–2424
Starr VP (1968) Physics of negative viscosity phenomena. McGrow-Hill
Tseskis AL (2008) On the properties of decaying two-dimensional turbulence: characteristic features of the spectra and coherent structures. Izvestiya Atmos Oceanic Phys 44(5):646–659
Vazaeva NV, Chkhetiani OG, Kurgansky MV, Kallistratova MA (2021) Spirality and turbulence in the atmospheric boundary layer // Изв. RAS. FAO. T. 57, № 1. C. 34–52

Chapter 3
Earthquakes

3.1 Statistics of earthquakes

Earthquakes, EQs, are among the natural phenomena that bring, along with floods, hurricanes, etc., a lot of damage annually, although in the whole world less than hydrometeorological phenomena, simply because they are rarer. EQs refer to the type of natural phenomena for which the connection with the ANK34 scales is not directly obvious, although here, as elsewhere, large events are much rarer than small ones, the degree of their heterogeneity is not the same, depending on external conditions, for example, on the place of the event. Below we will show that this is a consequence of the first ANK34 scale. This situation can be traced using the methods of similarity theory and dimensions. The vast majority, more than 90%, of strong EQs occur in certain areas confined to the boundaries of lithospheric plates. Convection in the mantle is non-uniform in space (and time over millions of years), entraining the plates, generates stresses at their boundaries. These stresses are relieved during the EQs by the formation of ruptures in the fragile earth's crust. At the turn of the 1940s, the law of repeatability of EQs depending on their strength (Boðvarsson et al. 1999; Fix 1972), called the Gutenberg-Richter 1942 law (Law G-R) was empirically established:

$$\log N(\geq m) = a - b \cdot m, \tag{3.1}$$

where $N(\geq m)$ is the cumulative number of EQs for the studied period with magnitude $\geq m$, a is a constant depending on the choice of units of measurement and the time interval, and, from the place of observation, $b \approx 1$. The magnitude m is related to the logarithm of the EQs energy and is estimated from the amplitude of surface or body seismic waves (Kasahara 1981).

The analysis of seismograms makes it possible to estimate the fracture parameters: length L, rupture area S, development time τ, the average displacement between the two sides of the fault u, and το calculate seismic moment, a measure of the EQs

G. Golitsyn and C. Varotsos, *The Stochastic Nature of Environmental Phenomena and Processes*, https://doi.org/10.1007/978-3-031-77015-9_3

energy in the form (see Kasahara 1981; Kagan 1994):

$$M = \mu \cdot S \cdot u, \tag{3.2}$$

where μ is the shear modulus of crustal rocks involved in the EQ. Between magnitude m and seismic moment, there is an approximate relationship justified by some theoretical considerations and statistics of observations (Golitsyn 1981).

$$m = 2/3 \cdot \log M - 6, \tag{3.3}$$

where the value of M is measured in the SI system, i.e., N•m. The value M is the moment of forces, acting in the system, i.e., it is a tensor (see Kasahara 1981). According to the dimension of the moment, this equals energy. In terms of the moment, the G-R law (3.1) is written with the exponent $b' \approx 2/3$. From (3.1) and hence it follows that

$$N(\geq M) \sim M^{-2/3} \tag{3.1'}$$

Kanamori and Anderson (1975) based on a simple cracking model and gap dynamics, as well as the postulation of scale similarity (scaling), accepted that the stress drop $\Delta\sigma$ (or strain drop) in the cortex (during the EQs) is approximately constant, obtained that the discontinuity area is S $\propto M^{2/3}$. This is true for EQs with $m \geq 6$ (Purcaru and Berkhemer 1982; Geller 1976).

This implies another interpretation of the magnitude (Golitsyn 2001)

$$m = \log(S/S_o), \tag{3.4}$$

where $S_o \approx 100\text{m}^2 = 0.01\,\text{ha} = 1\,\text{ar}$, i.e. one weave in common parlance. Thus, magnitude is a quantity related to the surface, and the seismic moment is bulk tensor quantity (Kasahara 1981; Kanamori and Anderson 1975) with modulus M_o. It is clear from this, that indicators (3.1) b and $b*$ further differ by a factor of one and a half. With a similar circumstance, we will meet more than once (see Sect. 11.2).

Special verification of the variability of the value of $\Delta\sigma$ on a material of about 200 EQs $m \geq 5$ in (Golitsyn 2001) showed that the stress drop is within 30–70 bar (1 bar = 1atm = 0.1 MPa = 10^5N/m^2) with a median value of 4.4 MPa at moment changes by 5–6 orders of magnitude.

Figure 3.1 presents the results of Golitsyn (2001) of checking relation (3.3) using the data (Purcaru and Berkhemer 1982). The determination coefficient r^2 between magnitudes calculated from seismograms and estimated by (3.3) is close to 0.8. Global data analysis catalogs showed that there is a value $M_{cr} \approx 1.6 \cdot 10^{20}$ N•m near which the magnitude m in (3.1) changes from b $\approx$ 1 to $b^* \approx 1.5$ (see Okal and Romanowitz 1994). This circumstance is associated with a rupture of the bottom of the cortex.

This assertion is supported by the fact that b $\approx$ 1.5 is also observed for the EQs near the mid-ocean ridges, where young crust, thin, ~ 1 km near these ridges with

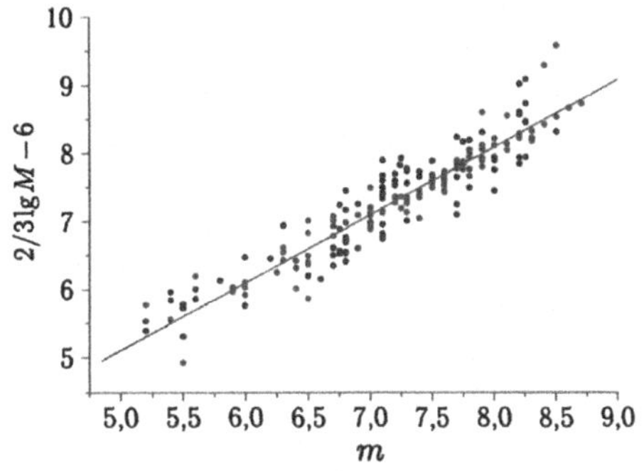

Fig. 3.1. Relationship between seismic moment (volume) and magnitude (fracture surface) in earthquakes (Golitsyn 2001).

thickness increasing as distance from them, proportional to the square root of their age, which characteristic of parabolic processes. The explanation for this change parameter *b*, as well as other regularities of the EQs process from the point of view of mechanics, was given by Grigoryan (1988), Kanamori and Anderson (1975) and later by Golitsyn (2001).

3.2 Similarity Theory for Earthquakes

An explanation of the EQs from the point of view of the theory of similarity was given in Golitsyn (1996) with the definition of numerical coefficients by comparison with seismogram processing data, which was not in Grigoryan (1988). These results will be reproduced here and then commented on from the point of view of probability theory by A. N. Kolmogorov 1934.

Material constants are chosen for the defining parameters of the EQ process bark rocks (Kasahara 1981): shear modulus $\mu = (3-7) \cdot 10^{10}\mathrm{N/m^2}$, rock density $\rho \approx 3 \cdot 10^3\mathrm{kg/m^{3\cdot}}$ and the value of the discharged stress $\Delta\sigma \approx 4\mathrm{MPa} = 4 \cdot 10^6\mathrm{N/m^2}$. Given that the properties of the crustal rocks and the value of $\Delta\sigma$ vary within relatively narrow limits, we will assume these quantities to be practically fixed. Value $\Delta\sigma/\mu$ order 10^{-4}, and this dimensionless similarity parameter will be considered little changing and further do not consider it. The thickness of the *h* plate also plays a role.

The root cause and measure of the power of all geodynamic processes is geothermal heat flux equal to $F \approx 4.5 \cdot 10^{13}$ W at an average density of 86 mW/m^2 (Schubert et al. 2001). The data on the EQs (Okal and Romanowitz 1994) show that the plate thickness *h* also plays a role. From selected four dimensional parameters—seismic moment, dip bark stress at rupture $\Delta\sigma$, geothermal flow *F*, and plate thickness *h*—it is possible to make scales of length and time (Grigoryan 1988).

$$L = (M/\Delta\sigma)^{1/3}, \tag{3.5}$$

$$T = M/F, \tag{3.6}$$

characterizing individual EQ. The magnitudes of the measured EQs vary from the strong in Chile in July 1960 with $m = 9.5$ occurring every many centuries or less often, up to $m = -1$ on the most dense and sensitive in the world in 1999 Icelandic measuring network Bodvarson et al. 1999). The seismic moment (3.3) then changes to 16 orders.

Of the four quantities selected above with three independent dimensions (mass, length, and time) one dimensionless criterion can be formed

$$\Pi = L/h = M^{1/3}/\left(\Delta\sigma^{1/3} \cdot h\right) \tag{3.7}$$

which we will consider as the main similarity criterion for EQs. At $M = M_{cr} = 1.6 \cdot 10^{20}\text{N} \cdot \text{m}$ (see above), $\Delta\sigma = 4 \cdot 10^6\text{N/m}^2$, $F \approx 4.5 \cdot 10^{13}\text{W}\left(\text{kgm}^2/\text{s}^3\right)$ and $h \approx 30\text{km} = 3 \cdot 10^4\text{m}$ (globe-average crustal thickness), we obtain $\Pi = 1.07$. This is the value of $\Pi \approx 1$ that separates the regions of the EQs spectrum, i.e., the G-R law, for the vast majority of EQs with $b \approx 1$ at $m \leq 7.5$ from the region of the spectrum with a steeper drop, where $b^* \approx 1.5$.

For dimensional reasons (since the cumulative number of events has a dimension of inverse time, i.e., frequency) can be written:

$$N(\geq M) = (F/M) \cdot f(\Pi), \tag{3.8}$$

where the left side is the number of events with moment $\geq M$ for a certain interval of time and f—is a function of the similarity criterion, which should be determined from observations. The works above present data on the global catalogs of EQ depending on the moment M for 828 events in 1977–1993, including those near mid-ocean ridges, Geller (1976). There, due to the small thickness of the crust for the values M within $10^{17} - 10^{20}$ N•m value $\Pi \geq 1$. At the same time $5.5 \leq m \leq 7.5$. Value $b = 1.05$ in the G-R law according to these data, i.e., $b \approx 1$, and then $N(M) \sim M^{-1}$, and the function $\Pi \rightarrow$ const $= 0.35$ (see Golitsyn 2001 and G12).

For values of the similarity parameter $\Pi < 1$, the function $f(\Pi)$ can be expanded into a Taylor series, and it must begin with a linear term (no excitation-no EQs):$f(P) = \text{c}_1\Pi = \text{c}_1 M^{1/3}\Delta\sigma^{-1/3}h^{-1}$.

The number of EQs with $m < 7.5$, which in the continental bark hundreds and thousands per year, has $c_1 = 0.35$. The definition of the desired coefficient gives $c_1 \approx 0.34 \pm 0.02$. The value of the constant coefficient in both cases at $\Pi \approx 1$ converges to 0.35 both for $b^* \approx 1$ and $b \approx 2/3$. Note that for $5 < m < 7.5$ in (Smirnov and Ispolinova 1995), has obtained $b = 0.65 \pm 0.02$. For the rarely used differential form of the G-R law for $\Pi < 1$ we will have

$$N(M) = [\text{d}[N(\geq M)]/\text{d}M] \propto M^{-5/3} \tag{3.8'}$$

Here the Kolmogorov Obukhov turbulence spectrum involuntarily recalls: The nature of the thirds here is purely geometric, and the game is between the magnitude of the seismic moment M, which is a measure of the energy of the EQs, and the volumetric elastic energy density of deformation, allowed by the stress jump $\Delta\sigma$, which gives the length scale (3.6), see Sect. 11.2 for details.

In G12, data are given for the fracture length L, its area S, and the time of development of the EQs along with the magnitudes and estimates $\Delta\sigma$, of the stress drop at EQ. This makes it possible to estimate numerical coefficients for length scales L, area $S = L^{2\cdot}$ and time of EQs development.

$$\tau = L \cdot (\rho/\mu)^{1/2}, \tag{3.9}$$

where $(\mu/\rho)^{1/2} = c_1$—is the scale of the propagation velocity of volumetric and surface seismic waves. As a result, we have for EQs with number i:

$$L_i = 2.3 \cdot L = 2.3 \cdot (M_i/\Delta\sigma)^{1/3};\ S_i = 0.34 \cdot L^2 = 0.34 \cdot (M_i/\Delta\sigma)^{2/3};$$
$$\tau = 1.4 \cdot L/c_1 = 1.4 \cdot (\mathrm{Mi}/\Delta\sigma)^{1/3} \cdot (\rho/\mu)^{1/2},$$

and the displacement size of adjacent blocks along the discontinuity is determined from (3.2) as

$$\mathrm{u} = 0.54 \cdot (M/S\mu) = 0.54 \cdot M^{1/3} \cdot \Delta\sigma^{1/3} \cdot \mu^{-1}, \tag{3.10}$$

whence even for EQs 1960 in Chile this shift is estimated to be about 20m. The presentation of this paragraph follows, mainly, the works of the author (Golitsyn 2001) with some recalculation of the numerical coefficients obtained there, as described in G12. There is also a description of the nature of the so-called induced EQs (see detailed article Volant and Grasso 1994). After the creation of large reservoirs, and production in large volumes of oil and gas, a series of small EZs are observed with magnitudes $\mathrm{m} \le 4$, i.e., with seismic moments $\mathrm{M} \le 10^{15}$N.m. For the above reasons, EQs in the cortex the previously established isostatic equilibrium is disturbed, and stresses arise due to an increase in pressure or the appearance of voids that are removed by ruptured rocks, i.e., by earthquakes.

3.3 Induced Earthquakes

The statistics of such events are satisfactorily described by the Gutenberg law—Richter, and the numerical coefficients in this statistics are close to those for natural EQ (Nikolaev and Galkin 1994). The driving force, i.e., the source of energy for forced EQs, is the change with time of the vertical gradient of the internal pressure in the system—the values of dp/dz. This gradient at equilibrium is:

$$\mathrm{dp/dz} = -\rho \cdot g, \tag{3.11}$$

where g is the gravitational acceleration. The change in this value with time, which has dimension $MT^{-2}L^{-2}$, leads to violation of isostasy, i.e., to the appearance of new and changing old stresses. This value replaces in the previous formulas geothermal flow F. Several such induced EQs are described in the collection (Nikolaev and Galkin 1994).

In Nikolaev and Galkin (1994) and G12, data on the statistics of not very large tsunami waves are also given at the posts of the Soviet Geophysical Service in the Far East. In the height range of waves from centimeters to a meter, the cumulative frequency of events turns out to be inversely proportional to the height of the waves, i.e., their energy. These waves are excited submarine EQs, most of which occur near mid-ocean ridges where the earth's crust is thin. For such STs, as we noted above, in the G-R law the value $b \approx 1$, which follows from (3.8).

What are the connections of the G-R law with the results of ANK34, interpreted as a structure function with zero initial conditions and frequency spectrum? Let's remember first the Kolmogorov scale $< u^2 >= \varepsilon t$, formula (1.31) for energy growth per unit masses for continuous processes. For discrete processes numbered index i, we should expect $< u_i^2 >= \varepsilon \cdot t_i$, whence for cumulative distributions with frequency dimension $1/t_i \sim \varepsilon / < u_i^2 >$ or in the case of EQ moment M, energy measure

$$N(\geq M)] \approx c \cdot \varepsilon_g / M\,, \tag{3.12}$$

where ε_g is the elastic energy generation rate per unit mass associated with the geothermal flow-numerical coefficient. On a logarithmic scale, this is Gutenberg-Richter's law. For large EQ, as noted above, the coefficient $c \approx 0.35$ and is close to the constant value for large values of the similarity parameter Π. For its small $c = c(\Pi)$ is also close to 0.35 and is a linear function of Π as the first expansion term in the Taylor series.

Thus, the ANK34 theory immediately explains the G-R law for strong EQ, i.e., similarity parameter $\Pi \gg 1$, and the use of methods of the theory of similarity and dimension at $\Pi \ll 1$, when the details of the internal processes of earthquakes are considered, explains more frequent events for smaller EQs. In both cases, predicted the correct exponents in the Gutenberg-Richter power law (3.1) for magnitudes are: $b^* \approx 1.5$ and $b \approx 1$, respectively, for strong and weaker events, i.e., for both asymptotic of the similarity parameter $\Pi \gg 1$ and $\Pi \ll 1$. A sufficiently detailed description of EQs in the process of evolution of their knowledge by the author can be found in (Golitsyn 2001), written when the author has not yet seen a close connection with the laws of the ANK34, he simply forgot about them, although many years before that he was the scientific editor of the book MV 75, where this law of the ANK34 discussed in connection with turbulence. The title of his work is characteristic: "Place Gutenberg-Richter law among other statistical laws of nature". Thus, the G-R law is the cumulative distribution of the number of EQs events (see Sect. 1.3).

Another unexpected manifestation of the connection with ANK34 gives a spectrum of seismic noise, a wave manifestation of processes associated with random displacements of rocks on the earth's crust. This spectrum from Fix (1972) is shown in Fig. 3.2 in the interval of periods recorded fluctuations from 0.1 to 2560 s.

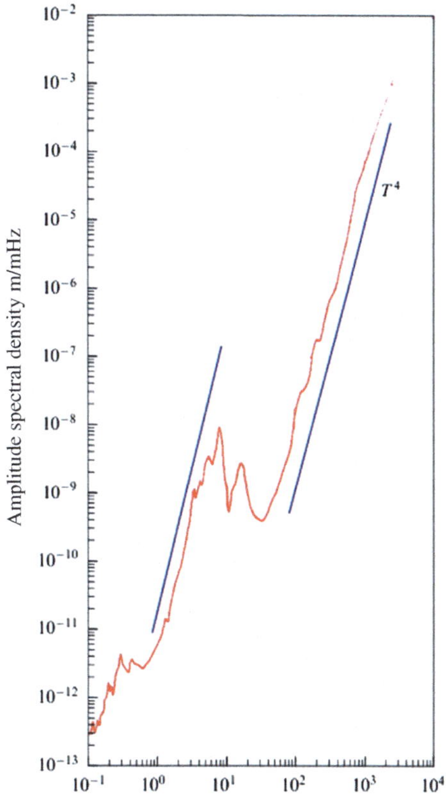

Fig. 3.2. Spectrum of microseisms in the interval of periods from 0.1 to 2400, Fix (1972) seconds.

It has two distinct power sections, both are proportional to the oscillation period T as T^4, i.e., the frequency of the amplitude spectral density changes as f^{-4}. The first plot for periods 1–10 s, and the second from 100 to 2400 s. The nature of the first site was described in 1950 by Longuet-Higgins (1950), who noted that these are periods of powerful wind waves when they break on the cliffs. However, in the middle XX century, spectral methods were rarely used in the study of natural phenomena and there was no idea about the frequency spectrum of sea waves (see Chap. 6).

The spectra $\sim \omega^4 = [f(2\pi)]^{-4}$ were theoretically obtained only in the 1960s. Figure 3.3 is a perfect illustration of signal flow from the collapse of sea waves through the earth's crust, where microseisms are caused. This is how the second invariant of A. N. Kolmogorov manifests itself in nature (1.33). This shows that wave excitation (and microseism) has a Markovian character, i.e., its spectrum is white noise. The same nature and excitation of seismic minute range, up to half an hour, as shown in Fig. 3.3. What exactly excites micro-oscillations in this range is still unclear. Probably here there are many uncorrelated causes, such as they have the Markovian nature which leads to the spectrum $f^{-4} \sim T^4$, according to (1.33) as was assumed in ANK.

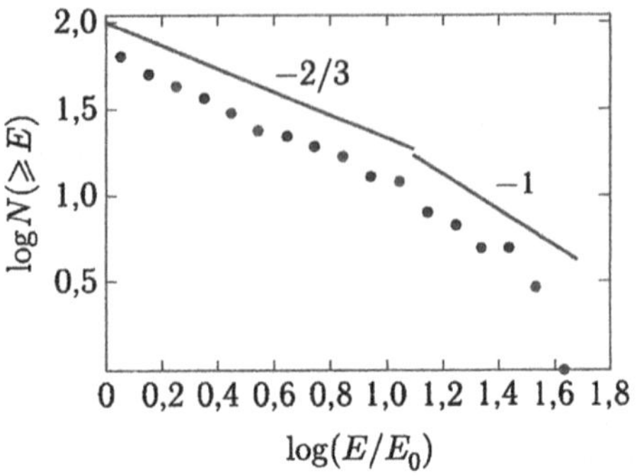

Fig. 3.3. Dependence of the cumulative number of stellar flares on their energy from G12.

3.4 Acoustic noise of stressed crystals

The work (Miguel et al. 2001) presents the measurement data of acoustic noise at viscoplastic deformation (creep) of crystalline materials at constant loading. It is known that the creep is explained by the movement of a large number of dislocations in the crystal lattice interacting with each other friend. In the experiments described, acoustic pulses are produced by an ice crystal that is compressed along one direction. Dislocations in the lattice under the influence of compression form a slowly changing configuration in which in a time-intermittent manner, rapid structural changes occur in a three-dimensional field of dislocations. These rapid changes are accompanied by radiation sound pulses of different intensity (energy E).

The measurements showed that the probability density distribution function of the number

pulses depending on their energy have a power-law form:

$$N(E) \propto E^{-n}, \quad n = 1.6 \tag{3.13}$$

A numerical dynamic model was built for a crystal, which describes the rearrangement of its dislocation structure during creep. This model is for the same $N(E)$ also gave a power law, but the exponent turned out to be equal to n = 1.8. The authors considered a satisfactory agreement between the results of laboratory and numerical experiments, although a decrease in the number of pulses low energy compared to measurements could be caused by insufficient spatial resolution of the numerical model. Due to the proximity of this research to the problems of the destruction of materials, in the end, to the mechanics of earthquakes, we will try to look at the results of this work from the point of view of theories of similarity and dimension.

As external parameters, we have the stress σ with the dimension $\mathrm{ML^{-1}T^{-2}}$, energy per unit volume (force per unit area), characteristic sample size h, and dynamic viscosity included in the equation of motion for dislocations. integrable numerically, $[n] = \mathrm{ML^{-1}T^{-1}}$, where M and T are the dimensions of the mass and time, $\eta = \rho v$, where v is the kinematic viscosity multiplied by the density substances ρ. The energy of the sound pulse E and their number in the kHz range in dependence on energy $N(E)$, $[N(E)] = T^{-1} \cdot E^{-1}$. In this equality on the right, the quantity E has the meaning of the dimension of energy. In the TEL system, the stress dimension $[\sigma] = \mathrm{E} \cdot \mathrm{L}^{-3}$ as pressure.

The given 4 quantities form a dimensionless similarity criterion, which we choose in the form:

$$\Pi_1 = (\sigma \cdot \rho)^{1/2} \cdot h \cdot \eta^{-1} = u_d \cdot h \cdot v^{-1}, u_d = (\sigma/\rho)^{1/2}, \tag{3.14}$$

Value $(\sigma/\rho)^{1/2}$ has the dimension of speed, and it can be considered which determines the speed of dislocation movement. Then our similarity parameter will make sense of the Reynolds number for the motion of dislocations. Π_1 value is fixed for a specific sample, and therefore the function $C(\Pi_1)$ is simply a dimensionless multiplier.

The measured value E and the set values σ and v enable uniquely organize the value with the dimension of the distribution function:

$$N(E) = C(\Pi_1) \cdot \sigma^{2/3} \cdot v \cdot E^{-5/3} \tag{3.15}$$

The exponent − 5/3 is 1/15 higher than according to laboratory data, and experiments, and 2/15 lower than in their numerical score. This agreement is somewhat better than at the authors themselves. Note that the exponent n = 5/3 is a consequence of ANK34, see Chap. 1.

The cumulative distribution of the number of acoustic pulses depending on their energy has the frequency dimension:

$$N(\geq E) = \int_E^\infty N(E) \cdot dE \propto E \cdot N(E) = (\sigma/E)^{2/3} v \tag{3.16}$$

in form and meaning it turns out to be similar to the Gutenberg-Richter law (3.8′) in seismic moment M for earthquakes with magnitude m < 7.5:

$$N(\geq M) = \frac{0.35F \cdot M^{-2/3}}{(\Delta\sigma)^{1/3} \cdot h}, \tag{3.8'}$$

where N(≥ M) is the number of earthquakes, EQs, 0.35 obtained from their global catalog, F=4.5•10^{13} W—total geothermal flow, $\Delta\sigma \sim 4\,\mathrm{MPa/m^2}$, discharged at EQ stress, a quantity that weakly depends on the EQs force, h is the thickness of the fragile earth crust, taken on average equal to 30 km, $M = \mu \cdot < u > \cdot S$ is the seismic moment, where μ is Young's modulus, $<u>$ is the average displacement of neighboring blocks along the discontinuity at the EQ, S is the break area. Note that in (3.16) the quantity $(\sigma/E)^{2/3}$ has the dimension reciprocal area, and v is the kinematic viscosity, a given value. The same formula (3.8′) can also be brought into the form, where the effective viscosity will be $v = M^{-1/3}(\Delta\sigma)^{-2/3}F$ and $S = (M/\Delta\sigma)^{1/3} \cdot h \equiv l \cdot h$ area.

The analogy is not entirely complete, as for the EQs, the effective viscosity depends on both the geothermal flow and seismic moment. Here the quantity l has the dimension of length and is the rupture length scale at EQs. Thus, the quantitative and physical analogy of acoustic emission of pulses of a loaded crystal and the process earthquakes are quite close to each other.

3.5 Starquakes

The first author has long heard about solar seismology, the manifestation of oscillations, registered in the photosphere of the Sun to obtain information about the internal structure of our luminary, about starquakes. At the end of 1996, an article appeared on the statistics of gamma-ray bursts at an object with coordinates SGR 1806-20. This statistics has a cumulative distribution for about a hundred events power characters with exponent $n = -0.66$. Sequential waiting times events, as well as EQs, have a log-normal distribution, a strong correlation between the waiting times of successive events, i.e., the trend to their groupings in time, and the lack of correlation between the intensity events and their waiting times. This work attracted the attention of V. I. Keilis-Borok, who used his experience in forecasting EQs, and successfully applied this methodology for forecasting gamma-ray flares on SGR 1806-20 (private communication by V.I.K.-B., April 2001).

I spent my summer 1997 vacation at the Institute for Astrophysics of the Max Planck Society near Munich. Several of my lectures on turbulence and convection justified my monthly stay. I entered the essence of the subject, the result of which was my article (Golitsyn 1998). The data themselves appeared by monitoring compliance with the International Treaty for banning nuclear tests on land, underwater, and in outer space, which concluded in 1963.

The objects under consideration are neutron stars with a radius of 10 km, detailed references in the paper (Golitsyn 1998). Their internal magnetic field reaches 10^{15} Gauss and its changes feed all processes on similar objects, magnetars. These are supernova remnants a special class of stars called plerions. Such objects are very rare in our Galaxy, and/or their lifetimes are short at this stage. Currently, there are 9 such objects (Wikipedia 2011) out of the total number of stars $4{\cdot}10^{11}$, and supernovae explode 2–3 times in 100 years (see Chap. 4).

The most powerful explosion at the object in question was recorded on 12/27/2004. The flare lasted 100–200 ms and was registered on the terrestrial surface and satellites. Its power is estimated at $1.5{\bullet}10^{39}\ J$, which is 3–4 orders of magnitude smaller than supernova explosions (see Chap. 4). In Golitsyn (1998), data on material parameters of magnetars, propagation velocities of ruptures in their crust. If such explosions use take the same numerical data of discontinuities as for the EQs, then the discontinuity length will be of the order 100 km, i.e., one hemisphere of the object will be completely separated from the other and shifted by tens of meters, G12, Sect. 6.9 in it.

We see because of this chapter that both EQs and ZVT processes appear with the same general rules.

References

Boðvarsson R, Rögnvaldsson ST, Slunga R, Kjartansson E (1999) The SIL data acquisition system—at present and beyond year 2000. Phys Earth Planet Inter 113(1–4):89–101

Fix JE (1972) Ambient Earth motion in the period range from 0.1 to 2560 seconds. Bull Seismol Soc Am 62(6):1953–1760

Geller RJ (1976) Scaling relations for earthquake source parameters and magnitudes. Bull Seismol Soc Am 66(5):1501–1523

Golitsyn GS (1996) Earthquakes from the point of view of similarity theory. Dokl RAS 346(4):536–579

Golitsyn GS (1998) Starquakes on SGR 1806-20 and other neutron stars. Lett Astron Mag 34(11):827–832

Golitsyn GS (2001) Place of the Gutenberg-Richter Law among other statistical laws of nature. Selected papers from volume 32 of Vychislitel'naya Seysmologiya 7:119–137

Grigoryan SS (1988) On the mechanism of earthquake occurrence and the content of empirical patterns of seismology. Dokl Acad Sci USSR 299(5):1094–1101

Gutenberg B, Richter CF (1942) Earthquake magnitude, intensity, energy, and acceleration. Bull Seismol Soc Am 32(3):163–191

Kagan YY (1994) Observational evidence for earthquakes as a non-linear dynamic process. Phys D 77:160–192

Kanamori H, Anderson DL (1975) Theoretical basis for some empirical relations in seismology. Bull Seismol Soc Am 65(5):1073–1095

Kasahara K (1981) Mechanics of earthquakes. Cambridge University Press (Russian translation: Kasahara K. Mechanics of earthquakes.—M.: Mir, 1985)

Longuet-Higgins MS (1950) A theory of the origin of microseisms. Philos Trans R Soc 243(857):1–35

Miguel M-C, Vespignani A, Zapperi S, Weiss J, Grasso JR (2001) Intermittent dislocation flow in viscoplastic deformation. Nature 405:667–671

Nikolaev AV, Galkin IN (eds) (1994) Induced seismicity.—M.: Nauka

Okal EA, Romanowitz BA (1994) On the variation of b-values with earthquake size. Phys Earth Planet Interior 87:1094–1101

Purcaru G, Berkhemer G (1982) Quantitative relations for large earthquakes. Tectonophysics 84:57–128

Schubert G, Turcotte DL, Olson P (2001) Mantle convection in the Earth and planets. Cambridge University Press, 940 pp

Smirnov VB, Ispolinova SI (1995) Size-frequency distribution for earthquakes. Dokl RAS 342(6):809–811

Volant P, Grasso J-R (1994) The finite extension of fractal geometry and power law distribution of shallow earthquakes: a geomechanical effect. J Geophys Res 99:21879–21889

Chapter 4
Cosmic Rays' Spectra

4.1 Cosmic Ray Spectrum

It was this topic, above all others, that attracted my attention to the statistics of natural phenomena in the first half of 1990. By that time, the energies of the rays were already measured in more than ten orders of magnitude and their spectrum was the number of particles measured per unit area per unit time, and even coming from a unit solid angle of space. Such a spectrum has a power-law form and is shown in Fig. 4.1. These are mainly galactic cosmic rays. Cosmic rays with energy less than 1 GeV = 10^9eV are created mainly in the solar corona. High energy rays have galactic and extragalactic origin. The main source of energy for them is supernova explosions. The most complete description of the processes associated with cosmic rays can be found in the book "Astrophysics of Cosmic Rays," published in 1990 in Russian and English edited by Ginzburg (1990), which has remained virtually unchanged since then.

The ideas about the mechanisms of acceleration of CR particles are based on E. Fermi in 1949 about acceleration on magnetic field inhomogeneities, now understood as fronts of collisionless shock waves, shock waves, arising in interstellar gas during supernova explosions. This general idea received quantitative formalization in the recent works of Malkov and Diamond (2001). In our Galaxy, the Milky Way, with a population of $4 \cdot 10^{11}$ stars and ages of order 10^{10} years, such explosions occur in random places two to three times a century. Therefore, the shock waves are also arbitrary in space, time, and in the directions of propagation and acceleration of CR particles, i.e., it is a random process with uncorrelated influences in the exact sense, i.e. Markovian.

The disk of our Galaxy has a thickness of the order of 200 pc, parsec, and a radius of 15 kpc, 1 pc = 3.26 light years $\approx 3 \cdot 10^{16}$ m, so its volume is about 10^{61} m^3. With such a geometry of the Galaxy and such a rate of formation of supernovae within it, shock waves of the order of a thousand explosions must simultaneously exist if shock waves propagate at the speed of light. This is clearly an estimate from below,

G. Golitsyn and C. Varotsos, *The Stochastic Nature of Environmental Phenomena and Processes*, https://doi.org/10.1007/978-3-031-77015-9_4

Fig. 4.1. Integral spectrum of cosmic rays (Wikipedia)

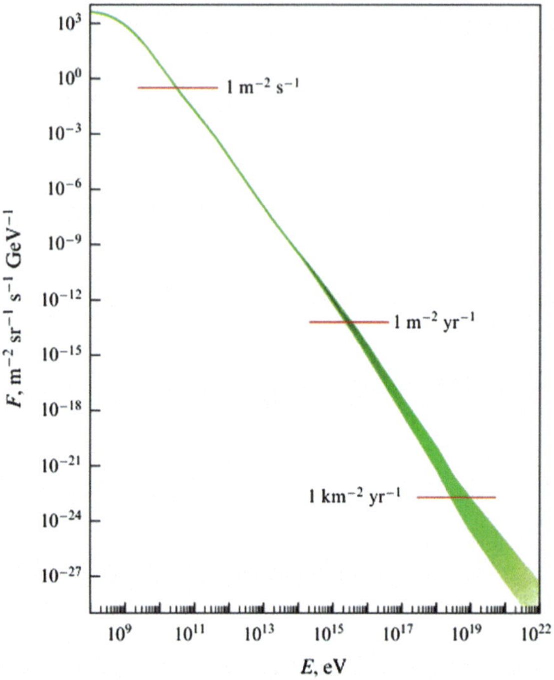

since their speed is an order of magnitude lower, which significantly increases the number of shock waves and objects for CR acceleration. From balance considerations (Ginzburg 1990), the volumetric energy density of CR particles is known, depending on their number and their energies:

$$w_o \approx 0.5\mathrm{e\,V/cm^3} \approx 10^{-13}\,\mathrm{J/m^3} \tag{4.1}$$

Thus, the total CR energy is of the order of W $\approx$ 10^{48} J. The explosion energy of an individual supernova is of the order of $10^{42} - 10^{43}$ J, i.e., on average about $3 \cdot 10^{42}$ J. With two to three explosions per century, this gives an energy production capacity of $G = 2 \cdot 10^{33}$W. This power is dissipated by photons, used to generate turbulent and other movements of interstellar gas, to maintain the galactic magnetic field, and to cosmic rays. The particles do not immediately leave the Galaxy, since they are twisted by a magnetic field with a strength of $H = 5 \cdot 10^{-6}$ Gauss. Its energy density $H^2/8\pi \approx 10^{-13}\mathrm{J/m^3}$, the same as for CR.

The description of CR uses the integral energy spectrum (in Sect. 1.1 called by then cumulative):

$$I(\geq E) = \int_E^\infty I(E)dE, \tag{4.2}$$

where differential spectrum $I(E)$ is the number of particles with energy in the interval $E \pm dE$, measured per unit area per unit time, i.e., per 1 m^2 per 1 second. Obviously, these units of measurement are not comparable to galactic scales.

Both spectra are associated with the energy distribution function of the number of particles, i.e., their energy concentrations:

$$n(\geq E) = \int_E^\infty n(E)dE = \frac{4\pi}{c}\int_E^\infty I(E)dE = \frac{4\pi}{c}I(\geq E), \tag{4.3}$$

where the particles are considered ultra-relativistic. The empirical spectrum of CR is approximated by two power-law sections (where ГэВ stands for GeV)

$$I(\geq E) = \begin{cases} 1\cdot[E(\mathrm{GeV})] & \text{for } 10 \leq E \leq 3\cdot 10^6\,\mathrm{GeV} \\ 3\cdot 10^{-10}\left[10^{-6}E(\mathrm{GeV})\right]^{-2.1} & \text{for } 3\cdot 10^6 < E < 3\cdot 10^9\,\mathrm{GeV} \end{cases} \tag{4.4}$$

(where ГэВ stands for GeV)

The presence of a break in the spectrum near $3 \cdot 10^{15}$eV, called the knee, is explained by the fact that the magnetic field no longer begins to hold all CR particles, since their Larmor radius becomes comparable to the thickness of the galactic disk. Particles with an energy of E > $3 \cdot 10^9$ GeV are practically not observed; they have so far been recorded in isolated cases. The relatively small rise in the spectrum at the very end is called the ankle, not yet explained. In the years, 2008–2014, European and Russian scientists carried out extensive and carefully prepared program "PAMELA" (Karelin et al. 2014). CR spectra were measured in a range up to the knee on our satellite *Resource 5*. They gave $I(E) \sim E^{-n}$, where $n = 2.67 \pm 0.02$, i.e., for the integral spectrum $n = 1.67 \pm 0.02$ and probably here the spread is smaller, because, according to (4.2) and our Sect. 1.1, the cumulative spectra are always smoother than differential ones.

However, long before these studies, the energy spectrum of cosmic rays was determined theoretically (Golitsyn 1997, 2005) and is very close to its empirical generalizations (4.4). First, this was done for the first section of the spectrum up to the knee, and then for the second section up to the ankle. It is obvious that the CR energy density is related to their concentration, i.e., it has a dimension inversely proportional to the cube of the distance between particles. Let us determine the CR spectrum using the method of dimensional theory.

For the system of measurement units, we choose time *T*, area *S*, and energy *E* (instead of mass multiplied by the square of velocity (Golitsyn 2008). To determine the integral spectrum with dimensions per unit area and unit time:

$$I(\geq E) = \mathrm{S}^{-1}\mathrm{T}^{-1}, [w_0] = \mathrm{ES}^{-3/2}, [G] = \mathrm{ET}^{-1}, [E] = \mathrm{E}.$$

From here it is immediately clear that the spectrum must be proportional to the generation rate G, then $w_o^{2/3}$, and the dependence is $E^{-5/3}$ provides the required dimension of the integral spectrum. Less formally: the value *G/E* gives the dimension of frequency, i.e., reciprocal time,

during which the process of CR acceleration on the shock wave blasts, and the unit area gives $(w_0/E)^{2/3}$, since this value is proportional to the number of particles per unit volume, i.e., their concentrations with dimension L^{-3}. Both quantities are random and independent, and their product gives the integral CR spectrum. Note that the dimension area is 2/3 of the volume dimension. Differential spectrum from here proportional $E^{-8/3}$.. The coincidence with the measurement data (Karelin et al. 2014) is surprising; one can it couldn't be better to say if you remember that $0.67 \approx 2/3$. Eventually:

$$I(\geq E) = a_1 \frac{G}{E}\left(\frac{w_0}{E}\right)^{2/3} \propto E^{-5/3}, \tag{4.5}$$

$$I(E) = \frac{2}{3} a_1 G\left(\frac{w_0}{E}\right)^{2/3} E^{-2} \propto E^{-8/3}, \tag{4.6}$$

where the numerical coefficient α_1 turns out to be of the order of 10^{-37}. The so small value results from the discrepancy between our standard units of measurement and galactic sizes and scales. However, if we take per unit area thickness of the galactic disk 200pc $\approx 6 \cdot 10^{18}$m, then such a unit of area will already be $\sim 4 \cdot 10^{37}\text{m}^2$. How can one not to recall Einstein's statement in 1911, quoted in Chap. 8 of BPW, that in the correct formulas obtained by dimensional considerations, the numerical coefficients should not be very small, nor very large, i.e. O(1)!

The spectrum after the knee is determined by CR particles moving into ever greater degrees from the Galaxy since the magnetic field no longer holds them. Next, the system of inequalities leads along this path:

$$n(\geq E) = \int_E^\infty n(E)dE = \int_E^\infty \frac{E}{E} n(E)dE < \frac{1}{E}\int_E^\infty En(E)dE = \frac{w(\geq E)}{E}. \tag{4.7}$$

Cumulative distribution function $n(\geq E)$ we know from relation (4.3), connecting it with the integral spectrum up to the knee, to estimate the density of the number of particles $n(\geq E)$ and their energies near the knee, we use inequality (4.7) between its first and last members. Knowing the spectrum according to (4.4) and (4.7) and taking (4.3) into account gives (Golitsyn 2005)

$$I(\geq E) = c_2 c^{-2/3} G^{5/3} w_0^{4/9} E^{-19/9} \propto E^{-19/9}; \tag{4.8}$$

the exponent in this interval turns out to be equal to[1] $19/9 = 2 + 1/9$, , and from observations (4.4) it is estimated as $2.1 = 2 + 1/10$!

What is the connection with Kolmogorov's laws of 1934? The equation of motion of particles is $dp/dt = f$, the derivative of the impulse concerning time is equal to the acting force, i.e., acceleration. These forces act on random shock waves, i.e., they exactly correspond to the conditions of AHK34. The integral spectrum $I(\geq E)$ is inversely proportional to time, as for all cumulative distributions (see Sect. 1.2). However, due to experimental methods for determining this spectrum, it also depends on the unit of area on which measurements are taken. Such an area is itself determined from balance considerations, giving the volumetric energy density w0 divided by the energy E, which estimates the volumetric concentration of particles N with a dimension of L^{-3}, where L—length. Therefore, the value $(w_0/\mathrm{E})^{2/3}$ estimates the area. Thus, the way the spectrum is determined separates it from several other natural processes, such as earthquakes. It was this circumstance that led the authors of the book [1] to write that some new physics may be needed to determine the CR spectrum. We see that novelty arises only from the methodological non-standardization of the concept of the CR spectrum, which requires the use of energy balance considerations to estimate its spatial density.

Note that when using similarity and dimension methods, it is convenient, for example, here, to use energy instead of mass as a unit of measurement [6].

4.2 Nowcasting Extreme Cosmic Ray Events

The development of a model for predicting extreme CR events, which can potentially cause significant issues, is crucial. The main goals of this model include:

(a) the ability to forecast extreme CR events promptly to prevent or mitigate the impact on telecommunication systems.
(b) estimating the duration and intensity of these CR events based on atmospheric conditions.

This model will aid in enhancing strategies to prevent, prepare for, and manage the consequences of extreme CR events. Varotsos et al. (2023) utilized daily measurements of cosmic rays intensity (CRI) from the neutron monitor station in Athens, Greece, from January 1, 2010, to January 14, 2022 (Fig. 4.2a). The station, known as Athens Neutron Monitoring Station (A.Ne.Mo.S), is located at 37.97°N, 23.78°E, with an altitude of 260 meters above sea level and an effective vertical cutoff rigidity of 8.53 GV (Mavromichalaki 2010). This station has been operational since November 2010 and is managed by the Faculty of Physics at the National and

[1] The first author recalls the late autumn of 1945, the result of the football games in UK of the then USSR champion Dynamo—19:9. There were 4 matches: Chelsea 3:3, Arsenal 4:3, Cardiff City 10:1, Glasgow Rangers 2:2. The whole country was incredibly excited and proud of this event, the first joyful one after the war.

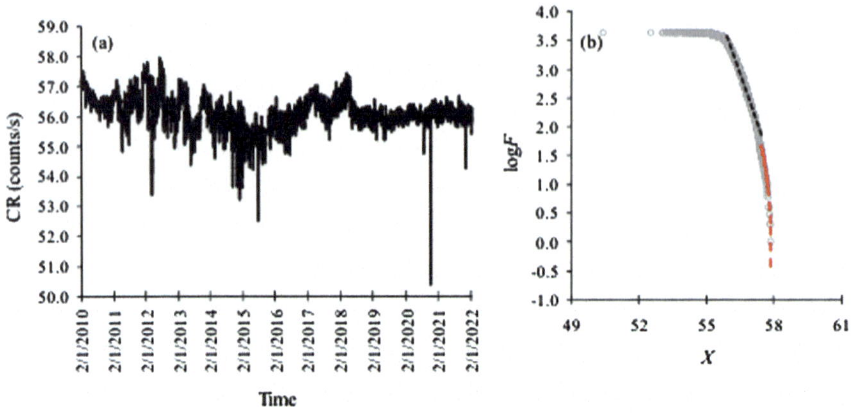

Fig. 4.2 **a** Daily measurements of CRI [in counts/s] during the period 1/1/2010 to 14/1/2022 obtained at A.Ne.Mo.S. **b** Semi-logarithmic plot of the cumulative number (F) of the CRI values greater than or equal to X versus X (gray circles). The dashed black line is the least-squares fit of the GR-scaling $\left(\log F = 65.6 - 1.11 \cdot X \text{ with } R^2 = 0.99\right)$, , while the dashed red line is the upper-truncated fit derived from Eq. (2) with $X_{max} = 57.84$

Kapodistrian University of Athens. Additionally, Varotsos et al. (2023) incorporated cosmic flux data and particle total energy data from the Cosmic Rays Database (CRDB) to further enhance their analysis. The CRDB utilizes various types of experiments and techniques to describe the composition and CR spectrum (Maurin et al. 2014).

Initially, the analysis indicated that the daily CRI values from 1 January 2010 to 14 January 2022 did not follow a Gaussian distribution, as determined by the Kolmogorov-Smirnov (KS), chi-square, and Anderson-Darling tests at a 95% confidence level. To identify the appropriate distribution for this dataset, we calculated the cumulative number (F) of daily CRI values that were greater than or equal to a specific X value of CRI. Subsequently, we plotted the logarithm of F against X (Fig. 4.2b).

Upon conducting regression analysis, we discovered a statistically significant linear relationship between log$F$ and X for high CRI values exceeding 55.86. This value was chosen to achieve the best linear fit:

$$\log F = d + m \cdot X. \tag{4.9}$$

The estimated values of $R^2 = .99(m \geq -1.11 \text{ and } d = 65.6)$ were found to be statistically significant based on the F-test (t-test) at a 95% confidence level. Consequently, it appeared that high CRI values followed a semi-logarithmic distribution similar to the Gutenberg-Richter (GR) law.

However, Fig. 4.2b indicated that the GR scaling accurately described CRI values only up to magnitude 57.42, beyond which a rollover occurred. To address this, we applied an upper-truncated GR fit to the data with $X \geq 57.42$, utilizing a model proposed by Burroughs and Tebbens (2002).

$$F' = 10^{d} \cdot \left(10^{mX} - 10^{mX_{\max}}\right) \tag{4.10}$$

The values of $m \geq -1.11$, $d = 65.6$ were derived from Equation (4.9), and $X_{\max} = 57.86$ was selected to ensure a more precise approach.

Subsequently, we tested the reliability of the aforementioned GR-fit by employing the KS test. The KS-statistic $D = .04$ confirmed the semi-logarithmic distribution for high CRI values greater than 56.039 at a 95% confidence level.

Further information on our new research results will be presented in Chap. 26.

References

Burroughs SM, Tebbens SF (2002) The upper-truncated power law applied to earthquake cumulative frequency-magnitude distributions: evidence for a time-independent scaling parameter. Bull Seismol Soc Am 92(8):2983–2993. https://doi.org/10.1785/0120010191

Ginzburg VL (ed) (1990) Astrophysics of cosmic rays.—M.: Nauka. English edn. North Holland Publications Co., Amsterdam

Golitsyn GS (1997) The spectrum of cosmic rays from the point of view of similarity theory. Astron Lett 23(1):127–132 (Pis' ma v Astronomicheskii Zhurnal, 23, p. 149, 23, 127–132)

Golitsyn GS (2005) Phenomenological explanation of the shape of the spectrum of cosmic rays with energies E > 10 GeV. Astron Lett J 31(7):500–505

Golitsyn GS (2008) Visualization for a number of problems of choosing energy as a unit of measurement instead of mass. Physics 178(7)

Karelin AV et al (60 authors) (2014) New measurements of the energy spectra of protons and helium nuclei of high-energy cosmic rays using a calorimeter in the PAMELA experiment. JETP 146(3):513–517

Malkov MA, Diamond P (2001) Modern theory of Fermi acceleration: a new challenge to plasma physics. Phys Plasmas 8(5):2401–2406. https://doi.org/10.1063/1.1345507

Maurin D, Melot F, Taillet R (2014) A database of charged cosmic rays. Astron Astrophys 569:A32. https://doi.org/10.1051/0004-6361/201321344

Mavromichalaki H (2010) Worldwide integration of neutron monitors. Eos TraNs Am Geophys Union 91(35):305–306. https://doi.org/10.1029/2010EO350001

Varotsos CA, Golitsyn GS, Efstathiou M, Sarlis N (2023) A new method of nowcasting extreme cosmic ray events. Remote Sensing Lett 14(6):576–584

Chapter 5
Turbulence and Rotation

5.1 Mesoscale Turbulence

Here we look at mechanical turbulence in rotating systems and show its connection with the laws of Kolmogorov of 1934. The rotation that introduces noticeable differences in the turbulence described by Kolmogorov and Obukhov in 1941. Laboratory measurements appeared at the very beginning of our century (Baroud et al. 2002, 2003), which confirmed the main conclusions of purely theoretical research and discovered new statistical patterns noticed before that but remained ununderstood. A systematic review of the state of the subject with some new results was made by Golitsyn (2007), and further development of this direction through numerical modeling was carried out by Gledzer (2008) with an analysis of the limits of applicability of the results obtained in measurements.

Rotation with angular velocity Ω sets the time scale in the system. Basic the quantity that excites turbulence remains the generation/dissipation rate kinetic energy of velocity fluctuations ε. This quantity can be excited either mechanically or by a system of sources and sinks in the vessel used. Dimensional analysis gives scales of length, acceleration, speed, fluctuations pressure, vortex (turbulent) mixing coefficient:

$$L_\Omega = \left(\varepsilon / \Omega^3\right)^{1/2}, \tag{5.1}$$

$$a_\Omega = (\varepsilon\, \Omega)^{1/2}, \tag{5.2}$$

$$u_\Omega = (\varepsilon / \Omega)^{1/2}, \tag{5.3}$$

$$\delta_P = \rho\varepsilon / \Omega, \tag{5.4}$$

$$K_\Omega = u_\Omega L_\Omega = \varepsilon / \Omega^2. \tag{5.5}$$

G. Golitsyn and C. Varotsos, *The Stochastic Nature of Environmental Phenomena and Processes*, https://doi.org/10.1007/978-3-031-77015-9_5

The velocity scale was indicated in Golitsyn (1980), the length scale in Browand et al. (1982), and the mixing coefficient in Baroud et al. (2002, 2003) and Boubnov and Golitsyn (1990), although the theoretical results were indicated in Zhou (1995) and Canuto and Dubovikov (1997) without comparison with measurement data. The scales of accelerations and pressure fluctuations were introduced in Golitsyn (2007). Experiments on convection in a rotating fluid give for horizontal movements speed estimation (Boubnov and Golitsyn, 1990):

$$u_\Omega = 2.4(\varepsilon/\Omega)^{1/2}$$

(or 1.7, if you take Coriolis parameter 2Ω). Length scales, speed, and vortex coefficient mixing have already been given by us in paragraph 1.3, but at a fixed time, specified here by the inverse rotation frequency Ω.

When considering convection, the rate of generation of kinetic energy in stationary is equal to the buoyancy flux density (see Chap. 11). Note also that in the 1941 theory, there is a single linear scale—the microscale of dissipation:

$$l_\nu = \nu^{3/4}\varepsilon^{-1/4}, \tag{5.6}$$

which is a consequence of the appearance of a new dimensional parameter ν, the kinematic viscosity of the medium. In this case, we have a new control parameter—the speed of angular rotation of the system Ω, which fixes random variables near their average values.

If kinematic viscosity and external linear scale L_0 are significant, then you can enter the usual Reynolds number:

$$\mathrm{Re} = \frac{u_\Omega L_0}{\nu} \tag{5.7}$$

and rotational Reynolds number:

$$\mathrm{Re}_\Omega = \frac{u_\Omega L_\Omega}{\nu} = \frac{\varepsilon}{\nu\Omega^2}. \tag{5.8}$$

In geophysical fluid dynamics, the role of rotation is characterized by the Rossby number:

$$\mathrm{Ro} = \frac{u_\Omega}{2\Omega L_0} = \frac{\varepsilon^{1/2}}{(2\Omega)^{1/2} L_0} = \frac{L_\Omega}{L_0}, \tag{5.9}$$

Therefore, the rotational Reynolds number is:

$$\mathrm{Re}_\Omega = \mathrm{Re}/\mathrm{Ro}. \tag{5.10}$$

Rotation is essential for $\mathrm{Ro} \ll 1$ for flow dynamics. At $\mathrm{Ro} \ll 1$ force the Coriolis flow is balanced by the pressure gradient, and this flow is called geostrophic. Let l

be the considered scale of movements, and when $L_\Omega \ll l \ll L_o$ and $l \gg lv$ from considerations of dimensionality and similarity for velocity structure functions of order p we can write:

$$S_p(l) = \left\langle [u(x+l) - u(x)]^2 \right\rangle = a_p u_\Omega^p f_p(l/L_\Omega), \tag{5.11}$$

where a_p is a numerical factor, f_p is an arbitrary function of its argument, type which can be found from experiments, laboratory and/or numerical, and has been defined by Golitsyn (2007). At $p = 2$, the structure-function S_2 is related to the energy density of spatial velocity fluctuations, i.e. with their spectrum (see Sect. 1.1):

$$E(k) = 2\int_0^\infty (1 - \cos kl) S_2(l) dl, \tag{5.12}$$

where $k = 2\pi/l$– spatial wave number. If $S_2(I) \sim I^n$, then the spectrum will also be power $E(k) \sim k^{-n-1}$. In Golitsyn (2007), using several assumptions, has shown that (in our notation):

$$E(k) \approx \frac{u_\Omega^2}{L_\Omega} k^{-2} = (\varepsilon\,\Omega)^{1/2} k^{-2} \tag{5.13}$$

and correspondingly:

$$S_2(l) \approx \frac{\varepsilon}{\Omega}\frac{l}{L_\Omega} = (\varepsilon\,\Omega)^{1/2} l = a_\Omega l. \tag{5.14}$$

The first formula was obtained by formally expanding the solution of the Navier-Stokes equation into an infinite series of perturbation theory (Zhou 1995; Canuto and Dubovikov 1997), but leaving only the first leading term of the expansion. For us, this is equivalent to expanding the function f_p in (5.11) into a Taylor series, leaving only the first linear term.

Experiments (Baroud et al. 2002, 2003) will confirm the power-law dependence of the magnitude of the spatial spectrum of velocity fluctuations on the wave number in the form k^{-n} с $n = 2.04 \pm 0.06$ (Fig. 5.1).

Velocities in Baroud et al. (2002, 2003) were measured spatially by fixed sensors, i.e., the signal was measured over time. Using Taylor's hypothesis about frozen turbulence, i.e., dispersive ratios $\omega = ku$, frequency spectra turned into spatial ones when $u = r\,\Omega$, where r is the coordinate of the sensor, i.e., spectra k^{-2} associated with the first scale ANK34 (1.31). The same is directly evident from measurements of temperature fluctuations in rotating vessels (see below). Mechanical turbulence there was created by the method of sources and sinks. Measurement data is also provided there. velocity structure functions (5.11) of various orders up to $p = 10$. Remarkably, $S_p(I) \sim I^{p/2}$ there it was convincingly established experimentally that distinguishes turbulence during rotation from the turbulence we are used to Kolmogorov–Obukhov,

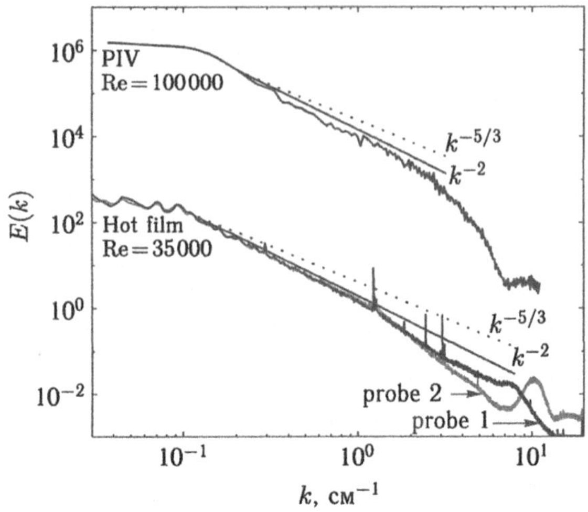

Fig. 5.1 Spatial spectrum of velocity fluctuations in a rotating vessel by Golitsyn (2007)

when the dimension follows $S_p(I) \sim I^{p/3}$ and where such equality is experimentally satisfied only for $p \leq 3$. Figure 5.2, taken from Baroud et al., (2003) and reworked by us in Golitsyn (2007), demonstrates these structure functions. They are constructed in the so-called extended self-similarity approximation when along the x-axis not the distance r that is plotted, but the ratio $5S_3 / 4S_2^{3/2}$, proportional to r in the case of fulfillment of the "law – 4/5" of A. H. Kolmogorov (see Chap. 2).

In Golitsyn (2007), based on the data published by Baroud et al. (2003), has estimated the dissipation rate $\varepsilon = 0.26 \pm 0.02$ cm^2/s^3 at a rotation speed of 11 rad/s = 1.75 revolutions per second. Figure 5.2 shows that for fixed values of l from 1 to 10 cm, the distance between adjacent curves $S_p(l)$ increases only slightly with their number p. For our values of ε and Ω, the numerical factor α_p in (5.11), which we estimated for different values of l, turns out to be equal to $\alpha_p = 5\ 10^{-3}p$ the accuracy of this estimate is better than 20%. As a result, we rewrite formula (5.11) as:

$$S_p(l) = 5 \cdot 10^{-3} p (2\varepsilon\,\Omega)^{p/4} l^{p/2}, \tag{5.15}$$

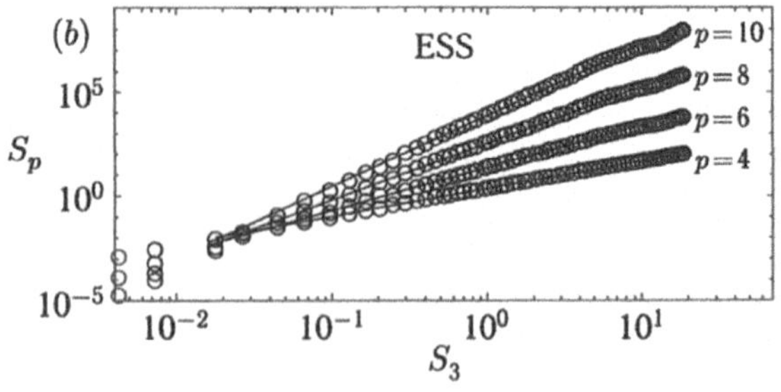

Fig. 5.2 Velocity structure functions of order p from 4 to 10 (Baroud et al. (2003)

which is true for $p = 2, 4, 6, 8$. For $p = 10$, the order of the moment S_{10} (l) is so high that the statistical significance of the measurement results for it may be insufficient, i.e., the ensemble of data is small.

An estimate of the dissipation rate ε allows us to get an impression of the nature of the phenomena in experiments. At $\Omega = 11$ rad/s we have $L_\Omega \approx 2\ 10^{-2}$cm $= 0.2$ mm.

The spatial resolution of the measuring equipment in the experiments described was an order of magnitude coarser. Thus, since $l >> L_\Omega$, all measurements were carried out in the Ro << 1 mode, i.e., rotation was decisive. For water, kinematic viscosity $\nu = 10^{-2}$ cm^2/s, i.e., Kolmogorov microscale $l_\nu \approx 4\ 10^{-2}$ cm $= 0.4$ mm, twice the rotational scale L_Ω.

Note that the moment $S_3(l)$, associated with the asymmetry of the probability distribution function, turned out to be positive in these measurements for most of the scale interval. This indicates a reverse cascade of energy transfer from small scales to large ones (see Chap. 2). The flow dynamics in Baroud et al. (2002, 2003), were quasi-two-dimensional, and the process of vortex merging, characteristic of an inverse cascade, was observed there.

The analysis of the structure of turbulence during rotation carried out by Golitsyn (2007) also allowed naturally two old measurements of convection in rotating liquids, i.e., thermal turbulence (Boubnov and Golitsyn 1990, 1991). One of them refers to measurements of the diffusion coefficient at different Rayleigh numbers and angular velocities. In these experiments, a dye was introduced to the bottom of the vessel, and then the change in spot area S was monitored over time, which turned out to be linear with t. The dS/dt value was identified with the mixing coefficient K_Ω from (5.5). According to these measurements, this value turned out to be many orders of magnitude greater than the molecular value but did not depend over time on the size of the spot. According to Taylor (1915), the eddy diffusion coefficient $K \approx ul$, and in (5.5)$\mathrm{k}_\Omega = u_\Omega\, l_\Omega = \varepsilon\, \Omega^{-2}$, whose factors do not depend on the scale of the phenomenon l. This contrasts with the Richardson–Obukhov turbulent diffusion coefficient, when the latter is proportional $\varepsilon^{1/3}\ l^{4/3}$, i.e., it strongly depends on the size of the spot (see Chap. 7). Then in 1990, for the authors of Boubnov and Golitsyn (1990, 1991), such a difference seemed mysterious.

In the last two references, the frequency spectra of temperature fluctuations $E_T(\omega)$ are also presented. Being normalized to a temperature dispersion of $\sigma_T{}^2$, all measured spectra merged into a thin rope (see Fig. 12 in Boubnov and Golitsyn (1995). According to dimensional considerations and measurement data, the temperature dispersion:

$$\sigma_T^2 = \frac{0.42\Omega f'}{\alpha g}, \tag{5.16}$$

where $\mathrm{f}' = \langle \mathrm{w}'\mathrm{T}' \rangle / (\rho \mathrm{c_p})$– kinematic heat flux, α—coefficient of thermal extensions, w′—fluctuation of vertical velocity, $\mathrm{c_p}$—heat capacity of the medium at constant pressure. In the frequency range from 1 to 0.01 Hz, the slope of the mentioned bundle is equal to – 2, i.e., $E_T = \sigma_T^2\omega^{-2}$. This corresponds to the spatial spectrum k^{-2} for the passive scalar when Taylor's frozen conjecture holds turbulence: $\omega = ku$. Here u is the speed of reverse rotation of the liquid in the vessel as a whole. This rotation occurs due to the law of conservation of the total rotational moment of the liquid in the measuring vessel. Convection in our experiments occurs during cooling from above due to the instability of the cold film through which the liquid gives up its enthalpy by cooling, evaporation, and radiant heat exchange. This creates a lattice vortex (see Chap. 12), transmitting cold thermals down. These thermals have a cyclonic rotation that coincides with the general rotation of the vessel. Conservation of the total momentum of the fluid in the vessel requires the occurrence of a slow general reverse rotation of the fluid as a whole.

This allows us to convert frequency spectra into spatial spectra using Taylor's hypothesis [see formula (1.5)] as a linear dispersion relation.

Applications of these concepts and ideas to real natural processes leave it for Chap. 9. Hurricanes, vortices in the atmosphere and ocean, etc. will be considered there.

5.2 The Process of Vortex Merging

Here it is appropriate to give a visual analytical description of such a process of reverse energy cascade from small to large scales, observed in many experiments. In the spirit of the ideas of Kolmogorov, we will use the Fokker–Planck equation for the probability density $p(\omega, t)$, where ω is the vortex component in a quasi-two-dimensional flow. It looks like:

$$\frac{\partial p}{\partial t} = \frac{D}{2}\frac{\partial^2 p}{\partial \omega^2}, \tag{5.17}$$

where the diffusion coefficient has the dimension of the cube of reciprocal time. It can be identified with the rate of enstrophy generation described in Sect. 2.3. In a random two-dimensional flow, the statistical characteristic of vorticity is the mean square of the vortex, called enstrophy $\Omega_e = \langle\omega^2\rangle$, and its generation rate will be the vorticity diffusion coefficient:

$$D = \frac{d\Omega_e}{dt}. \tag{5.18}$$

Equation (5.17) is the Fokker–Planck equation for the probability density function p in the absence of systematic motions when it is expanded in a Taylor series to a quadratic term in the vortex variable ω (Lifshitz and Pitaevsky 1979). It has a second moment:

$$\langle\omega^2\rangle = 2Dt \tag{5.19}$$

in full accordance with the description of Brownian motion or random walk, when the mean square coordinates in n-dimensional space:

$$\langle x^2 \rangle = 2nDt. \tag{5.20}$$

As there, the mean square of vorticity increases with time, this occurs through the merging of vortices. Their total number decreases, and one central vortex may remain in a closed vessel. This pattern was indeed observed several times in a 10-meter Coriolis vessel operating at the University of Grenoble in France (after working for many hours).

References

Baroud CN, Plapp BB, She ZS, Swinney HC (2002) Anomalous self-similarity in a turbulent rapidly rotating fluid. Phys Rev Lett 88:14501–14504

Baroud CN, Plapp BB, Swinney HC, She ZS (2003) Scaling in three-dimensional and quasi-two-dimensional rotating turbulent flows. Phys Fluids 15(8):215–239

Boubnov BM, Golitsyn GS (1990) Temperature and velocity field regimes of convective motions in a rotating fluid layer. J Fluid Mech 219:215–239

Boubnov BM, Golitsyn GS (1991) Diffusion in liquid under geostrophic convection regime. Dokl Acad Sci USSR 317(3):602–605

Boubnov BM, Golitsyn GS (1995) Convection in rotating fluids. Kluwer, 232 pp

Browand FK, Hopfinger EL, Gagne V (1982) Turbulence and waves in a rotating tank. J Fluid Mech 125:505–534

Canuto VM, Dubovikov MS (1997) Physical regimes and dimensional structure of rotating turbulence. Phys Fluids 9:2132–2135

Gledzer E (2008) Rotation and helicity effects in cascade models of turbulence. Doklady Phys 53(4)

Golitsyn GS (1980) Geostrophic convection. Dokl RAS 251(6):1356–1359

Golitsyn GS (2007) Turbulence during rotation: scales, regimes, spectra, structural functions. Dokl RAS 415(3):325–329

Lifshitz EM, Pitaevsky LP (1979) Physical kinetics. See 2.—M.: Fizmatlit, 528 pp

Taylor GI (1915) Eddy motion in the atmosphere. Phil Trans R Soc Lond 215A:1–26

Zhou Y (1995) A phenomenological treatment of rotating turbulence. Phys Fluids 7:2092–2102

Chapter 6
Sea Wind Waves

6.1 Characteristics of Waves and Similarity Criteria

Waves on the surface of the sea are one of the most fascinating natural phenomena. Against the background of a clear period of the main peak, the random nature of the disturbance is clearly visible. And here the laws of the ANC34 and its schools are very clearly manifested. However, a quantitative understanding of the development of waves under the influence of wind began only in the late 1950s. As long as the wind overtakes the waves, you can think that they are developing.

This is how the concept of age of excitement arose:

$$\Omega = U/c_{\phi}, \tag{6.1}$$

where U is the wind at some level, usually considered at z = 10 m, c_{ϕ} is the phase velocity of the main peak.

The propagation of water surface disturbances taking into account surface tension is described by the dispersion equation:

$$\omega^2 = \left[g + (\sigma/\rho)k^2\right]k \text{ th } kh, \tag{6.2}$$

where g is the acceleration of gravity, h is the depth of the liquid layer, $k = 2\pi/\lambda$ is the wave number. The second term in parentheses is compared with the first when λ = 1.7 cm (20 °C). At $k\lambda >> 1$ excluding capillary ripples

$$\omega^2 = gk \tag{6.3}$$

for deep water. For simplicity, we will limit ourselves to this case. Then the phase and group velocities are equal to:

G. Golitsyn and C. Varotsos, *The Stochastic Nature of Environmental Phenomena and Processes*, https://doi.org/10.1007/978-3-031-77015-9_6

$$c_\phi = \frac{\omega}{k} = \frac{g}{\omega} = \left(\frac{g}{k}\right)^{1/2}, \quad c_{gr} = \frac{d\omega}{dk} = \frac{1}{2}\left(\frac{g}{k}\right)^{1/2} = \frac{1}{2}c_\phi. \tag{6.4}$$

The wave age is the first similarity parameter for wind waves:

$$\Omega = \frac{U\omega_p}{g} = \frac{Uf_p}{2\pi g} = \left(\frac{2\pi U^2}{\lambda g}\right)^{1/2}, \tag{6.5}$$

where f_p is the peak frequency in Hertz. The wave energy per unit area is given by (Komen et al. 1994)

$$E = \rho_w g \frac{h_s^2}{16} = \rho_w g \frac{U^4}{g^2}\frac{\tilde{h}_s^2}{16}, \tag{6.6}$$

where ρ_w—density of water, h_s is the so-called significant wave height, equal on average to 1/4 of its full height at the peak (Komen et al. 1994). The tilde sign denotes the dimensionless value of the corresponding variable, and the height is normalized to U^2/g. The momentum is defined as E/c_ϕ, action—how E/ω_p. The main parameter of nonlinearity is the steepness of the wave:

$$s = \frac{1}{2}kh_s = \pi\frac{h_s}{\lambda} = \frac{h_s g \Omega^2}{2U^2}. \tag{6.7}$$

In practice, the reference wind speed at a height of 10 m is significantly determined by the stratification of the atmosphere. The atmosphere is stable when it is warmer than water, and convectively unstable when it is colder. The age of waves also affects the wind in the lower layer of the atmosphere through the coefficient of wind resistance on the water surface (Kraus and Businger 1999). The problem is complex, which is most clearly characterized by the fact that waves develop in the middle latitudes much more slowly in the spring when the atmosphere usually has a warmer water surface (Komen et al. 1994) than in autumn.

The state of excitement depends on the place and time of action of the wind on the water. This is determined by the external similarity parameter—for example, the distance e^χ from the lee shore. Combined with wind and acceleration g is determined dimensionless acceleration (fetch):

$$F = \frac{gx}{U^2}. \tag{6.8}$$

Note that the product of two similarity parameters (6.5) and (6.8) is equal to $\Omega F = \omega_p/U = \omega_p xT$, where $T = x/U$ is the time of influence of the wind on the developing wave.

6.2 Fetch Law

The beginning of planned and systematic field studies of the laws of wind wave development was laid out by Hasselman et al. (1973), which were continued in several other projects. Research before the early to mid-1990s is summarized in Komen et al. (1994). Dependences of the frequency of the main wave peak and its height in Badulin et al. (2007) on accelerations are given for 23 field experiments. The results are described in Gagnaire et al. (2011) extensive numerical experiments on the development of waves depending on their ages that remain unsurpassed. The most complete generalization of these results by 2017 is given in Zakharov (2017), Golitsyn (2010) and Golitsyn et al. (2021). The resulting empirical patterns are called fetch laws, which have the form:

$$\frac{Uf_r}{g} = AF^{-\alpha} \text{ or } \frac{gT}{U} = \frac{F^{\alpha}}{A} \tag{6.9}$$

$$\frac{g^2\varepsilon}{U^4} = BF^{\beta}, \quad \varepsilon = \frac{h_s^2}{16}, \tag{6.10}$$

where A and $\alpha > 0$, B and $\beta > 0$ are parameters determined over a large series of observations. Excluding acceleration F from these two formulas gives a relationship between height and period at peak:

$$16\varepsilon = h_s^2 = \frac{U^4}{g^2}\left(\frac{AgT_p}{U}\right)^{\frac{\beta}{\alpha}}. \tag{6.11}$$

The evolution of wave characteristics in real time is determined using the acceleration laws (Golitsyn 2010) by the relations:

$$\frac{dE}{dt} = \frac{dE}{dx}\frac{dx}{dt} = c_{zp}\frac{dE}{dx}, \tag{6.12}$$

whence it follows that the derivative of acceleration concerning time is the group velocity, i.e., the speed of movement of the energy of the main peak in space. Then, using (6.4) and (6.9), we obtain:

$$\frac{dx}{dt} = c_{gr} = \frac{U}{2\Omega} = \frac{g}{4\pi f_p} = \frac{U}{4\pi A}\left(\frac{gx}{U^2}\right)^{\alpha} \tag{6.13}$$

or in the dimensionless form:

$$\frac{dF}{d\tau} = (1-\alpha)F^{\alpha}, \quad \tau = (1-\alpha)\frac{t}{T_0} = \frac{(1-\alpha)gt}{4\pi AU}. \tag{6.14}$$

The solution of this nonlinear equation with a zero initial condition gives the desired relationships between fetch, physical time, wave age, and their height, steepness, and action depending on the indicators α and β (see Golitsyn 2010).

According to Badulin et al. (2007), Gagnaire et al. (2011) and Zakharov (2017) values α are in the range from 0.23 to 0.33, and β—from 0.7 to 1.0. The values of these coefficients are smaller when the water is colder than the air (Komen et al. 1994). When the atmosphere is colder than water, convection occurs in it, while the heat transfer from the water cools its upper layer, and convective mixing with deeper waters also occurs there. All this facilitates the transfer of momentum and wind energy to the water. In the tropics, this usually happens at night because the atmosphere cools faster than water, and cumulus clouds form over the ocean at nighttime. With stable stratification of the atmosphere, the exchange between ocean and atmosphere is complicated, and the values of the indicators α and β are decreasing. However, an often almost approximate relationship is always preserved (Gagnaire et al. 2011; Zakharov 2017).

$$3\alpha \approx \beta. \tag{6.15}$$

For example, in the most extensive and detailed project (Komen et al. 1994): $\alpha = 0.33$, $\beta = 1.0$. At lowest values $\alpha = 0.23$ in Golitsyn (2010) and $\beta = 0.7$. In intermediate cases (Kraus and Businger 1999) relation (6.15) is satisfied to within a few percent, for example, K. Melville (private communication with the author, February 2009) reported $\alpha = 2/7$ and $\beta = 4/5 = 0.8$. Zakharov (2017) provides data on 23 experiments, six of them with $\alpha = 0.33$ and $\beta = 1.0$. In Golitsyn et al. (2021), an analysis of all available data gave $\beta = 3.25\ (1 \pm 0.07)\ \alpha$.

The index α is responsible according to (6.9) for the connection between fetch and the period of the wave peak, and β according to (6.10) connects the square of the wave height with fetch. It immediately follows that:

$$h^2 \sim T^3. \tag{6.16}$$

This ratio, namely $h_s \sim T^{3/2}$, was empirically established by Toba (1978) and received the name of the law named after him. It is shown in Fig. 6.1. Surprisingly small scatter of empirical points relative to this law, which is a reflection of relation (6.15), but this relation itself reflects in reality A. N. Kolmogorov's law of random moves (1.33), which states that the average square of displacements in an ensemble of particles:

$$\langle x^2 \rangle = \varepsilon t^3. \tag{6.17}$$

A detailed examination of this issue with new additional experimental data is given in Golitsyn et al. (2021).

Figure 6.1 contains a lot of the own data of the last two authors of the article Golitsyn et al. (2021) on measurements at the Gorky Reservoir about 30 km wide. The straight thick line corresponds to the indicator $n = 1.47 \pm 0.03$ in dependence

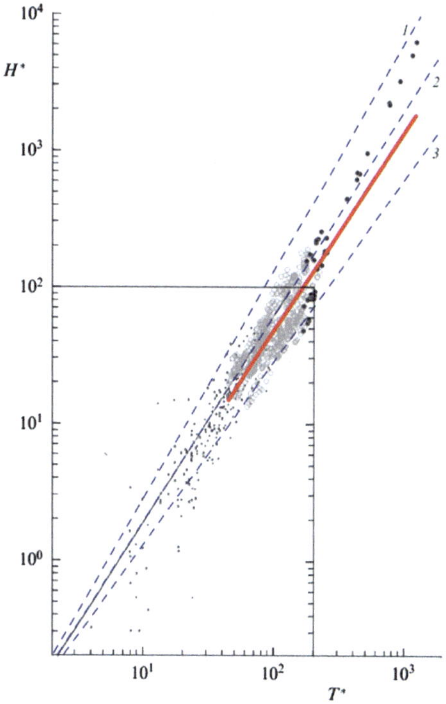

Fig. 6.1 Statistical substantiation of Toba's law $h^2 \sim T^3$, where h is the height sea wave crest, T—its period, dotted lines to the left of the straight line the lines correspond to young waves, and on the right-old ones, according to Kitaygorodsky (1962)

of h, $T^{n,}$ and to the left of it the points correspond to young waves with a steeper spectrum [see further (6.20)] up to $n = 5/3$, and to the right are more established or damped waves $3.5 > n > 4/3$ with independence (6.20).

Thus, the value of wave age is justified as the main characteristic of the degree of wave development when it is believed by means that the average age in the ocean Ω is close to 1.2. The figure under discussion describes the evolution of the development of the wave structure: from steeper waves to steady waves with decaying, breaking waves up to $\Omega \geq 0.83$. Based on the acceleration equations and formula $1.2 \leq \Omega < 2$ (6.12), for characteristic ages Ω—2, 1.2, and 0.83—acceleration times and lengths, wave heights, their lengths and periods are calculated for an average wind over the ocean of 9 m/With. All these quantities are functions of the parameters α and A, and acceleration times are hours, acceleration lengths are tens of kilometers, average wave heights are on the order of 1–2 m, their peak lengths are tens of meters, and periods are on the order of 5 s. An increase in wind above 10 m/s noticeably changes the given figures. Simple algebraic forms for the wave characteristics listed above make it possible to evaluate them.

6.3 Wave Frequency Spectra

This connection can be considered as a structure function of displacements with zero initial conditions (see Chap. 1). The connection of such a structure function with stationary random increments of the second order with its spectrum is carried out by a Fourier-type transformation (see Sect. 1.3)

$$D_h(\tau) = \langle [h(t+\tau) - h(t)]^2 \rangle = 8 \int_0^{\infty} (1 - \cos \omega t)^2 E(\omega) d\omega. \tag{6.18}$$

If $D(t) \sim t^k$ is a power structure function, then the spectrum will also be power ω^{-k-1}. For $k = 3$, as in (6.17), the spectrum will be ω^{-4}.

Such a frequency spectrum of sea waves was obtained from considerations of similarity and dimension by Kitaigorodsky (1962) and Zakharov and Filonenko (1966) as a solution to the kinetic equation for the Fourier wave components of a free surface and confirmed experimentally in 1973 by Toba (1973). Consequence laws of random moves by A. N. Kolmogorov 1934 (ANK34) are confirmed like Toba law $h \propto T^{3/2}$, and by frequency spectrum of sea waves $S \sim \omega^{-4}$. This power-law portion of the frequency spectrum is observed for approximately one-third of a decade after the peak frequency and carries approximately 95% of the wave energy. In Zakharov (2017), this section is called the "Hasselmann Sea," and the next section, where $n = -5$, is called the "Phillips Sea." The ratio of these two spectra is determined by the age of the disturbance. Phillips spectrum obtained in 1958 from dimensional considerations for the high-frequency part, assuming that, due to the nonlinearity of the process, the wind no longer plays a role, is equal to (Phillips 1958): $S_h(\omega) \approx g^2\omega^{-5}$.

Measurements from flyby vehicles also make it possible to measure the spatial spectrum of waves. Using the dispersion relation (6.3), we transform the frequency spectrum into a spatial one $S_h(k)$, and we get a dependence $k^{-5/2}$. This form was confirmed by measurements (Baranovsky et al. 1992), where it was found that $S_h(k) \sim k^{-n}$ and $2.4 < n < 2.6$, at least up to wave components approximately an order of magnitude smaller than the main peak.

As mentioned at the beginning, the main similarity parameter in the process of wave development by wind is age (6.1), which determines whether the wind overtakes the phase speed. It has long been visually noticed that at the beginning of development young waves are steeper than in the middle of their development, then they collapse, and when the wind subsides, the waves turn into swell. In Golitsyn (2010) it is estimated that with average wind $U_{10} = 9$ m/s and neutral stratification, the development process takes up to half a day, and with a stable atmosphere and the same average wind speed, the time for wave development increases by 2–3 times. Since the steepness of the waves decreases as fetch $F^{-1/6}$, then their frequency spectrum is slightly flattened. The most complete and consistent development of wave spectrum depending on its age was traced to the initiative of Badulin et al. (2007). It was shown there that the spectrum:

$$S_h(\omega) \sim \omega^{-4} f(\Omega), \tag{6.19}$$

where the dimensionless age function can be represented as:

$$f(\Omega) = \begin{cases} \Omega^{-1/3}, & \text{if}\,\Omega > 2 \\ 1, & \text{if}\,1.2 < \Omega < 2 \\ \Omega^{1/3}, & \text{if}\,0.83 < \Omega < 1.2 \end{cases}. \tag{6.20}$$

Open ocean waves have an age of about 1.2 (Komen et al. 1994; Golitsyn 2010) and an average height (Golitsyn 2017) $h_s = 2.7$ m. At the same time, in the Southern Hemisphere, the waves are higher than in the Northern Hemisphere due to greater fetch because of a larger area of open water. The process of development of waves, as asymptotics, is based on the third scale of Kolmogorov (1.33), but the process of development itself is transformed due to the evolution of waves with age, and this leads to small corrections to the spectrum in the form of a function $f(\Omega)$, whose asymptotics are given by (6.20).

The theory of ANC34 provides a statistical model for many natural processes and phenomena and, thereby, a rationale for using considerations of similarity and dimensionality in concluding using the quantities involved in them. Such conclusions should always be accompanied by comparisons with measurement data obtained under specific and controlled natural conditions. These conditions can introduce new dimensional quantities that form other dimensionless similarity parameters. These parameters can influence the numerical coefficients obtained when comparing experimental data with theoretical ones (see B02). In the case of wind waves, this is expressed by the appearance of formulas (6.19) and (6.20). This leads to a modification of the Toba relation: $h^2 \sim T^3$ which will now be:

$$h^2 \sim T^n, \quad n = 3 \pm 1/3, \tag{6.21}$$

which is in good agreement with the data on the diffusion of impurity spots of various sizes and with the data collected by Golitsyn et al. (2021) on determining the wave spectra. Figure 6.1, taken from Golitsyn et al. (2021), shows that relation (6.21) is related to age waves, and real points reveal a fairly large statistical scatter, apparently also due to inaccuracies in measurements of both the height and period of the wave peak. However, in general, the tendency towards the implementation of Toba's law can be traced with confidence. The data in Sect. 7.2 testify in favor of formula (6.20).

References

Badulin SI, Babanin AN, Zakharov VE, Resio D (2007) Weakly turbulent laws of wind-wave growth. J Fluid Mech 591:339–378

Baranovsky VD, Bondur VG, Kulakov VV et al (1992) Using the brightness spectrum and polarization characteristics of outgoing radiation in remote sensing of the sea surface. Issl Earth Space 10(2):235–243

Gagnaire E, Benoit M, Badulin SI (2011) On weakly turbulent scaling of wind sea in simulation of fetch-limited growth. J Fluid Mech 669:178–213

Golitsyn GS (2010) Energy cycle of wind waves on the ocean surface and acceleration laws. Izv RAS FAO 46(1):10–18

Golitsyn GS, Troitskaya YI, Baidakov GA (2021) Analysis of frequency spectra of sea waves and acceleration laws from the point of view of the probabilistic laws of AN Kolmogorov and his school. Izv RAS FAO 56(1):67–71

Hasselmann K, Barnett TP, Bows E et al (1973) Measurement of wind-wave growth and swell decay during the Joint North Sea Wave Project (JONSWAP). Ergänzung zur Deut Hydrogr Z Reihe A 12(8):1–95

Kitaygorodsky SA (1962) Application of similarity theory to the analysis of waves excited by wind. Izv Acad Sci USSR Ser Geophys 1:73–82

Komen GJ, Cavaleri L, Donelan M, Hasselmann K, Hasselmann S, Janssen PAEM (1994) Dynamics and modelling of ocean waves. CUP, Cambridge

Kraus EB, Businger JA (1999) Atmosphere–ocean interaction, 2nd edn. Oxford University Press, Oxford

Phillips OM (1958) The equilibrium range in the spectrum of wind-generated ocean waves. J Fluid Mech 4(3):426–434

Toba Y (1973) Total balance in the air-sea boundary processes: III. On the spectrum of wind waves. J Oceanogr Soc Japan 29(3):209–229

Toba Y (1978) Stochastic form of the growth of wind waves in a single-parameter representation with physical implications. J Phys Oceanogr 8(5):494–507

Zakharov V. E., Filonenko N. N. (1966). Energy spectrum for stochastic oscillations of a liquid surface. Dokl. Academy of Sciences of the USSR. 170 (6). 1292-1298.

Zakharov V (2017) Analytic theory of a wind-driven sea. In: Proceedings of IUTAM Symposium Wind Waves, London, UK

Chapter 7
Turbulence Eddy Mixing in the Atmosphere and on the Sea Surface

Since the time of Charles Fourier (since 1821), diffusion has been studied based on the parabolic equation that he proposed for the propagation of heat. Adolph Fick used it to study diffusion in 1855, and only 70 years later L. F. Richardson realized that diffusion occurs much faster in the atmosphere. Probably the first to understand this was Taylor (1915), who to evaluate diffusion coefficient used the formula:

$$K \approx r\delta u \approx \frac{1}{2}\frac{dr^2}{dt} \sim \frac{dS}{dt}, \tag{7.1}$$

where r is the size of the spot, S is its area, δu is the velocity difference at its edges. The next paper by Taylor (1921) showed how diffusion is accelerated in the presence of a vertical velocity gradient. Its elegant and important generalization was given in 1958 in a short note by Novikov (1958). Relation (7.1) is very general and is widely used below.

7.1 Atmospheric Diffusion

Richardson (1926), found some universality, most clearly formulated by him in 1929 (Richardson 1929). When considering the diffusion of a cluster with size l (air balls, dust or smoke), he introduced (mentally) a vertical plane through the center of the cluster and considered the standard deviation σ cluster particles from this plane. With Fick's diffusion dl^2/dt does not depend on spot size r. Richardson (1926) showed that this value increases with diffusion from a point source approximately as $l^{4/3}$. His graph, illustrating this dependence, contains seven points for distance intervals from 1 cm to thousands of kilometers.

In a 1929 paper, rarely cited, Richardson (1929) presented new results determination of the diffusion coefficient for several small distances from 30 cm to 10

G. Golitsyn and C. Varotsos, *The Stochastic Nature of Environmental Phenomena and Processes*, https://doi.org/10.1007/978-3-031-77015-9_7

m and corrected his calculations in Richardson (1926), for large distances. These experiences are so unconventional and witty that they should be at least briefly described. The relative diffusion coefficient of two markers initially separated by distance l_o, perpendicular to the direction of the average wind, and released into the air simultaneously, was determined by the formula:

$$K(l) = \frac{1}{n}\sum \frac{\left(l_0 - n^{-1}\Sigma l\right)^2}{2t}, \tag{7.2}$$

where l—parallel l_o component of the distance between markers at time t, the expression in parentheses is a small value compared to l_o, n—number of experiments. Initial separation of particles 1 cm. Dust grains in a sunbeam in a darkened room. Rough estimates are given by eye $K \sim 0.2\ \text{cm}^2/\text{s}$, which is on the order of the molecular diffusion coefficient. To do this, we must assume that the relative speeds of the order of 1 cm/s at distances of the order of 1 cm, and then the formula for the square of the displacement Brownian particle $\langle \Delta\chi^2 \rangle = 2nDt$, where $n = 3$—dimension of space, with t ~ 1 s gives the desired order for the diffusion coefficient.

The initial separation is 30 cm. The tracing of two dandelion seeds was released simultaneously. The observer stood 6 m downwind and assessed the distance between seeds by eye. 18 such experiments gave an average value of $K = 38\ \text{cm}^2/\text{s}$.

The initial distances between markers are 154 and 181 cm (at different times in different weather and different places). Markers—tobacco smoke from two cigarettes. For the first cases on average over 40 experiments $K \approx 43\ \text{cm}^2/\text{s}$. For the second—based on 71 experiments—$K \sim 800\ \text{cm}^2/\text{s}$.

The initial separation 5 m: hydrogen-inflated baby rubber balls, green and yellow cost one penny. A very complex measurement procedure with three participants was carried out on February 19 and 20, 1928, and included 8 experiments. It was received on average $K(l) = 0.7\ \text{m}^2/\text{s}$.

The initial separation is 100 m. Many observers stood in the direction of the wind with 4 pairs of balls were released. The score received $K(l) = 10\,\text{m}^2/\text{s}$.

The initial separation is 11.5 km. Pairs of standard pilot balloons results observations of their speeds were published officially and made it possible to make corresponding assessments. The average value for altitudes 600–1800 m is estimated at 120 m^2/s, which is probably a lower estimate (Richardson 1929). The initial separation is 28.5 km. Received 160 m^2/s.

All these data are shown at Fig. 7.1 black circles along with seven estimates from the 1926 article (Richardson 1926), corrected in Richardson (1929). The hollow circles show the values of the turbulent diffusion coefficient that were calculated using the structure functions published by Lindborg (1999). The first author conducted a systematic analysis of data from the European MOZAIC project, in which hundreds of commercial airline aircraft were equipped with g-force and temperature sensors. At an average flight speed of U ~ 250 m/s (900 km/h), Taylor's hypothesis of "frozen turbulence", $x = Ut$, is a fully applicable justification, i.e., the temporal structure of the recorded signal can be considered as the spatial structure of turbulence.

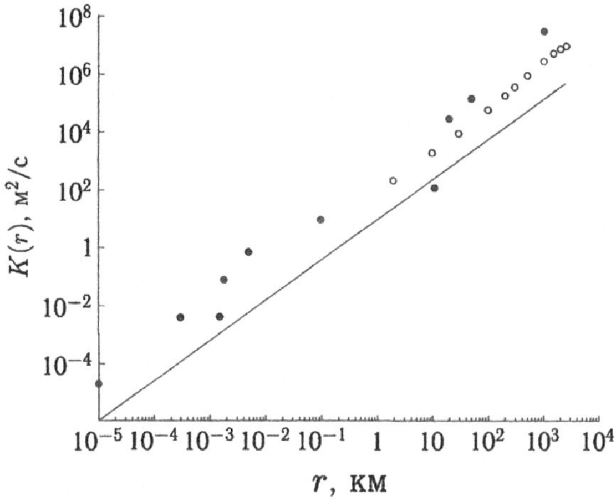

Fig. 7.1 Values of the coefficient of vortex mixing, and diffusion, according to Richardson (1926, 1929), hollow circles according to Lindborg data from Golitsyn and Fortus (2020)

Integration of overloads in time gives the time structure of flow velocity fluctuations averaged over the size of the aircraft. A similar project was previously carried out in the USA (Gage and Nastrom 1986). In both projects, the spectra for both speed and temperature reveal a clear pattern $k^{-5/3}$ in the scale range from 1 km to 500 km. For scales larger than 500 km, in both projects, the spectra approach the k^{-3} dependence up to 3000 km. A discussion of these results is given in Sect. 2.4. In Lindborg (1999), Lindborg also calculated the structure functions of horizontal velocity for distances of 2–2500 km. Their form for the longitudinal and transverse velocity components of the flight is as follows:

$$D_{ll}(r) = a_1 r^{2/3} + b_1 r^2 - c_1 r^2 \ln r, \tag{7.3}$$

$$D_{nn}(r) = a_2 r^{2/3} + b_2 r^2 - c_2 r^2 \ln r, \tag{7.4}$$

where the numerical coefficients are determined based on statistics of more than 30 thousand hours of flight time and are as follows:

$$a_1 = 3.6 \cdot 10^{-3}, b_1 = 2.4 \cdot 10^{-9}, c_1 = 0.16 \cdot 10^{-9},$$
$$a_2 = 4.0 \cdot 10^{-3}, b_2 = 6.5 \cdot 10^{-9}, c_2 = 0.43 \cdot 10^{-9}.$$

All values are in SI units, i.e., meters and seconds. Hence the variance a relative speeds is equal to:

$$\sigma_u = [D_{ll}(r) + D_{nn}(r)]^{1/2} = \left[(a_1 + a_2) r^{2/3} + (b_1 + b_2) r^2 - (c_1 + c_2) r^2 \ln r\right]^{1/2}, \tag{7.5}$$

The first term in brackets on the right corresponds to the spectrum $k^{-5/3}$, the second—to the spectrum k^{-3}, a logarithmic term appears from the Kraichnan (1970) corrections, taking into account nonlocality of interactions of spectral components in the enstrophy transfer interval.

The relative diffusion coefficient is determined by (6.1) and (6.4)

$$K(r) = r\sigma_u \sim r^{4/3}, \tag{7.6}$$

which is reflected by the hollow dots in Fig. 7.1. It shows how close he was Richardson to modern ideas almost a hundred years ago: a comparison of the latter formulas with (7.2) gives $\delta u \sim r^{1/3}$! This figure combines data obtained in the surface layer of air, with different average winds, and in different areas, the first 6 points—in the lower and middle troposphere, the last 4 black points—in the upper troposphere, hollow circles are ours, G12. There seems to be a slight spread of such a diverse set of data regarding addiction $r^{4/3}$ is related to which according to A. M. Obukhov 1/ $3K = c_{RO}\varepsilon^{1/3}r^{4/3}$, and the cube root of the speed dissipation ε is a primary factor in reducing this variation in the upper troposphere, according to Lindborg (1999), $\varepsilon \sim 10^{-4}\text{m}^2\text{c}^{-3}$. Then from the first three hollow points figure, where the structural functions ~$r^{2/3}$ constant in the Richardson-Obukhov law can be estimated at 0.2, MY75. This law corresponds to scale (1.32) in the ANK34 time scale system. The proximity of classic and modern results is evidence of the constancy of the laws of nature and their correct understanding and stability of their assessments by previous generations of scientists and a hundred years before us.

Subsequent work by Lindborg and his colleagues (Brethouwer et al. 2007; Brethouwer and Lindborg 2008) explains the presence of spectral interval $K^{-5/3}$ for horizontal scales 1÷500 km specific turbulence in a stably stratified environment. In this case, the velocity fields are homogeneous and isotropic only in horizontal layers, and then the straight line is a cascade of turbulent energy from large eddies to smaller ones.

Vertical turbulent exchange in the atmosphere is determined by its stratification. With unstable stratification, i.e., with convection, for small scales, for example, in the boundary layer, one can use the general formula (7.1). Then for vertical speed, the Deardorff formula holds:

$$w \approx (bz)^{1/3}, \tag{7.7}$$

that is, the coefficient $K(r)$ will be proportional $b^{1/3}z^{4/3}$, in agreement with Richardson and Obukhov, since in the quasi-stationary case the buoyancy flow $b \approx \varepsilon$, order rate of dissipation of kinetic energy of turbulence.

Exchange with strong stability was studied by Monin and Obukhov back in 1953 (see MЯ71, Chap. 11.5), and more recently by Lindborg and Fedina (2009), when the coefficient of vertical turbulent diffusion does not depend on height but is determined by stratification parameters.

In the recent work of the first author with Fortus (Golitsyn and Fortus 2020), calculations of the turbulent coefficient diffusions were re-run with a composite spectrum corresponding to approximations (7.2), (7.3) and (7.4). Methodological issues of numerical calculations are also discussed there. Actions with composite spectra when the 1941 spectra are at distances up to 500 km begin to be overpowered in energy by the spectra of geostrophic turbulence (Charney 1971; see Chap. 2.4).

7.2 Coefficient of Horizontal Eddy Mixing at the Sea Surface in Dependence on Wave Age

Description of the distribution of impurities over the water surface is an important practical problem. The presence of wind waves makes this a statistical problem. This can be addressed by detailed operational surface wave forecast models. However, it is always important to know the fundamental theoretical laws of this process. In 1950–1970 in connection with applied aspects problems and the lack of adequate wave forecast models, much effort has been directed to field measurements of the diffusion of surface impurities.

This work was started in the late 1940s by Richardson and Stommel (1948) and continued by Stommel (1949). The first, carried out on January 6, 1948, from a pier off the west coast of Scotland, studied the mutual movement of clearly visible flat white pieces of cut turnips. The size of the cluster of these pieces on the surface of the water was about three meters or less. The distances between particles and their changes over time were visually monitored. Processing of measurement data was carried out according to the method described in Sect. 7.1 by formula (7.2).

It was found that the relative diffusion coefficient K, just like in the atmosphere, is a power function of the spot size with the same exponent (in our reprocessing 1.32 $\approx 4/3$ according to tabular data (Richardson and Stommel 1948):

$$K(r) \sim r^{\beta}, \quad \beta = 4/3, \tag{7.8}$$

The same value of 4/3 was obtained by Stommel (1949) for a scale interval up to one kilometer with Stommel's ingenious use of various tracers. Figure 7.2 shows the results of our data reprocessing Stommel (1949), where a straight line corresponds to a slope of 4/3. These results do not differ significantly from those in Stommel (1949). An overview of research results up to the mid-1980s can be found in the book Ozmidov (1986). Okubo (1971) analyzed a large amount of measurement data in both the spatial and temporal evolution of sea surface. Both types of dependencies are close to power-law ones when both r and time t change over at least three orders of magnitude: r^{β} and t^{γ}. For the diffusion coefficient (7.8) magnitude $\beta \approx 1.15$ within dimensions of up to many hundreds of kilometers, which is depicted in Fig. 7.3. The corresponding sizes of spots change over time in the same power-law manner as t^{γ} with $\gamma = 2.33$. For times from an hour to a month, this dependence, formula (3) from (Okubo 1971), has the form $S(t) = 0.0108\, t^{2.34}$ (recalculation in Golitsyn 2007).

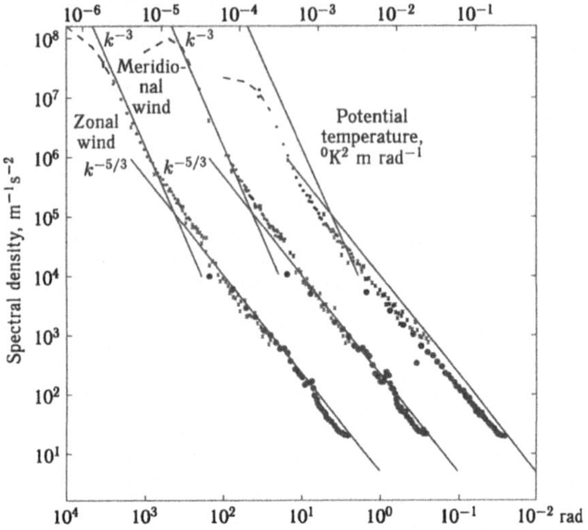

Fig. 7.2 Spectra of horizontal components of velocity and potential temperature (Gage and Nastrom 1986)

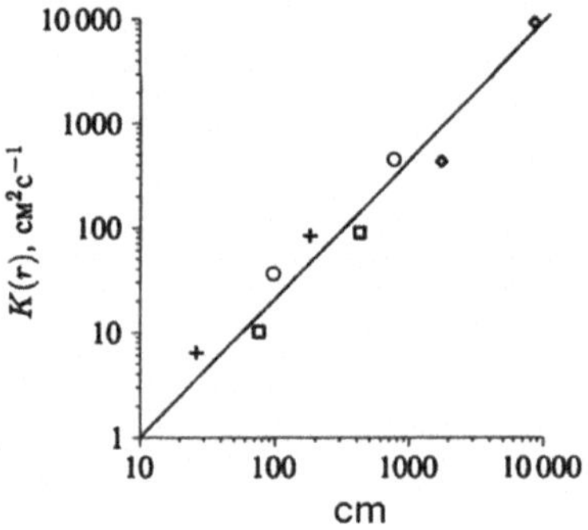

Fig. 7.3 The diffusion coefficient of markers on the water surface at scales from 1 to 100 m according to Stommel (1949)

The first author's work (Golitsyn 2007) indicated the relationship between the energy frequency spectrum of wave elevations on the water surface and the studied value of the horizontal turbulent mixing coefficient $K(r)$, determined by (7.1), where δu is the average velocity difference in the fluid on scale r. This difference should be estimated as the square root of the velocity structure function:

$$\delta u = [D_u(r)]^{1/2}, D_u(r) = \langle [u(x+r) - u(x)]^2 \rangle, \tag{7.9}$$

where, for simplicity, we abstract from the vector nature of the velocity field, from the angular spectrum of waves, and consider its component in the direction of wave propagation, i.e., wind. The structural function $D_u(r)$ is associated with the spatial spectrum energy of velocity fluctuations by the relation:

$$D_u(r) = 2\int_0^\infty (1 - \cos kr)E_u(k)dk, \tag{7.10}$$

where $k = 2\pi/r$—wave number. In reality, the frequency spectra of surface elevations are measured—for wind waves $Ez(\omega)$. Knowing them, we can estimate the frequency spectrum of fluctuations of the vertical velocity component $w = dz/dt$:

$$E_w(\omega) = \omega^2 Ez(\omega).$$

Due to the incompressibility of the liquid in the wave, the frequency spectrum of horizontal velocity fluctuations during waves should have the same frequency dependence:

$$E_u(\omega) = aE_w(\omega) = a_1\omega^2 E_z(\omega), \tag{7.11}$$

where α_l is the numerical coefficient that must be determined experimentally or from some detailed models. Its value may depend on the similarity parameters for waves, e.g., the wave age $\Omega = U/c_\varphi$, which for waves in deep water is equal to $\Omega = U_{10}\omega_p/g$. The wind speed is taken at the standard level of 10 m, and ω_p—angular frequency of the wave peak, which is a function of acceleration—see Chap. 6. 2. Further for simplicity, we take the quantity α_1 constant. Also, for simplicity, we will limit ourselves to the case of waves in deep water, when the dispersion equation $\omega^2 = \text{kg}$. In this case, simple analytical results can be obtained. The case of a general dispersion relation for waves in a layer of arbitrary depth was studied numerically in Golitsyn (2007).

The relationship between frequency and spatial spectra is given using group velocity by the relation:

$$E_u(k) = E_u(\omega)\cdot c_{gr}, \quad c_{gr} = \frac{d\omega}{dk} = \frac{g}{2\omega} = \frac{1}{2}\left(\frac{g}{k}\right)^{1/2}. \tag{7.12}$$

As a result, we obtain for the spatial spectrum of horizontal velocities in wave:

$$E_u(k) = 2a_n g^{1/2} k^{-5/2} E_z(\omega(k)). \tag{7.13}$$

Next, using transformation (7.10), we find the structure function of velocities, after which formula (7.9) gives the dependence of the diffusion coefficient on distance r:

$$K_n(r) = a_2(a_1 a_n)^{1/2} r^\beta,\ \beta = \frac{n+1}{4}, \tag{7.14}$$

where n is the exponent in the energy-carrying part of the frequency spectrum of elevations immediately to the right behind the peak frequency ω_p for approximately half an order of magnitude frequencies or an order in wave numbers K, the so-called Hasselmann sea (see Chap. 6). Numerical coefficients α_2 and α_n a subject to experimental determination.

Note that the structure functions of velocity fluctuations, as follows from definitions (7.9) and (7.10), are equal for the spectrum $E_z(\omega) \sim \omega^{-n}$:

$$D_u \cong a_1 a_n r^{\frac{n-3}{2}}. \tag{7.15}$$

with the spectrum of KZT (Kitaygorodsky, Zakharov, Toba), when $n = 4$, we have $K(r) \sim r^{5/4}$, $D_n(r) \sim r^{1/2}$, for the Phillips spectrum describing high-frequency part, $n = 5$ and $K(r) \sim r^{3/2}$, a $D_5(r) \sim r$.

With these classical spectra, the early results of Richardson and Stommel (1948) with $\beta = 4/3$ are similar to atmospheric values. From (7.14) we find that this value of the exponent β corresponds to $n = 13/3$. This value was obtained in a numerical model (Gagnaire-Renou et al. 2011) for young waves with age $\Omega > 2$, i.e., at low accelerations. Considering that the measurements (Richardson and Stommel 1948; Stommel 1949) were made close to the coast, the explanation for the value $\beta = 4/3$ can be based precisely on the young age of the waves. The same indicator $n = 13/3$ was first obtained by Hasselman (1974) under the condition of a constant momentum flux from the wind to the water surface.

Small Okubo exponents (1971) all refer to large scales of hundreds of kilometers and large diffusion times. In the same model (Gagnaire-Renou et al. 2011), old waves with indicators $\Omega < 1.2$ up to $\Omega \approx 0.83$ when saturated, waves have a spectral slope with an exponent n $\approx$ 11/3. Our formula (7.14) gives this $\beta = 7/6$, which is very close to the Okubo value of 1.15. The work by Okubo (1971) also provides tables of primary data for both the diffusion coefficient and the growth of spot sizes over time. Therefore, we can count the uncertainties exponents, which was done with a security of 95% and gave $\beta = 1.15 \pm 0.05$; $\gamma = 2.33 \pm 0.11$, and in Ozmidov (1986): $\beta = 1.15$, $\gamma = 2.34$ (Fig. 7.4).

All that remains is to admire the accuracy of Okubo's eye as he drew straight lines among the scattering of points in logarithmic coordinates.

For the age of excitement within $1.2 < \Omega < 2$ according to Gagnaire-Renou et al. (2011) $n = 4$, and then $\beta = 5/4$, although experiments confirming such dependencies could not be found. Let us dwell on the dependence of the spot area on time $S(t)$. From the parabolic diffusion equation with a mixing coefficient depending on the coordinate as $K(r) \sim \beta$ it follows, taking into account (7.14), that

$$r^2 \approx S(t) \sim t^{\frac{2}{2-\beta}} = t^{\gamma}, \gamma = 8(7-n)^{-1}. \tag{7.16}$$

At constant diffusion coefficient $\beta = 0$ we obtain $S(t) \sim t$, as in Brownian motion (or as in convection in a rotating fluid, Chap. 9). For $n = 13/3$, young waves, we have $S(t) \sim t^3$, as in the atmosphere (and as in Sect. 1.3), and for $n = 11/3$, old waves, we have $S(t) \sim t^{12/5}$. The last value is very close to the empirical value and describes diffusion at times from an hour to a month: $\gamma = 2.34$ (empirics Golitsyn 2011, 2.33±

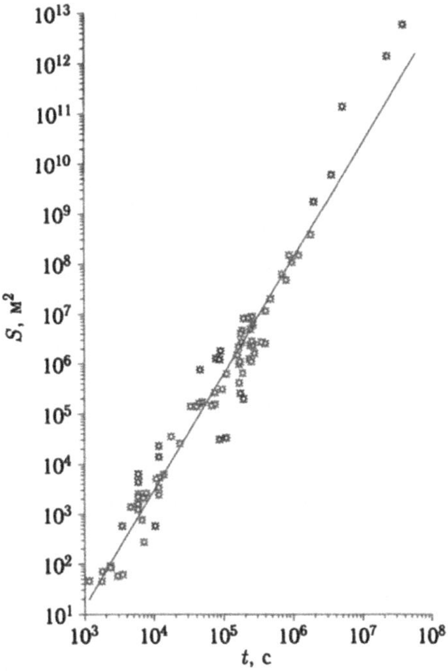

Fig. 7.4 The area of the pollutant stain depends on time (Ozmidov 1986) up to several months (G12)

0.11), obtained in Okubo (1971), collected in Fig. 7.5. If $\beta = 1$, then $S(t) \sim t^2$, as in ballistic expansion, when r ~ t, apparently, if this stage is achieved, it is rare, since such data could not be found. In the same way, for $n = 5$ it turns out that $S(t) \sim t^4$, and this stage is not traced in real data. We remind you that our above processing gave $\gamma = 2.33 \pm 0.11$.

Our formulas (7.14) and (7.15) connect the age of excitement with the real physical time of its development. In medium winds $U_{10} = 9$ m/s (see Chap. 6.3 and Golitsyn 2011) it is possible (G12) to obtain that age $\Omega = 2$ achieved in one and a half hours, age 1.2—in four hours, and saturation $\Omega = 0.83$—in half a day. The synoptic period in the atmosphere, i.e., a constant wind, lasts two to three days, then it is not surprising that Okubo's data (Okubo 1971) correspond to the values $0.83 < \Omega < 1.2$, i.e., developed excitement. A more detailed description of these results can be found in Golitsyn (2011) and Golitsyn and Chkhetiani (2014).

Vertical turbulent diffusion in stably stratified flows, especially important for the ocean with its thermocline, highly stably stratified, for a long time, did not lend itself to a fairly simple theoretical description. During the war, Obukhov (1946) showed that with strong stability, the coefficient of vertical turbulent mixing is related to the inverse Richardson number. Only recently has a new approach to this problem began to be developed by Lindborg and Brethower (2008) and Lindborg and Fedina (2009), where the formula was derived.

$$K_z = \varepsilon_p N^{-2}, \quad (7.17)$$

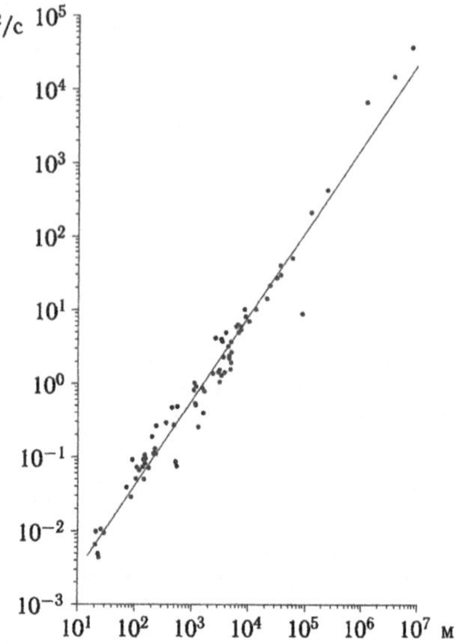

Fig. 7.5 The diffusion coefficient according to Okubo (1971) is up to thousands of km

where ε_p is the rate of generation (dissipation) of potential energy (buoyancy) in a layer with a gradient of potential temperature and humidity, N^2 is the square of the Brunt–Väisälä frequency, which can be associated with the Richardson number. Detailed numerical experiments confirm the theoretical conclusions in Lindborg and Brethower (2008) and Lindborg and Fedina (2009), although detailed comparisons with field data have not yet been made.

References

Brethouwer G, Lindborg E (2008) Passive scalars in stratified turbulence. Geophys Res Lett 35:L06809

Brethouwer G, Billant P, Lindborg E, Chimaz J-M (2007) Scaling analysis and numerical simulations of strongly stratified turbulent flows. J Fluid Mech 585:343–368

Charney J (1971) Geostrophic turbulence. J Atmos Sci 29(6):1087–1095

Gage KS, Nastrom GD (1986) Theoretical interpretation of atmospheric spectra of wind and temperature observed by commercial aircraft. J Atmos Sci 43:729–740

Gagnaire-Renou E, Benoit M, Badulin SI (2011) On weakly turbulent scaling of fetch limited growth. J Fluid Mech 669:178–213

Golitsyn GS (2007) Waves and diffusion on the sea surface. Izv RAS FAO 43(1):93–97

Golitsyn GS (2011) Coefficient of turbulent diffusion of impurities on the water surface depending on the stage of wave development. Izv RAS FAO 47(3):426–432

Golitsyn GS, Chkhetiani OG (2014) Effect of viscosity on the horizontal diffusion of impurities in the field of wind waves. Izv RAS FAO 50(6):547–553

Golitsyn GS, Fortus MI (2020) Random processes with stationary increments and composite spectra. Izv RAS FAO 56(4):418–427

Hasselmann K (1974) On the spectral dissipation of ocean waves due to white capping. Bound Layer Meteorol 6(2):107–127
Kraichnan RH (1970) Inertial-range transfer in two- and three-dimensional turbulence. J Fluid Mech 47:525–535
Lindborg E (1999) Can the atmospheric kinetic energy spectrum be explained by twodimensional turbulence? J Fluid Mech 338:259–288
Lindborg E, Brethower G (2008) Vertical dispersion by stratified turbulence. J Fluid Mech 2008(614):303–314
Lindborg E, Fedina E (2009) Vertical diffusion in stably stratified flows. Geophys Res Lett 36:L01605
Novikov EA (1958) On turbulent diffusion in a flow with a transverse velocity gradient. Appl Math Fur 22(5):412–414
Obukhov AM (1946) Turbulence in a temperature-inhomogeneous atmosphere. Proc Inst Theor Geophys USSR Acad Sci 1:95–115
Okubo A (1971) Oceanic diffusion diagrams. Deep-Sea Res 18:789–802
Ozmidov RV (1986) Diffusion of impurities in the ocean. L Gidrometeoizdat
Richardson LF (1926) Atmospheric diffusion shown on a distance-neigbour graph. Proc R Soc Lond A110(756):709–737
Richardson LF (1929) A search for the law of atmospheric diffusion. Beitr Phys Frei Atmos 15:24–29
Richardson LF, Stommel H (1948) Note on eddy-diffusion in the sea. J Meteorol 5(5):238–240
Stommel H (1949) Horizontal diffusion due to oceanic turbulence. J Mar Res 8(3):199–225
Taylor GI (1915) Eddy motion in the atmosphere. Philos Trans R Soc A215:1–26
Taylor GI (1921) Diffusion by continuous movements. Proc Lond Math Soc 20:196–211

Chapter 8
Statistical Structure of the Relief of Celestial Bodies—Kaula's Rule

The space age and its technologies led to the establishment of new natural patterns that remained ununderstood for many decades after their discovery. These include statistical patterns of gravity fluctuations, and then relief, first for our planet, and then for other celestial bodies (Turcotte 1997). In its original form, when it was noticed by Kaula (1966), soon after the names of the author called "Kaula's rule," which stated that fluctuations of the Earth's gravitational field, being expanded into a spatial spectrum into spherical harmonics, have an amazing property: the coefficients of these harmonics, starting from number $n \geq 4$, decrease as n^{-2}. Later it was found that Turcotte (1997) and the surface relief, considered as a random field, have property or the Moon, Mars, Earth (Rexer and Hirt 2015), Venus (Turcotte 1997) for the asteroid Vesta (Konopliv et al. 2014) with a size slightly less than 300 km, and in 2016 for a celestial body several kilometers in size (McMahon et al. 2016). In (Rexer and Hirt 2015) there was a special method for calculating harmonics up to $n = 43200$ was developed and such calculations for the Moon were carried out up to linear scales of 120 m, and for the Earth—up to 700 m. For Mars, the number of harmonics in Rexer and Hirt (2015) is half as much.

All this time, the origin of Kaula's rule remained unclear, and it was even called the "rule of thumb"[1], and only in 2019 the physical and mathematical origin and meaning of this rule were explained (Gledzer and Golitsyn 2019a, 2019b]. The explanation is based on the rules of A. N. Kolmogorov, described here in Ch. 1.3.

In fact, this is a consequence of the shortened equation (1.3) in the form of a simple Fokker–Planck equation, although it was the ideas of A. N. Kolmogorov inspired the authors Gledzer and Golitsyn (2019a) to deeply analyze the "Kaula rule" from the point of view of the distribution function for the probability of the relief field $p(y, h)$, where y is the meridional coordinate, h is the vertical coordinate, i.e. the height of a given surface area. The dimensions of such a section are determined by the resolution

[1] In ancient Rome, at gladiatorial games, the fate of the defeated gladiator was decided, without explanation, by the emperor: thumbs up - life, thumbs down - death. Nowadays, a thumbs-up is a sign of approval.

G. Golitsyn and C. Varotsos, *The Stochastic Nature of Environmental Phenomena and Processes*, https://doi.org/10.1007/978-3-031-77015-9_8

of the equipment producing star border relief measurements. The absence of a latitude coordinate in the distribution function p is associated with the features of spherical harmonics (their numbers in longitude $n \le m$) and the practice of presenting the results obtained with their help. Standard the form of the Fokker–Planck equation is:

$$\frac{\partial p}{\partial t} = \frac{D}{2} \frac{\partial^2 p}{\partial h^2}, \tag{8.1}$$

and it is in this form that a random temporary recording is interpreted by the device at a flyby vehicle on a satellite or airplane. With a known flight speed u, the time in the record is transformed into the horizontal coordinate $y = ut$, and then equation (8.1) takes the form:

$$\frac{\partial p}{\partial y} = \frac{D_1}{2} \frac{\partial^2 p}{\partial h^2}, \quad D_1 = \frac{D}{u}. \tag{8.2}$$

The diffusion coefficient D_l has the dimension of length. But if vertical and horizontal coordinates have their dimensions, then $D_1 = L_h^2/L_y$, which reflects the diffusion nature of the relief formation process. Last thing the equation gives the scale and structure function with zero initial conditions in the form of a known solution to this equation, namely, the second moment of the function probability distribution *p(y, h):*

$\langle h^2(y) \rangle = 2D_1 y$ with appropriate spectrum:

$$S(k) = \frac{D_1}{\pi} k^{-2}, \quad k = \frac{2\pi}{\lambda_y}, \tag{8.3}$$

where λ_y is the horizontal wavelength. This equation in this form applies to small areas. Thus, in Turcotte (1997), spectra are presented for $2 < \lambda_y < 60$km, measured in the 1980s in Oregon for flat, hilly, and mountainous areas. All of them have a power-law form k^{-n}, with $n = 2.03 \pm 0.04$ over 24 meridional and latitudinal spans, as estimated in Gledzer and Golitsyn (2019a).

The derivative of the relief along the horizontal coordinate is the slope angle, and the spectrum of these angles for $n = 2$ is equal to

$$S_\varsigma(k) = k^2 S(k) = D_1/\pi = \text{ const }, \tag{8.4}$$

i.e., this is a white noise spectrum, which corresponds to the δ-correlation of horizontal angles, and this in probability theory is a Markov process. This is, in our terms, the main position in the theory of A. N. Kolmogorov in 1934 (see Ch. 1.3):

$$\langle \varsigma(y_1) \varsigma(y_2) \rangle = \varsigma^2 \delta(y_1 - y_2). \tag{8.5}$$

In the book Turcotte (1997), the spectrum index $n = 2$ was proposed along with other statistical characteristics, such as the Hausdorff measure, and the fractal index, which are related to n, but without explaining their physical meaning, in connection with the spectrum of slope angles. Gravity acts along the relief slope, water flows, rocks fall, and the slope resists the wind—all these relief-forming factors act uncorrelated (up to certain not very large scales) and lead to a relief spectrum k^{-2}, that is, angles are a factor $g \sin\theta$, forming a relief.

Celestial bodies are finite formations, and their spectral harmonics are discrete. Let us give the basics of the spherical term used here. Harmonic analysis. Consider a random relief surface $h(\varphi, \theta)$ on a sphere with radius r, where φ is longitude, $0 \leq \varphi < 2\pi$, θ is the complement to latitude: $0 \leq \theta \leq \pi$. Then the derivative of the relief is equal to the angle θ.

$$\frac{\partial h(\phi, \theta)}{r\partial\theta} = \varsigma_\theta(\varphi, \theta)$$

Let us denote $y = \cos\theta$ and represent the relief as a sum of associated Legendre polynomials $P_n^{|m|}(y)$:

$$\frac{h(\varphi, \theta)}{r} = -\sum_{n=1}^{\infty}\sum_{m=-n}^{n} a_m^n \exp(im\varphi)\Phi_n^{|m|}(y),\ \Phi_n^{|m|}(y) = N_n^{|m|}P_n^{|m|}(y),$$

$$\int_{-1}^{1} \Phi_n^{|m|}(y)\Phi_k^{|m|}(y)dy = \delta_{nk},\quad N_n^{|m|} = \left(\frac{2n+1}{2}\frac{(n-|m|)!}{(n+|m|)!}\right)^{1/2}. \tag{8.6}$$

Then, using (8.6), the angle of inclination of the relief is expressed by its derivative:

$$\zeta_\theta(\varphi, \theta) = \sum_{n=1}^{\infty}\sum_{m=-n}^{n} a_m^n \exp(im\varphi)\sin\theta\frac{d\Phi_n^{|m|}(y)}{dy}. \tag{8.7}$$

Having performed all the necessary (and tedious) actions described in Gledzer and Golitsyn (2019a), for the spectral components of the relief we obtain:

$$S_n = \frac{\alpha^2}{n(n+1)},\quad E_\theta = 2\pi\sum_{n=1}^{\infty} S_n, \tag{8.8}$$

where α is the average angle of the slope in the relief, E_θ—total "energy" of inclination angles, obtained after summation over n. The total "energy" of the relief itself (in the root mean square sense after summing the amplitudes of all harmonics) turns out to be equal:

$$E = 4\pi\sum_{n=1}^{\infty}\frac{r^2a^2}{n(n+1)} = 4\pi r^2\alpha^2 = 4\pi rD_1, \tag{8.9}$$

where D_l is the horizontal diffusion coefficient in equation (8.2), and at the same time it was it is considered that:

Fig. 8.1 Amplitudes of spherical relief harmonics for the Earth, *a*, and Venus, *b*, from (Turcotte 1997).

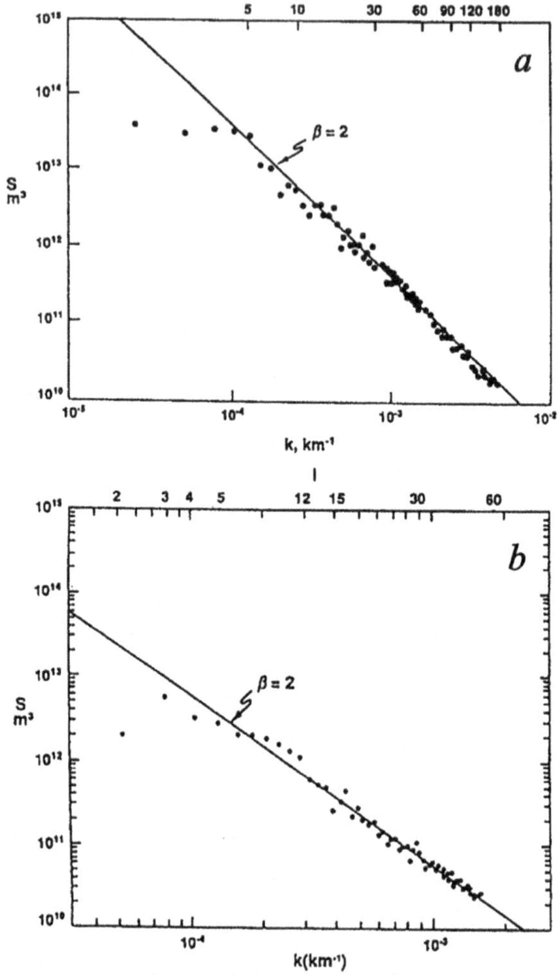

$$\sum_{n=1}^{\infty}\frac{1}{n(n+1)}=1, \text{ due to } \frac{1}{n(n+1)}=\frac{1}{n}-\frac{1}{n+1}$$

and the denominators of neighboring terms differ by one.

The resulting spectrum (8.8) differs from the "rule of thumb" spectrum Kaula n^{-2} since the amplitude of the harmonics here drops as $[n = (n+1)]^{-1}$. With increased harmonic numbers n this difference from n^{-2} decreases as $[n^2 = (n+1)]^{-1}$, but for small n the difference is quite noticeable, for example, for $n = 1$ the spectral component is half that of Kaula, however, harmonics with small n are determined by internal geodynamics of celestial bodies and, as can be seen in Fig. 8.1, amplitudes only starting from n $\geq$ 4 fits more or less satisfactorily on the line n^{-2}.

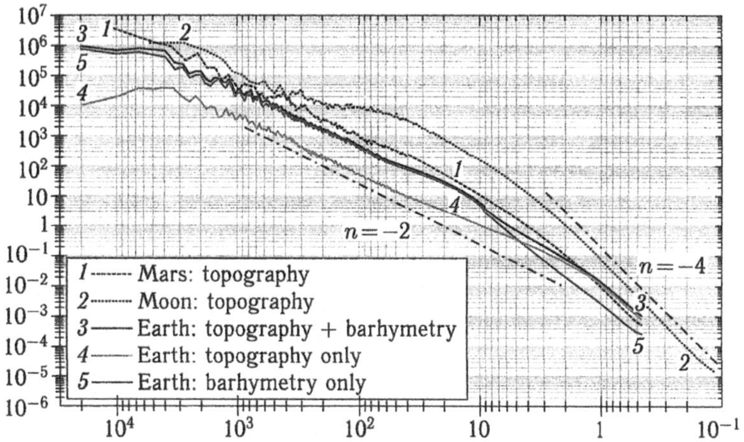

Fig. 8.2 Spatial spectra for the Earth, Mars, and the Moon according to Rexer and Hirt (2015), where asymptotes k^{-2} and k^{-4} were added in Gledzer and Golitsyn (2019a), from Gledzer and Golitsyn (2019b).

This figure, taken from the book (Turcotte 1997) makes it possible to estimate the diffusion coefficient D_1 for the Earth and Venus. According to (8.9), the spectral harmonic is equal to:

$$S_n = \frac{4\pi r D_1}{n(n+1)}. \tag{8.10}$$

To estimate the horizontal diffusion coefficient D1 for the Earth, we selected harmonics with numbers $n = 5, 10, 20, 30, 40, 60$, and 90. This gave $D_1 = 1.3 \pm 0.3$m. For Venus numbers 5, 10, 15, 20, 30, 40, and 50 were taken, which resulted in $D_1 = 0.2 \pm 0.03$m. The accuracy of determining D_l values is about 20%, the accuracy of determining values by eye we estimate the component with Fig. 8.1, but there is also a natural scatter. The difference in the values of the spectral coefficients for the two planets was noted by Turcotte (1997), with the remark that on Venus the data relate only to fairly smooth equatorial regions. D_l value is an analog of the numerical constant in Kaul's law n^{-2}.

Similar estimates can be obtained with much greater accuracy and statistical power from the data of Rexer and Hirt (2015), but the authors of Gledzer and Golitsyn (2019a), with a total age of more than 160 years, left this for younger colleagues with the hope that such work will yield a lot of new and interesting things. Data from Rexer and Hirt (2015), are presented in Fig. 8.2.

They are supplemented by two lines corresponding to the spectral dependences k^{-2} and k^{-4}. The transition to a steeper spectral dependence of the relief can be explained by the appearance at short distances of a correlation between slope angles (Gledzer and Golitsyn (2019a):

$$B_\theta(\theta(y_1)\theta(y_2)) = \exp(-\beta y), \tag{8.11}$$

where $\beta = 1/y_o$ is the reciprocal value for the wavelengths where such a correlation appears. The spectrum for such a correlation function is (see Ch. 1.1)

$$S_0(k) = \frac{\beta}{\pi\left(\beta^2 + k^2\right)}, \tag{8.12}$$

and if we know the spectrum of the derivative, here of the angles, then the spectrum of the quantity itself—the relief—will be:

$$S(k) = k^{-2} S_0(k) = \frac{k^{-2}\beta}{\pi\left(\beta^2 + k^2\right)}. \tag{8.13}$$

This spectrum has corrected asymptotics at both ends at $k^2 \ll \beta^2$ it is proportional to k^{-2}, and at small distances $k^2 \gg \beta^2$ spectrum k^{-4}. The gravity on the Moon is six times less than on Earth, and there is a high-frequency part of the spectrum manifests itself much earlier than on Earth and Mars, as can be seen in Fig. 8.2. Determination of the physical nature of the constant β and its connection with material constants of planets and their surfaces can be found in Golitsyn (2021), where it is shown that $\beta_0 \propto g$, which also follows from visual analysis of Fig. 8.2. It is obvious that the values of β are determined by the properties of the material of the surface rock, and the latter are related to the speed of sound in them. Eventually $\beta_0 = g/c_o^2$, $c_0^2 = \mu/\rho$, where μ is the modulus of uniform compression, i.e. Young's modulus. The work (Gledzer and Golitsyn (2019a) is an expanded version of the preliminary publication (Gledzer and Golitsyn (2019b).

References

Gledzer EB, Golitsyn GS (2019a) Kaula's rule—a consequence of probability laws by A. N. Kolmogorov and his school. Russ J Earth Sci 19:ES6007. https://doi.org/10.2205/2019ES000651

Gledzer EB, Golitsyn GS (2019b) Structure of the relief and gravitational field of planets: Kaula's rule as a consequence of the probabilistic laws of A. N. Kolmogorov and his school. Dokl. RAS. 485(4):391–394

Golitsyn GS (2021) Features of the spectrum of the surface of the Moon and planets. Astron Messenger 55(1):34–37

Kaula WM (1996) Theory of Satellite Geodesy. Bleinsdell, Waltham. Ma– pp 124

Konopliv AS et al (2014) The Vesta gravity field. Icarus 240:103–117

McMahon JW et al. (2016) Understanding the Kaula's rule for small bodies. In: Proceedings 47-th lunar and planetary science conference, AGU, Washington, D.C.. https://www.hou.usra.edu/meetings/lpsc2016/pdf/2129.pdf

Rexer M, Hirt C (2015) Ultra-high spherical harmonics analysis and application to planetary topography of earth. Mars and Moon Surv Geophys 36(6):803–830

Turcotte DL (1997) Chaos and fractals in geology and geophysics. 2nd ed. Cambridge, Cambridge University Press, pp 398

Chapter 9
Stochastic Motions at the Prescribed Rotation (Hurricanes et al)

9.1 The Scale of Events and Similarity Parameters

As noted in the Introduction and Chap. 5, random motions during rotation are characterized by the fact that in them, time is specified by the period of rotation, which determines the speed and reactions of emerging objects. This increases the stability hydrodynamic, and simply mechanical, systems and movements (for example, bike). This is clearly seen in the example of convection, the onset of which tightens with increasing rotation, which was first shown analytically Chandrasekhar (1961). The process is characterized by three similarity numbers: Rayleigh number:

$$\mathrm{Ra} = \frac{\alpha g h^3 \Delta T}{k\nu}, \tag{9.1}$$

where α is the coefficient of thermal expansion of the liquid, h is the thickness of the considered layer across which the gravitational acceleration g acts, ΔT is the difference temperatures between the lower and upper boundaries of the layer, k and ν are molecular thermal diffusivity and viscosity coefficients.

The second similarity parameter is the Taylor number:

$$\mathrm{Ta} = \frac{4\Omega^2 h^4}{\nu^2} = \mathrm{Ek}^{-2}, \tag{9.2}$$

the square of the ratio of the Coriolis force to the viscous forces, Ek is the Ekman number. In Chandrasekhar (1961) graphs of stability curves are presented in a linear approximation for different types of boundary conditions for velocity in a flat layer of liquid. For not very large Ta the value of $\mathrm{Ra_{cp}}$ increases slowly, but for $\mathrm{Ta} > 10^6$ it grows like $\mathrm{Ta}^{2/3}$. The third similarity parameter—Prandtl number $\mathrm{Pr} = \nu/k$.

Chandrasekhar (1961) established that convection occurs when the Rayleigh number (9.1) $\mathrm{Ra} > 10^6$ reaches a critical value proportional to:

G. Golitsyn and C. Varotsos, *The Stochastic Nature of Environmental Phenomena and Processes*, https://doi.org/10.1007/978-3-031-77015-9_9

$$\mathrm{Ra_{cr}} = k_i \mathrm{Ta}^{2/3} = k_i \mathrm{Ek}^{-4/3}, \tag{9.3}$$

where k_i—numerical coefficient determined by boundary conditions. From (9.1), (9.2), (9.3) we can obtain an expression for the reduced force of gravity for sufficiently large Taylor numbers:

$$g' = g\alpha\Delta T = 2.83 k_i \mathrm{Ta}^{2/3} k v h^{-3}.$$

The role of rotation is significant, and the movements are geostrophic when the dimensionless number Kibel–Rossby

$$\mathrm{Ro} = U / l_c d << 1,$$

where U is the speed of movement, $l_c = 2\Omega\sin\theta$.– Coriolis parameter on the sphere, θ—addition to latitude. This number is the ratio of the rotation period to dynamic period $T_o = d/U$, where d is the diameter of the phenomenon.

In convection, the main driving role is played by the buoyancy flux density $b = \alpha g f / \rho c_p$, where f is the heat flux per unit area, c_p is the heat capacity at constant pressure. Under quasi-stationary conditions, on average equal to the speed of energy generation/dissipation. The amount of kinetic energy in the system is equal to incoming energy multiplied by the shortest time characteristic of the system, which for Ro << 1 is estimated by the inverse Coriolis parameter l_c^{-1}. As a result, from dimensional considerations, as well as from the Kolmogorov scale (1.31), we obtain for layer

$$U = c(b/l_c)^{1/2}, \tag{9.4}$$

where $c_1 \approx 1.7$ is a numerical coefficient with an accuracy of several percent (Boubnov and Golitsyn 1995), according to numerous laboratory measurements. Similar considerations lead to the horizontal scale of area irregular vortices arising in the flow after reaching Ra > 30 $\mathrm{Ra_{kp}}$. (Ch 13). Identity with the ANK34 scales at a fixed Coriolis time is obvious: the square of the speed in (9.4) is proportional to time l_c^{-1}, and the area scale (see Ch 1.3).

$$S = c_1 b l_c^{-3}, \tag{9.5}$$

where $c_1 \approx 10$ (Boubnov and Golitsyn 1995) is necessary at comparison with experiments. Equation (9.5) is the moment $< x^2 > = \varepsilon t^3$. Fixed time at a given excitation allows the development of a certain size of vortices and currents. This is clearly manifested in the comparison of polar and tropical hurricanes: the first ones at latitude θ = 70° are several times smaller than tropical ones at latitude θ = 20°. Indeed, (sin 20°/sin70°)$^{3/2} \approx 0.22$.

The fixation of the rotation period is clearly manifested in horizontal diffusion (Boubnov and Golitsyn 1991, 1995), when the area of the marker spot grows linearly with time. Then the diffusion coefficient is constant and equal in size (Ch. 5)

$$K = b\Omega^{-2} \sim bT_r^2,$$

where T_r—period of rotation, constant value. It is instructive to compare this expression with the second scale (1.31), indicated by Obukhov (1962).

$$\langle ux \rangle = \varepsilon t^2, \tag{9.6}$$

where t—already current time. Of course, both scales are the same by dimension. The setting of the rotation period makes the current more stable compared to the cases of its absence, introducing at the same time some specificity. A series of results obtained in laboratory experiments (Bubnov and Golitsyn 1990, 1991) demonstrates a regular and irregular mode of convection in a rotating vessel with a diameter of 17 cm (see Figs. 13.6, 13.7, and 13.9).

Sir James Lighthill (1923–1998), a foreign member of the USSR and of many other academies, headed in 1990–1999. International Scientific Committee at the UN decade (1990–1999) to combat natural disasters. In his lectures in 1995, he noted that the fact remains unclear why tropical hurricanes reach kinetic energy of the order of $10^{18} \div 10^{19}$ J, which is equivalent to the explosion of thousands of hydrogen bombs. The strength and power of hurricanes have always amazed people, bringing terrible calamities and destruction. The hurricane of 1265 destroyed the Mongol fleet, which went after the conquest of China to conquer Japan. Every year in the world of about 80 hurricanes are observed, and the total effect from them and other meteorological disasters noticeably surpasses all other natural disasters according to the UN.

9.2 Hurricanes

Systematic observations of hurricanes began in World War II, especially over the Pacific Ocean during the US war with Japan. They received systematic development at the beginning of the satellite era. Understanding that hurricanes arise as a result of thermodynamic disequilibrium between the ocean and the atmosphere, which came only from the middle of the twentieth century (Riehl 1954). No wonder its hurricane season characterized by conditions when the surface of the ocean water is warmer than the air (Palmen and Newton 1969). Then a flow of sensible and latent heat (evaporation) arises, forming a buoyancy flow:

$$b = -\frac{g}{\rho_0}\langle \rho' w' \rangle, \tag{9.7}$$

where ρ_0 is the density of the ambient air, the dash marks the fluctuations of the vertical components of velocity and density. The state is determined not only by temperature fluctuations, but also by water vapor, which is 0.622 times lighter than air.

We remind you that under quasi-stationary conditions, the value of b is equal to ε, the rate of generation/dissipation of kinetic energy of turbulence. We remember A.N. Kolmogorov 1934 (Sect. 1.3), that the scale of the mean square velocity is equal to $U^2 = bt$, and the area scale $< x^2 > = S = bt^3$. The time scale is set by the rotation of the planet, on the sphere—by the Coriolis parameter $l_c = 2\omega \sin\theta$, where θ is the complement of latitude. Rising convective columns, and thermals, suck the air surrounding them, concentrating its angular momentum. To obtain the total kinetic energy, you need to multiply the product of these two scales per mass of a single atmospheric column $M = 10^4$ kg/m^2. Eventually:

$$K_e = Mb^2 l_c^{-4}. \tag{9.8}$$

This compact formula, previously obtained by dimensional analysis (Golitsyn 2008), gives the correct order of magnitude for both tropical and polar hurricanes. So, for latitude 20° $l_c = 0.5\ 10^{-4}\ c^{-1}$ and $b = 1$ kW/m^2 we get $K_c \approx 10^{18}$ J, and if we consider the factor c_1 in (9.7), then we get 10^{19} J. Note that 1 Mt of explosive trinitrotoluene is equal to 1.6 $10^{15} J$. At latitudes $\theta = 70°$ the Coriolis parameter is $l_c = 1.37\ 10^{-4}\ c^{-1}$, and polar hurricanes arise in them when cold air invades seas open from ice, the buoyancy parameter arises mainly due to the huge temperature differences between air and water, where $T \approx 0°$, so their sizes are several times smaller than the size of tropical hurricanes, (Fig. 9.1). Figure 9.2 gives a photo of a polar hurricane, and Fig. 9.3—a hurricane of similar formation over the Black Sea.

A detailed consideration of the formation of heat and moisture flows, considering the emerging winds, made it possible to obtain the necessary conditions for the occurrence of hurricanes (Golitsyn 2008, 2009). Let us briefly describe the process of

Fig. 9.1 Tropical cyclone-hurricane "Ivan" 09/15/2004 (https://www.nesdis.noaa.gov/search/content/Ivan)

Fig. 9.2 Polar hurricane (mesocyclone). G12

obtaining them. The interaction of the ocean and the atmosphere, and their exchange of momentum, heat, and moisture is described by the so-called bulk formulas (Gray 1968; Kitaygorodsky 1970; Fairall et al. 2003):

$$\tau = \rho\langle u'w'\rangle = \rho u_*^2 = c_d \rho U^2, \tag{9.9}$$

$$F_{SH} = \rho c_p\langle w'T'\rangle = c_T \rho c_p U \Delta T, \tag{9.10}$$

$$F_{LH} = \rho\langle q'w'\rangle = c_E \rho U \Delta q, \tag{9.11}$$

where u', w'—fluctuations of the horizontal and vertical components, T' and q'—fluctuations of temperature and water vapor mixing ratio, c_d, c_T, c_E—resistance coefficients for impulse, sensible and latent heat, U—speed wind at a standard height of 10 m above sea level, $\Delta T = T_1 - T_a$ temperature difference between the water surface and the atmosphere at z = 10 m, Δq is the difference between the mixing

Fig. 9.3 Hurricane-like whirlwind over the Black Sea 09/25/2005 (Efimov et al. 2007)

ratios for water vapor at the sea surface and the same height of 10 m. At the surface, the relative humidity is assumed to be 100%. The mixture ratio is related to the specific density *e* (*T*), i.e., water vapor pressure, ratio (Riehl 1954; Palmen and Newton 1969):

$$q = 0.622\frac{e}{p},$$

where p—atmospheric pressure, $0.622 = \mu_w / \mu_\alpha$—molecular weight ratio for water vapor and air. Relative humidity $r = e/e_s$, where e_s is pressure saturation. According to the Clayperon–Clausius equation, the value of e_s depends exponentially on temperature. There are various approximations of this dependence (Golitsyn 2008, 2009). With an accuracy of several percent, the heat of condensation of water vapor is $L = 10^6 L_1$ J/kg, and the value of L_1 in the observed range of sea surface temperatures from—2° to 31° varies from 2.506 to 2.430.

An essential parameter for describing convection is the buoyancy flux, which, considering the equations of state of air and water vapor, is equal to:

$$b = -\frac{g}{T}\langle w'T'\rangle - 0.622\langle w'q'\rangle$$

Using formulas (9.10) and (9.11), this expression is reduced to the form:

$$b = -g'U, \quad g' = c_T \frac{\Delta T}{T} g \left(1 + 0.378 \frac{\Delta e}{\rho \Delta T}\right), \quad \Delta e = e_s(T_s) - r e_s(T_s) \quad (9.12)$$

where the value g' is the reduced acceleration of gravity in heated and humidified air, $0.378 = (\mu_a - \mu_w)/\mu_a$ and also accepted the following (Kitaygorodsky 1970; Fairall et al. 2003) the equality of the coefficients c_T, $c_E \approx 1.3 \cdot 10^{-3}$.

Equation (9.12) determines the buoyancy flux before moisture condensation. Calculations and observations for hurricanes show that at typical relative humidity values r = 0.75–0.8 per 10 m, cloudiness forms at an altitude of 300–400 m. TH, tropical hurricanes, penetrate through the entire troposphere up to 15–18 km (Riehl 1954; Palmen and Newton 1969), and PH, polar hurricanes, reach 5–6 km. It wouldn't be a big mistake if you neglect the thin sub-cloud layer and assumed that latent heat, i.e., condensation begins to act directly from the surface. Then the reduced acceleration (9.12) can be rewritten as:

$$g' = c_T \frac{\Delta T}{T} \mathrm{Bo}^{-1}, \quad \mathrm{Bo}^{-1} = \frac{\mu_w}{\mu_a} \frac{L \Delta e}{p c_p \Delta T} = 0.614 L_s \frac{\Delta e}{T}, \quad (9.13)$$

where the Bowen number Bo is the ratio of sensible and latent heat fluxes, and it is accepted that $p = p_s = 1.013 \cdot 10^5$ Pa—average pressure at sea level.

From our parameters l and l_c one can construct the square of the velocity (9.4) and the area scale (9.5), which gives the formula for the kinetic energy of a hurricane (9.8). Formulas (9.9–9.11) for the buoyancy flow are linearly proportional to the wind speed U, and the velocity scale (9.4) is proportional to $b^{1/2}$. This makes it possible to express the speed following the thermodynamic disequilibrium between the ocean and the atmosphere through the reduced force of gravity:

$$U \approx 3 g' l_c^{-1}, \quad (9.14)$$

where $3 \approx 1.7^2$ from (9.4). For $T_s = 300\text{K} = 27°\text{C}$, $\Delta T = 1\text{K}$, r = 0.8, and $\theta = 20°$ latitude we get $U = 34$ m/s, typical speed in specifications. For PH at $\theta = 70°$, $Ts = 275\text{K}$, $r = 0.7$, to get hurricane winds $U \geq 33$ m/s, must have $\Delta T \geq 22$ °C. These are temperature differences in good agreement with observations and numerical experiments (Rasmussen and Turner 2003; Efimov et al. 2007). The latest work reproduces a hurricane in the Black Sea western part on 09/25/2005, which arose during $\Delta T \approx 12$ °C.

To describe the development of hurricanes, a theory of penetrating convection is needed (Turner 1973; Zilitinkevich1987). The basic formula for the dependence of the height h of the penetrating layer in a stable stratified atmosphere looks like:

$$h(t) = N^{-1}(2bt)^{1/2}, \quad (9.15)$$

where N is the Brunt-Väisälä frequency, a characteristic of atmospheric stability, defined as:

$$N^2 = -\frac{g}{\rho}\Gamma, \quad \Gamma = \left(\frac{d\rho}{dz}\right)_a - \frac{d\rho}{dz}, \quad \left(\frac{d\rho}{dz}\right)_a = -\frac{\rho g}{c^2}, \tag{9.16}$$

where c is the adiabatic speed of sound. Observations indicate that the hurricane is in the tropics it can develop in a day, and in high latitudes—in a few hours, since its upper limit is lower. Based on formulas (9.12) and (9.16), works (Golitsyn 2008, 2009) constructed graphs in coordinates ΔT and Ts for the necessary development conditions for tropical hurricanes per day (Fig. 9.4) and 5.5 h in polar conditions (Fig. 9.5). In addition to necessary conditions, we also need very diverse sufficient conditions on wind shear with height, on the proximity of vertical thermal and humidity gradient to adiabatic values and their homogeneity, etc., (Gray 1968; Golitsyn 2008, 2009), otherwise condensation of water vapor may not occur. For all this, there are not yet (and unlikely to be soon) the necessary measuring instruments with the required vertical resolution. Therefore, we will not have to expect operational predictions of the emergence of new hurricanes tied to specific places for a long time (if ever possible), although, when an object has already arisen, its further evolution and trajectory are already predictable, and the accuracy of such forecasts are improving.

In Fig. 9.1, 9.2, 9.3, 9.4 examples are presented of tropical and polar hurricanes and a hurricane-like vortex over the Black Sea (see Munk et al. 2000; Rasmussen and Turner 2003; Efimov et al., 2007). Nomograms for calculating the evolution of hurricanes, where c is the adiabatic speed of sound. Observations indicate that a tropical hurricane can develop in a day, and in high latitudes—in a few hours since its upper limit is lower. Based on formulas (9.12) and (9.16), works (Golitsyn 2008, 2009) constructed graphs in coordinates ΔT and Ts for the necessary development conditions of tropical hurricanes per day (Fig. 9.4) and 5.5 h in polar conditions (Fig. 9.5).

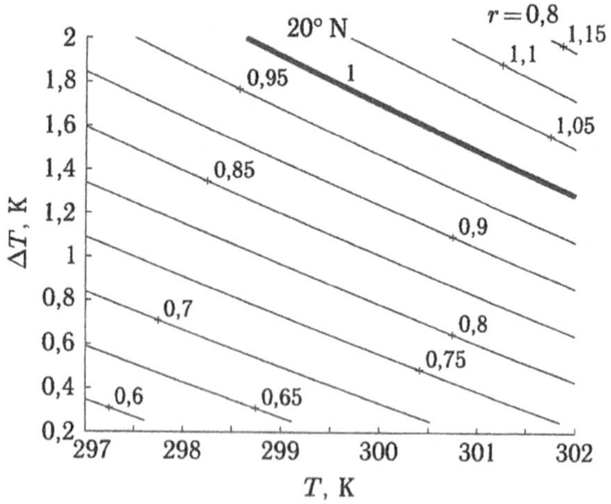

Fig. 9.4 Nomogram for the conditions for the onset of development of a tropical hurricane in one day before reaching $h = 18$ km, horizontally—ocean surface temperature, vertically—the difference in atmospheric temperature at specific humidity $r = 80\%$. (Golitsyn 2008, 2009)

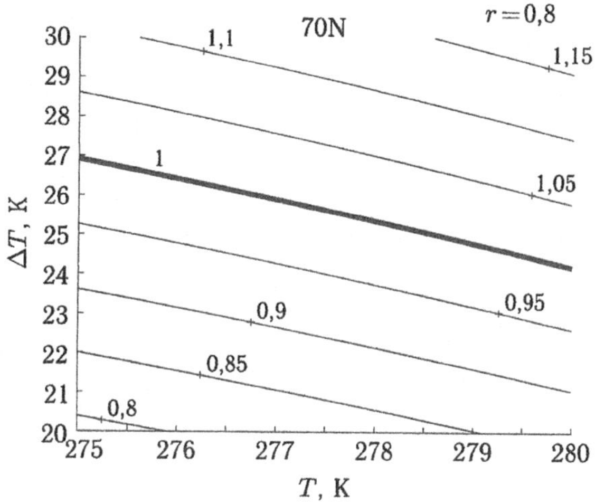

Fig. 9.5 Nomogram for the development of a polar hurricane in 5.5 h. (Golitsyn 2008, 2009)

In addition to necessary conditions, we also need very diverse sufficient conditions on wind shear with height, on the proximity of vertical thermal and humidity gradient to adiabatic values and their homogeneity, etc. (Gray 1968; Golitsyn 2008, 2009), otherwise, condensation of water vapor may not occur. For all this, there are not yet (and unlikely to be soon) the necessary measuring instruments with the required vertical resolution. Therefore, we will not have to expect operational predictions of the emergence of new hurricanes tied to specific places for a long time (if ever possible), although, when an object has already arisen, its further evolution and trajectory are already predictable, and the accuracy of such forecasts are improving.

Figures 9.1, 9.2, 9.3, 9.4 give examples of tropical and polar hurricanes and a hurricane-like vortex over the Black Sea (see Munk et al. 2000; Rasmussen and Turner 2003; Efimov et al. 2007). Nomograms for calculating the evolution of hurricanes were first published in Golitsyn (2008, 2009) with a detailed discussion of their use.

The surface of the ocean in the absence of strong waves gives a remarkable picture of spiral vortices, as shown in Fig. 13.8—marine satellite image surface area of about a thousand km^2 (cf. Figure 13.7). An analysis of the nature of these vortices is given in Golitsyn (2012). The sea gives up its heat to the atmosphere when it is colder. With winds up to 3 m/s, waves do not yet prevent the visibility of spiral vortices from space. With such weak winds and temperature differences of several degrees, the sum of sensible and latent heat released into the atmosphere will (Kitaygorodsky 1970; Fairall et al. 2003; Rasmussen and Turner 2003) be less than 100 W/m^2. In this case, the buoyancy flow in water is directed inward, and at water parameters, it will be $\leq 5\ 10^{-8}$ m^2 s^{-3}, which is five orders of magnitude less than in atmospheric hurricanes.

At the same time, according to (9.4) the speed in water will be of the order of 5 cm/s, and the diameter will be of the order of 5 km. Such vortices are never observed in strip $\pm 5°$ along the equator, which clearly confirms (Golitsyn 2012) their hurricane-like hydrodynamic nature in water, in which the orders of magnitude are much different from those typical for air. Thus, the vortices of Fig. 9.3 can be called micro-hurricanes on the sea surface. The uniformity of processes and phenomena in nature is obvious.

9.3 Hurricane-Like Vortices

Hurricanes and specific hydrodynamic formations arise in the presence of rotation in the system and energy sources in it. The latter leads to convection and rise of heated volumes, and this causes convergence of surrounding masses, which concentrates the angular momentum introduced and associated with the general rotation of the Earth. Vortices arise, the size and intensity of which depend on the sources of buoyancy b and the properties of the medium.

The closest, fortunately, and rare, analog is fire tornadoes that occur during long-term fires in large cities, as was the case in 1943 in Hamburg, and 1945 in Dresden and Hiroshima. Calculations show (Andrianov et al. 2003) that a few hours after the start of a fire, a cyclonic vortex with winds of up to 70 m/s and more is formed with a large release of energy. Another example are *vortices*, which almost all have a cyclonic rotation. They occur in the rear part of extensive thunderstorm mesocyclones with dimensions of the order of 100 km, where, according to Doppler radar observations, downward movements exist. The mesocyclones themselves rotate as a whole with angular speeds, 10 times faster than planetary rotation. There are several reasons for downward movements: cooling of the upper boundaries of the clouds by thermal radiation escaping into space, which gives negative buoyancy to descending tornadoes along with raindrops, which, evaporating, also cool the air. The convergence of ambient air into a tornado is uneven in height, the speed of the surrounding air increases as it descends, which leads to compression of the jet (like water falling from a faucet is compressed).

9.4 Energy of Tornadoes and Landspouts

As we have already mentioned Kolmogorov's seminal work, "Random Motions," published in 1934, established a significant relation between random velocities and the spatial extent of motion in natural events and phenomena. This groundbreaking study was based on the Fokker–Planck–Kolmogorov equation, which describes the probability distributions of a six-dimensional vector, $p(t, u_i, x_i)$. The second moments of this vector with respect to velocity and coordinates were later determined by A.M. Obukhov in 1958 and further validated in subsequent studies. Building upon Kolmogorov's ideas, various natural processes and phenomena were elucidated in a book, providing insights into the behavior of atmospheric vortices like tornadoes and landspouts. By estimating the moments of these small-scale vortices, it becomes feasible to calculate the rate of energy generation, denoted as ε:

$$< u^2(t) > = c_1 \varepsilon t$$
$$< x^2(t) > = S = c_2 \varepsilon t^3 = r^2 \quad (9.17)$$

The expressions for these moments involve dimensionless constants, c_1 and c_2, which are determined through theoretical and experimental comparisons to ascertain the rates of kinetic energy generation per unit mass. These constants, assumed to be of the order of unity, are crucial for understanding the dynamics of such phenomena.

Golitsyn et al. (2023) utilized empirical data from 164 tornadoes and landspouts in Russia since the early twenty-first century. These events were documented by reliable observers and included information on whirlwind intensity measured on the Fujita scale and the maximum width of the whirlwind (horizontal size *D* of the vortex). The diameters of these vortices range from 15 to 1750 m, with average velocities ranging from 25 to 100 m/s (Fig. 9.6a) (Table 9.1). Assuming axial symmetry of the vortex, the area *S* is calculated based on the vortex radius *R* and diameter *D as*: $S = \pi D^2/4$.

The intensity of a whirlwind on the Fujita scale can be converted into a range of horizontal wind velocity *u* typical for a specific tornado intensity. In the absence of additional data, a standard average wind velocity is accepted for further analysis. However, this may result in some variations compared to more precisely measured parameters, such as the estimation of the forcing value of ε obtained from the Eqs. (9.17) as: $\varepsilon = \; < u >^3/S^{1/2}$

If the measurement errors are not known, all of these equations will produce the same results. However, when the measurement errors are known, the above calculations will yield a smaller error in estimating the forcing due to its lower power. For the whirlwind sample being analyzed, the values of the forcing ε range from 7×10^3 m^2/s^3 (Fig. 9.6b, Table 9.2), which exceeds the typical value of 5×10^4 m^2/s^3 observed in estimates within the Earth's atmosphere by up to seven orders of magnitude. The consistency of these estimates with common sense can be attributed to Kolmogorov's law of random motions. The characteristic estimates of insert variable] range from 0.4 to 23.5 s (Fig. 9.6c). It is worth noting that the median values of the characteristic impact time t_0 only change weakly (by a factor of three, Table 9.2) when transitioning from tornadoes of intensity F_0 to tornadoes of intensity F_4. The time t_0 is closely related to the time of vortex rotation around its axis, and the reciprocal value of t_0 represents the vertical vorticity. It is higher for weaker vortices and relatively lower for the most intense vortices.

Let t_0 be the duration of the vortex's impact on the environment, then:

$$t_0 = (S/ < u >^2)^{1/2} \quad (9.18)$$

We can calculate the final energy of an individual vortex by assuming that the dimensionless constants c_1 and c_2 are approximately equal to unity, knowing the height of the vortex, and multiplying Eqs. (9.17) by the mass of the atmospheric column M. This allows us to estimate the kinetic energy E involved in the motion or, more practically, in the destruction as $E = M\varepsilon^2 t^4$.

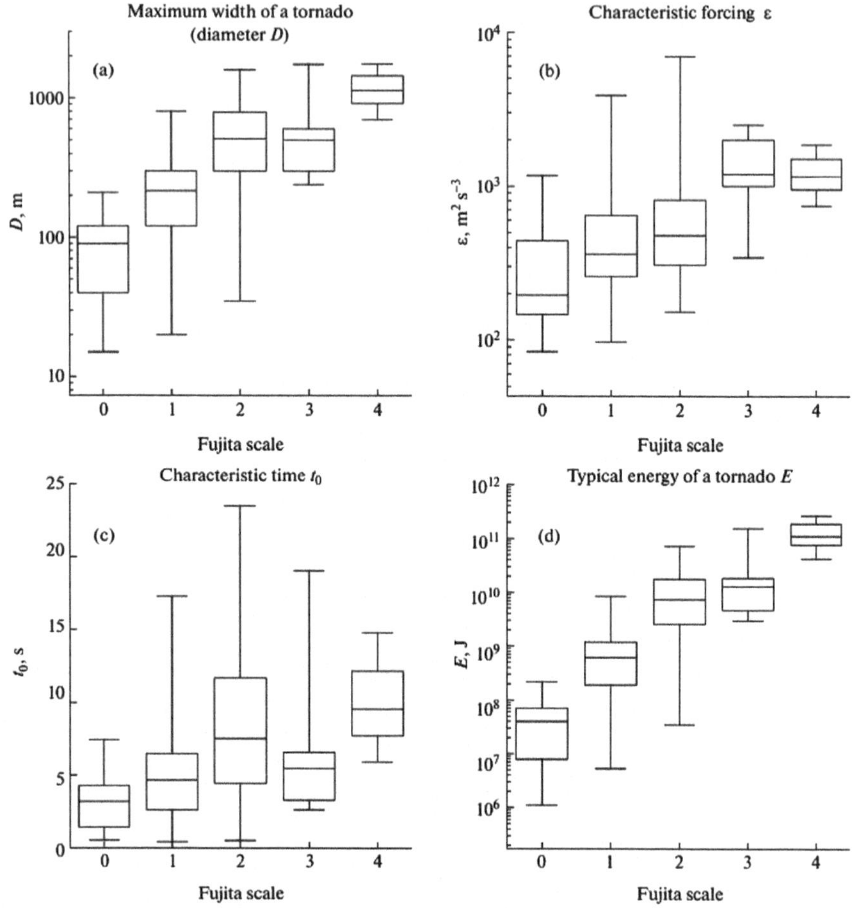

Fig. 9.6 Distribution of characteristics of tornadoes and lanspouts with respect to intensity on the Fujita scale: **a** the maximum width of a whirlwind (diameter D), as well as **b** the forcing values (the rates of energy generation) ε calculated in Golitsyn et al. (2023), **c** the characteristic time of impact $t_0 = (S/ < u >^2)^{1/2}$ and **d** the typical energy E

Table 9.1 Sample of tornadoes and landspouts used in Golitsyn et al. (2023)

Intensity on the Fujita scale	Speed range for this intensity, m/s	Mean speed in the range, m/s	Number of tornadoes of a given intensity in the sample
F0	18–32	25	21
Fl	33–50	41	66
F2	51–70	60	61
F3	71–92	81	13
F4	93–116	105	3

Table 9.2 Median values of tornado characteristics for different intensities (Golitsyn et al. 2023)

Intensity on the Fujita scale	Forcing ε, m^2s^{-3}	Characteristic time t_0, s	Tornado energy *E*, J	Energy in equivalent of TNT, tons of TNT	Tornado energy ***E'*** for a typical size of a house, J
F0	195.9	3.2	3.98×10^7	9.5×10^{-3}	7.72×10^5
F1	361.9	4.6	6.11×10^8	0.1	2.13×10^6
F2	477.9	7.5	7.35×10^9	1.8	3.60×10^6
F3	1199.4	5.5	1.29×10^{10}	3.1	6.56×10^6
F4	1156.0	9.5	1.11×10^{11}	26.4	1.10×10^7

The expression of the characteristic time of impact is given by Eq. (9.18), with a height of 10 m considered for the layer involved in the destruction, known as the Fujita index. It should be noted that the strongest wind is typically observed in the lower part of the tornado. In this case, the mass per unit area (*M*) is assumed to be 10 kg/ m^2, which corresponds to a temperature close to 300 K. Based on these assumptions, the calculated values for the energy (*E*) are on the order of 10^9 J (ranging from 10^7 to 10^{11} J) for the analyzed tornadoes (Fig. 9.6d). These values are eight orders of magnitude lower than the typical energy values of tropical hurricanes.

If we express the energy of whirlwinds in terms of TNT equivalent, the median values range from 9.5 kg TNT for F_0 intensity tornadoes to 26.4 tons of TNT for F_4 intensity tornadoes (Table 9.2). Additionally, the characteristic values of energy (*E'*) for the volume of 10^3m^3, which is similar to the volume of rural structures most frequently subjected to destruction, are estimated to be $E' = E \times 100/S$. The values of *E'* (the energy per unit area) vary from 4.4×10^3 to 3.18×10^8 J, increasing with an increase in tornado intensity (Table 9.2).

It is important to emphasize that the approximations provided pertain to a height of approximately 10 m, where rural constructions with a single-store are situated. The velocities are indicated based on the Fujita scale, which corresponds to the extent of destruction, with the highest speeds in tornadoes typically occurring near the surface as indicated by Golitsyn et al. (2023). This phenomenon is a result of the conservation of angular momentum, akin to water flowing out of a rotating faucet. In line with the principle of conservation, the observed alteration in the stream's diameter leads to an escalation in its horizontal velocities. The dynamic interplay between the vortex and the underlying surface significantly contributes to amplifying the surface wind speed in a tornado. The computations enable the estimation of the correlation between the width and duration of the vortex (expressed through the vortex path length) and its intensity. The estimates of whirlwind energy we have derived can be utilized for evaluating potential damage linked to tornado passage, including for adapting to climate variations concerning the overall intensification of hazardous convective phenomena observed in Russian regions (Golitsyn et al. 2023).

References

Andrianov SA, Vasilchenko II, Golitsyn GS, Zabavin VN, Shcherbin MD (2003) Numerical modeling of the formation processes of a "fire tornado." Izv. RAS. FAO. 39(1):3–13

Boubnov BM, Golitsyn GS (1995) In: Convection in rotating fluids. Dordrecht, Kluwer Academy Publications, pp 224

Boubnov BM, Golitsyn GS (1990) Temperature and velocity field regimes of convective motions in a rotating fluid layer. J Fluid Mech 219:215–239

Bubnov BM, Golitsyn GS (1991) Diffusion in liquid under geostrophic convection regime. Dokl Acad Sci USSR 317(3):602–605

Chandrasekhar S (1961) In: Hydrodynamic and hydromagnetic stability. Clarendon Press, Oxford University Press, pp 652

Efimov VV, Shokurov MV, Yarovaya DA (2007) Numerical modeling of a quasi-tropical cyclone over the Black Sea. Izv Atmos Ocean Phys 43(6):723–733

Fairall CW et al (2003) Bulk parametrization of air-sea fluxes: updates and verification for the COARE algorithm. J Climate 16(3):572–591

Golitsyn GS (2009) Tropical cyclones and polar lows: velocity, size and energy scales, and relation to the 26 °C cyclone origin criteria. Adv Atmos Sci 26(3):585–598

Golitsyn GS (2012) On the nature of spiral vortices on the surface of seas and oceans. Izv Atmos Ocean Phys 43(3):391–395

Golitsyn GS, Chernokulsky AV, Vazaeva NV (2023) Energy of Tornadoes and Landspouts. Phys Atmosphere and Hydrosphere Doklady Earth Sci 513(1):1211–1214. https://doi.org/10.1134/S1028334X23601554

Golitsyn GS (2008) Polar and tropical hurricanes: their energy, sizes and quantitative criteria for their occurrence. Izv Atmos Ocean Phys 44(5):579–591

Gray WM (1968) Global view on the origin of tropical disturbances and storms. Mon Wea Rev 96(6):669–700

Kitaygorodsky SA (1970) Physics of interaction between the atmosphere and the ocean. – L.: Gidrometeoizdat, pp 284

Munk W, Armi L, Fischer K, Zachariasen F (2000) Spirals on the sea. Proc Roy Soc London 456(A):1217–1280

Obukhov AM (1962) Description of turbulence in terms of Lagrangian variables. In collection. Atmospheric diffusion and air pollution. M.: IIL, pp 138–140

Palmen E, Newton CW (1973) Atmospheric circulation systems. Ch. 5. New York, Acadamy Press, 1969. pp 606. (Russian translation: Palmen E., Newton C.W. Atmospheric circulation systems. - M.: Mir, - 614 p

Rasmussen E, Turner J (eds) (2003) In: Polar lows. Mesoscale weather systems in polar regions. Cambridge University Press, vol. 612, pp 108

Riehl H (1960) Tropical meteorology. McGraw-Hill, New York, 1954. pp 392. (Russian translation: H. Riehl. Tropical meteorology. - M.: IIL, 370 p.)

Turner JS (1973) In: Convection of buoyant fluids. Cambridge University Press, pp 367

Zilitinkevich SS (1987) Theoretical model of penetrating turbulent convection. Izv RAS FAO 23(6):593–610

Chapter 10
Size Distributions for Lakes and Rivers. Flood Damage

10.1 Distributions for Rivers and Lakes

As in the previous sections, here there are many small rivers and lakes, but few large ones. Their size and quantity reflect the relief of the earth's surface; they should be in its valleys and lowlands, and in what proportions began to become clear only in the last quarter of the twentieth century, which is described in Chap. 8. Floods are also closely related to topography (and precipitation), so it is natural to consider all this in one point. The areas of lakes, of course, should depend on the level of precipitation and processes of evaporation, which is determined by the climate of the region. Size distributions for lakes and rivers are presented in Turcotte's book (1997) in figures 8.3 and 8.4. They are well described thereby power laws. For rivers, the cumulative length distribution $N\ (\geq l)$ proportionally l^{-n}, where $n = 1.9$, and for lakes cumulative distribution by area

$$N(\geq S) \propto S^{-n}, \quad n = 0.95, \tag{10.1}$$

and according to the average linear size, defined as $S^{1/2}$ corresponding to the indicator is also 1.9 (cf. Chap. 8). Equality of indicators in cumulative distributions for river lengths and average linear size of lakes indicates the unambiguous role of relief in determining these distributions: in both cases, the catchment areas, their distributions according to sizes. A slight difference in the exponent in the distribution of lakes by area from one, or two for river lengths, may characterize some increased role of large-scale processes (or faster erosion of small areas, for example, due to faster evaporation of small and shallow lakes) compared to the purely "white noise" of processes that form the relief (see Sect. 1.3). If $n = 2$ or 1, then these are purely Kolmogorov, not processed purely by the nature distribution of ANK34.

Ryanzhin (2005) also calculated the cumulative distribution functions of the number of lakes by area. The areas were normalized to 386,400 km^2, the area of the largest lake—the Caspian Sea in 1980. The exponent of the cumulative number

G. Golitsyn and C. Varotsos, *The Stochastic Nature of Environmental Phenomena and Processes*, https://doi.org/10.1007/978-3-031-77015-9_10

of lakes depending on their normalized area was found to be 0.925, one-four less than given above. The Ryanzhin database includes 8.65×10^6 lakes with area ≥ 1 ha $= 10^4$ m^2 $= 10^{-2}$ km^2, which is twice as much as in Turcotte (1997). The increased number of small lakes leads to a slight decrease in this indicator.

Let's try to give a physical meaning to these power laws. First of all, about the accuracy of determining the value of the exponent using empirical data. It is determined by the least squares method for deviations of the number of objects from the power law dependence. The correlation coefficient, r, of this number with the specified dependence must be the greatest, which determines the value exponent n. The dependence $r = r(n)$ usually has a very blunt maximum, i.e., a large radius of curvature, therefore values of n close to $n = 2$ can also approximate the desired statistical pattern quite well. Unfortunately, authors of scientific articles seldom give statistical limits depending on the size of the base and the degree of connection (this is determined by the number of degrees of freedom in the random series under study), in which the value of the exponent n in power distributions should be located with a given degree of probability. Therefore, to simplify the interpretation of formula (10.1), we can first approximately assume $n = 0.95 \approx 1$, in accordance with Sect. 1. 3 and Chap. 8, where for a spatial spectrum of relief $n = 2$.

Most rivers and lakes in the middle and high latitudes of Europe and North America were formed after the melting of ice sheets that formed the topography of the earth's surface over the last million years. In Asia and other continents, the relief was formed by tectonic processes and erosion, and there the role of glaciers is not clearly identified. Therefore, the number of lakes at present should be viewed as the number of events, and processes that led to their appearance, for some characteristic period. Just to be certain, let's take ten thousand years for such a period, the moment when the last ice sheets of North America and Fennoscandia disappeared, $T \approx 10^4$ years $= 3\ 10^{11}$ s. Then we must assume that the value of the cumulative distribution of the number of lakes over areas $N (\geq S)$ belongs to this period, i.e., has the dimension of inverse time or frequency T^{-1}. In this case, we will write formula (10.1) as equality for $n = 1$:

$$N(\geq S) = DS^{-1}, \tag{10.2}$$

where the factor D has the dimension of the diffusion coefficient. According to the data Fig. 8.4 from Turcotte's book at $n = 2$ the value of the coefficient D turns out to be of the order of 10^3 km^2/year, i.e., 30 m^2/s, which refers to the area of all lakes in the world. This value can be interpreted as the coefficient of broadening of the total area of lakes. It should be considered an upper estimate since many lakes are older.

Ryanzhin (2005) in his database has volumes for several thousand lakes. For them, the dependence of the volume of lakes on their area was determined:

$$V = S^{-n}, \quad n = 1.137. \tag{10.3}$$

The similarity parameter can be selected $\Pi = S^{1/2}/h$, where h is the average depth of the lake. If the similarity parameter (10.3) was the same for all lakes, then the exponent n should have been equal to 3/2, but here we must write:

$$V = S^{3/2} f_1(\Pi), \tag{10.4}$$

where is the function f_1 (Π) must be proportional $\Pi^{-0.363}$, so that formula (10.3) is satisfied. Note that formula (10.3) characterizes a typical fractal dependence, similar to the dependence of the length of the coast of Great Britain on the size of the measurement unit found by Richardson. Examples of such dependencies are given in Turcotte (1997) for Norway, Australia and other areas. Such dependencies do not lend themselves to simple physical interpretations (or, in the spirit of B02, their numerical coefficients depend on several similarity parameters in the form of intermediate asymptotics).

We see that formula (10.2) is a formula of the general type (1.31). This is a typical formula for the Brownian process, when a particle is acted upon by random impulses, the correlation time of which is much less than the reaction time of the system, and then the time correlation function of the impulse action process can be represented as a delta function of time, energy whose spectrum is frequency white noise intensity $2\sigma_v^2\tau_o = D$, where σ_v^2—dispersion of the impulse process, τ_o—internal time of this process. In such interpretation, for the area S we take the area covered by the trajectory of the Brownian particle during the time $\tau = \tau_o \geq (S)$. The reciprocal of such time will be the cumulative frequency N ($\geq S$) according to Sect. 1.3.

It is necessary to conclude this section by noting that the energy spectrum of the surface relief, described in Chap. 8, for regional sections is proportional to k^{-2}, where $k = 2\pi/l$—wave number, l—wavelength of the spatial harmonic. The dependence of the spherical harmonics of the spatial relief of the earth's surface on the number of these harmonics is similar, Chap. 8. There is an obvious connection between these results and statistical properties of the earth's relief with the results outlined in this paragraph: the number of depressions in the relief, depending on their size l, should fall with height l, like $l^{-2} = S^{-1}$, i.e., the larger the area, the smaller the number of such lakes. From here $N (\geq l) \sim l^{-2} \sim S^{-1}$.

10.2 Number of Floods in Dependence on Their Damage Values

Figure 10.1 from the book of Smith and Ward (1998), shows the cumulative number of cases (about 100) of floods depending on the damage, U. With a coefficient of determination close to 0.7, a cumulative dependence was drawn there,

$$N(\geq Y) \propto Y^{-n}, \quad n = 0.65, \tag{10.5}$$

i.e., the greater the damage, the smaller the number of such cases to a degree close to 2/3.

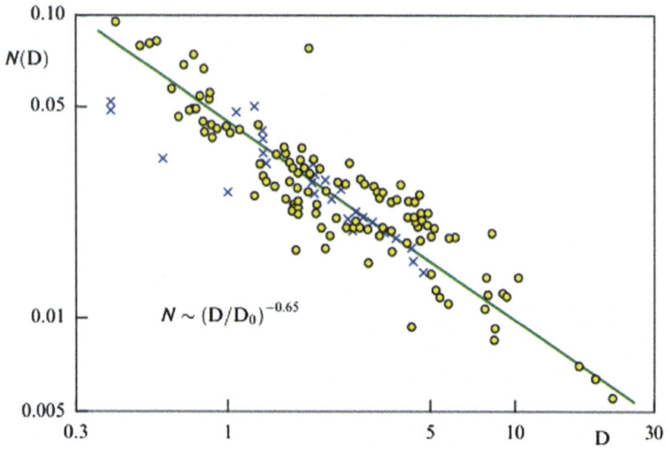

Fig. 10.1 Cumulative distribution of the number of floods versus their damage (Smith and Ward 1998)

Damage can be assessed by the flooded area, depending on its topography, and the volume of precipitation is a random variable:

$$V = p\tau S_0, \tag{10.6}$$

where p is the average intensity of precipitation during the time of its occurrence τ over the river catchment area S_o (or part of it). The hypothesis naturally arises that the exponent in (10.6) is determined by the relationship between area and volume, as is the case for the cases of cumulative distributions of the number of earthquakes and volcanic eruptions (see Chap. 3).

Heavy and prolonged precipitation is a process that, fortunately, is quite rare and irregular for each specific area. The process of rain formation itself is very complex and depends on many factors. It is known (Palmen and Newton 1973) that the temporal and spatial correlation radii for precipitation are almost an order of magnitude smaller than for other synoptic processes. This serves as a basis for considering the processes leading to the formation of precipitation to be delta-correlated in time. Then the general theory of such processes, set out in Sect. 1.3, gives that the precipitation itself must be cumulatively distributed over their volume as follows:

$$N(\geq V) \sim AV^{-1}, \tag{10.7}$$

where N ($\geq V$)—number of cases over a certain period (for example, a century) with the volume of precipitation $\geq V$, A—quantity with dimension VT^{-1}, where V is the dimension of volume, T is time, i.e. A is the rate of precipitation generation. It can only be determined from observational data.

We will consider the damage, Y, proportional to the volume of precipitation:

$$Y = \alpha V, \tag{10.8}$$

where α—proportionality coefficient with dimension equal to $[\alpha] = YV^{-1}$, where Y—the amount of damage, for example, in money. We estimate the flood area as

$$S_y = h\beta^{-1}L, \tag{10.9}$$

where h is the depth of flooding, proportional to the amount of precipitation, $h = h\ (V)$, β—the average angle of the side slope of the river valley, L—length of the flooded valley, proportional value h/β_1, where β_1—slope coefficient for river flow.

Specific damage per unit area (with average watershed exploitation terrain) on average will be proportional to the depth of flooding with the coefficient proportionality α_1, with dimension $[\alpha_1] = YL^{-3}$.

$$y = \alpha_1 h. \tag{10.10}$$

Now we estimate the volume of water flooding the area, taking into account (10.10) as:

$$\gamma V = hS_y = yS_y/\alpha_1 = V/\alpha_1, \tag{10.11}$$

where γ is the coefficient of water loss from its evaporation and seepage into the soil, which is further omitted for simplicity. Considering (10.9) and the remark that on average $L \propto h/\beta_1$, we find from (10.11) the volume of water in the form:

$$V \sim hS_y \sim \alpha_1 h^3/\beta\beta_1. \tag{10.12}$$

The area and damage are thus determined by geomorphological parameter—the average product of the angles of the lateral slope and the slope of the valley $\beta\beta_1$. It varies for different regions and serves as a natural parameter set by external conditions. Using formulas (10.8) and (10.12), it is also expressed through the internal variables of the problem as:

$$(\beta\beta_1)^{-1/3} = h^{-1}(y/\alpha_1)^{1/3} = \Pi_1, \tag{10.13}$$

which can serve as a natural similarity parameter for river floods.

The general theory, Sect. 1.2, for cases of weakly correlated influences, gives dependence (10.7) for the cumulative distribution of the number of precipitation events with volume $\geq V$. Let us rewrite it for damage considering (10.8) dependence Y (V):

$$N(\geq y) = c(\Pi_1)\alpha y^{-1}, \tag{10.14}$$

where for the numerical proportionality coefficient $c(\Pi_1)$, as in cases of earthquakes and volcanoes, it turns out to be possible to depend on the parameter similarities Π_1 in (10.13).

With increasing damage values Y_B in (10.13), other things being equal, our parameter similarities Π_1 also grows, and with small damages—as $Y^{1/3}$. Numerical coefficient in (10.14), $c(\Pi_1)$, for small Π_1 can be expanded near zero in a Maclaurin series $c = c_o + c_1\Pi_1$. Constant coefficient c_o here must be equal to zero, since no precipitation and no flooding, i.e., no damage. At $c_1 \neq 0$ now the general ratio (10.14) taking into account (10.13) gives:

$$N(\geq Y) \propto V^{-2/3}, \tag{10.15}$$

in good agreement with empirics (10.5). This relationship can be understood as follows: that it reflects the relationship between damage in the flooded area and volume of precipitation: the larger this volume, the larger the area, the greater the damage and, namely, in the proportion given by the last formula. When deriving this formula, it was the same technique of expansion of the unknown function from the similarity parameter is used, as was the case in the Monin–Obukhov theory MY65 for a stratified boundary layer, as in the derivation of the cumulative distribution function for the frequency of earthquakes and volcanic eruptions in Chap. 3.

An additional relation to this subparagraph is the statistical connection between the maximum water flow module during flood q_p and the catchment area of river A, given in Robinson and Sivapalan (1997): $q_p = KA^{\theta}$, where $\theta = -0.33$. This dependence is statistical in nature with a coefficient of determination of about 0.9. The modulus of water flow in a river is the ratio of flow q to the catchment area A, so the latter relation can be written as:

$$q_{\max} = kA^{2/3}. \tag{10.16}$$

Having graphs of frequency probability distributions for individual rivers, the inverse value of the distribution of the function $q_{\max}$ can be represented as a cumulative probability, namely frequency N ($\geq q_{\max}$). Then formula (10.16) will give:

$$N(\geq q_{\max}) = k^{-1}A^{-2/3}. \tag{10.17}$$

Let us note that both of the last two formulas obey the first formula (1.34), in which the cumulative distribution of the quantity α is inversely proportional to the quantity itself α, for which the distribution is sought. This explains the indicators in Sect. 10.3 in formulas (10.18) and (10.19) and shows greater generality for the processes spill, expressed (10.15)–(10.17). In all these formulas the exponent b is close to 2/3.

The damage in (10.15) is related to the severity of the flood, which depends on the catchment area. Thus, (10.17) serves as a justification for the cumulative dependence of the number of cases of damage on the magnitude of the damage itself. Simply, the damage is determined by the total volume of precipitation that falls on the area, and the dimension of this area is equal to $V^{2/3}$, i.e., the area of the spill is thus related to the volume of precipitation.

10.3 Statistics of Numbers of Mud Mushrooms on the Ocean Surface Near the River Mouths

Physically similar patterns are shown by the statistics of mushrooms (plumes), and areas of the sea surface occupied by turbid river runoff. They are clearly visible on satellite images, and even with the naked eye from the window of an airplane flying along the coast. Their good statistics are given and analyzed in Warwick and Fond (2004). Cumulative distributions have a power-law form over a wide intervals scale:

$$P = cA^b, \quad (10.18)$$

$$N(\geq P) = aP^{-\beta}, \quad (10.19)$$

where P is the area of the mushroom on the water surface, A is the catchment area of the corresponding river, a and c are coefficients. For mushrooms from seven rivers in Morocco $b = 0.60 \pm 0.09$ with the coefficient of determination $r^2 = 0.89$ in the range of their catchment areas from 21 to 1100 km^2. Data on 110 rivers are also provided in Warwick and Fond (2004) for California, flowing into the Pacific Ocean. In the range of their catchment areas from 1 to 230 km^2 a meaning has been found for them $b = 0.63 \pm 0.15$ at $r^2 = 0.63$.

Greater variation for California's small rivers is evident due to the diversity of their drainage conditions in the state's coastal mountains. For large rivers of the world with catchments of 10 to 10^7 km^2 (the last figure refers to Amazon) magnitude $b = 0.70$ with $r^2 = 0.94$, according to Warwick and Fond (2004). The closeness of b values to the 2/3 indicator in formula (10.5) demonstrates the relationship between the volume of precipitation in catchment areas and their areas.

The number of mushrooms themselves off the coast of California depending on their area, and their cumulative distribution, follows the formula (10.19) with the value $\beta = 1.02 \pm 0.03$. This is in excellent agreement with our theoretical formula (1.31), where $\beta = 1$, for the cumulative number of events given in Sect. 1.3, which describes the results of A.N. Kolmogorov for random movements and their consequences, in which there is no dependence on the function of any dimensionless parameter, and clearly follows the results of ANK34. Thus, all the results of this section are determined by the relationships between the areas of watersheds and the volume of precipitation on them. A more complete summary of this material can be found in Golitsyn (2018, 2024).

References

Golitsyn GS (2018) Power-law distributions of spill areas in hydrology. Water Resour 45(4):380–384

Golitsyn GS (2024) A N Kolmogorov's 1934 paper is the basis for explaining the statistics of natural phenomena of the macrocosm. Phys Usp 67(1):80–90. https://doi.org/10.3367/UFNe.2023.05.039355

Palmen E, Newton CW (1973) Atmospheric circulation systems.—M.: Mir,—614 pp

Robinson JS, Sivapalan M (1997) An investigation into the physical causes of scaling and heterogeneity in regional flood frequency. Water Resour Res 33(5):1045–1059

Ryanzhin SV (2005) Are there many lakes on Earth? Nature 4:18–25

Smith K, Ward R (eds) (1998) Floods: physical processes and human impacts. Ward, Wiley, Chichester, 382 pp

Turcotte DL (1997) Chaos and fractals in geology and geophysics, 2nd edn. Cambridge University Press, Cambridge, 398 pp

Warwick JA, Fond DA (2004) Dispersal scaling from the world rivers. Geophys Res Lett 31:L 04301. https://doi.org/10.1029/2003GL019114

Chapter 11
Additions and Comments to Previous Sections

The material presented here, and its understanding were formed by the first author over many years. The main thoughts were about the essence of these results and how they add up to the overall picture of nature. Give a general idea of the ones discussed here processes was my need, and I hope this picture will be interesting and useful for other researchers as well. Nowadays, scientists are working on some very specific topics, which are just small parts of the entire universe.

The first part of the additions concerns the rule for the fastest response of natural systems to external influences, which came to mind in the late 1970s in connection with studies of convection and atmospheric circulation. The second part, which has occupied almost the entire scientific life of the first author, is how and why in statistical characteristics, such as spectra or structure functions, third parts of powers of natural numbers occur so often and in a variety of ways. In the third part, several cases are mentioned, briefly describing examples of the consequences of the laws of Kolmogorov 1934 for cumulative distributions.

11.1 The Rule of the Fastest Response to the External Forcings

When considering most natural physical processes and phenomena, we, as a rule, know the power of energy supplied to them, called forcing. If we somehow know how to estimate the reaction time to this "forcing," then their product gives us the acquired energy of the process. It turns out that if the system under consideration has several time scales, then the response to the impact occurs within the shortest time. This statement arose from the analysis of several processes. Its complete justification is obvious for linear systems of ordinary differential equations of order higher than the first when there are several different magnitude increments, but the largest increment

G. Golitsyn and C. Varotsos, *The Stochastic Nature of Environmental Phenomena and Processes*, https://doi.org/10.1007/978-3-031-77015-9_11

acts, and this corresponds to the shortest reaction time. It turns out, G12, that reaction times can be found from similarity criteria for the processes being studied.

Consider the Reynolds number Re = UL/v, Where U—speed flow, L is its linear scale, and v is kinematic viscosity. Let's multiply and divide this number by L and introduce a dynamic time scale $\tau_d = \mathrm{L/U}$ and the viscous scale is $2\,\mathrm{L}/\tau_\nu = L^2/\nu$. Then our similarity criterion will be presented as the ratio of two time scales:

$$\mathrm{Re} = \tau_\nu/\tau_d.$$

If the Reynolds number is much greater than one, then this means that $\tau_d << \tau_\nu$. In this case, one must keep in mind the existence of a critical Reynolds number O(10^3), to which the situation under consideration should be attributed. If we know forcing—ε, the rate of generation (and dissipation in the statistical equilibrium case) by unit mass, then the kinetic energy is estimated as:

$$U^2 \sim \varepsilon\tau_d = \varepsilon L/U.$$

This relation, up to a numerical coefficient and sign, is reduced to the expression for the third moment of velocity in Kolmogorov's "law—4/5" (see Sect. 2.1): $< U_1^3 >= -4/5U\varepsilon r$, and it also reveals the "2/3 law" for the structure function.

If we are dealing with a small scale, so, Re $\leq$ 1, That $\tau_\nu \leq \tau_{d.}$, and then

$$U^2 \sim \varepsilon\tau_\nu = (\varepsilon/\nu)L^2.$$

And again, up to a numerical coefficient, this is the formula Kolmogorov for the structure function of velocity in the scale domain, smaller than the internal viscous scale of turbulence (Sect. 2.1).

There are examples when Re << 1, which can be on a small scale or very high viscosities, such as in the Earth's mantle, where the viscosity $\nu = 10^{17}$ m^2/s. The lowest velocities of terrestrial natural processes are the speeds of convective movements in the mantle. The earth's crust is a few hundred kilometers thick continent to a few kilometers near mid-ocean ridges, from which mantle material with a volume of about 3 km^3 per year emerges. Convection in the mantle is spatially inhomogeneous; it contains inhomogeneous ascending plumes of mantle matter that tear the earth's crust into lithospheric plates. The relative velocities of these plates are observed by GPS and GLONASS systems from 1 up to 18 cm/year. The Pacific Nazca plate has the highest speed, and on average they are about 5 cm/year.

Let us turn to the earth's mantle, which under the lithosphere occupies a depth of up to 3000 km, and below is a liquid core, the fluid speeds of which are estimated at 5 km/year, which is consistent with the drift of some components of the geomagnetic field, and such speeds are sufficient for a geomagnetic dynamo (Golitsyn 2008). Total geothermal flux from the earth's interior W consists of its average density in 86×10^{-3} W/m^2 and total area of the earth in 510×10^6 km^2 = 5.1×10^{14} m^2, where the total power W = 86×10^{-3} W/m^2 $\times$ 5.1×10^{14} m^2 = 4.4×10^{13} W. The term ε is the rate of energy generation/dissipation per unit masses. The total mass of the planet $M = gr^2/G$, where $G^{-1} = 1.5 \times 10^{10}$ kg s^2/m^3—reverse the value of the gravitational constant. Substituting terrestrial parameters into it, $g = 10$ m/s^2, $r = 6370$ km, received $M_{\oplus} = 6 \times 10^{24}$ kg, which coincides with ~ 1% accuracy with astronomical mass of our planet.

The mass of mantle material, extending up to 3000 km deep, we estimate at 4×10^{24} kg, from where the forcing for the entire mass will be $4.4 \times 10^{13} \times 5 \times 10^{-4} = 2.2 \times 10^{10}$ W, where is the multiplier 5×10^{-4}—an estimate of the share of the total flow heat used to generate geodynamics and seismic processes. From here to unit of mass, i.e., $\varepsilon = W/M \cdot \left(2.2 \times 10^{10}/4 \times 10^{24}\right) \approx 5.5 \times 10^{-15} \mathrm{m}^2/\mathrm{s}^3$. Time $\tau_v = L^2/\nu = 9 \times 10^{12}/10^{17} = 10^{-4}$ second. This time does not physically correspond to anything, but let us accept it as formal asymptotic consequence for a viscous time in the mantle at $L = 3 \times 10^6$ m and viscosity $\nu = 10^{17}$ m^2/s.

Then: $\mathrm{U} = (2\varepsilon\tau_v)^{1/2} \approx \left(2 \times 5.5 \times 10^{-15} \times 10^{-4}\right)^{1/2} \approx 10^{-9} \mathrm{m/s} = 10^{-7} \mathrm{cm/s}$

For 1 year $= 3 \times 10^7$s, then the speed we obtained is 10^{-7} cm/s 3×10^7 cm/year will be 3 cm/year. The authors did not expect to obtain such agreement with the average estimated by GPS a value of 5 cm/year.

Our order-of-magnitude estimate of the mass of mantle matter may turn out to be overestimated. If you take it 3×10^{24} kg, then the speed will be 3.55 cm/year, in addition, ~ 3% adds exact value 1 year $= 3.16 \times 10^7$ s, generally 4 cm/year. However, such clarifications are unnecessary. Estimation of velocities and cumulative distribution of lithospheric plates across areas indicate the reality of the magma parameters used and the energy cycle of geodynamics and seismic.

The authors have repeatedly heard that such unimaginably low speeds are impossible to imagine. But this is the speed at which our nails grow. In the year 52 weeks, and a week or two the nails grow by about a millimeter, i.e., 3–5 cm/year. In 1995, the first author presented his results on earthquakes at Cambridge University. A well-known specialist who attended the seminar in geodynamics Prof. Dan McKenzie said that he gave the speed comparison with nail growth back in the 1980s. Prof G. Golitsyn apologized that he didn't know this and came up with it himself in the 1990s. After that, McKenzie called him to a famous pub in the city, and they drank beer at the table where sat Watson and Crick, who discovered 40 years earlier, the DNA double helix structure.

A non-trivial example here is provided by the general circulation of the atmosphere slowly rotating Earth, Mars, Venus, and Titan, the satellite of Saturn. The first author have brought out in 1970, using considerations from similarity theory and dimensionality, the formula for kinetic energy of the atmosphere in the form:

$$K = \sigma^{1/8} q^{7/8} c_p^{-1/2} r^3.$$

Remembering the Stefan–Boltzmann radiation law $q = \sigma T_e^4$, Where T_e—effective temperature of outgoing thermal radiation arriving at the planet's disk, $\pi g r^2 = Q$,, where $(c_p T_e)^{1/2}$ is of the order of the speed of sound in the atmosphere, then the formula for kinetic energy can be reduced to the following form $K = Q\tau_e$, where $Q = \pi r^2 q$, the arrival of energy to the disk,

$$\tau_e \approx r/c_e \approx r\left(c_p T_e\right)^{-1/2}$$

τ_e—time establishing local thermodynamic equilibrium in the atmosphere with characteristic horizontal size—planet radius r. This time—the smallest for macro-processes in the atmosphere. This example was initiated for the "rule of the fastest reaction", the actions of which were also revealed in other processes with external influence. For rapidly rotating planets the numerical coefficient will be a function of the square of the rotational Mach number $M = \omega r/c$. In the case of Jupiter, where the American probe Galileo descended into its atmosphere up to pressures of 20 atm., recorded speeds of 100–200 m/s in complete agreement with G12, where a speed estimate as 130 m/s is given.

A new and characteristic example of the rule is convection. This has two similarity parameter—Rayleigh and Prandtl numbers:

$$\text{Ra} = \frac{\alpha g \Delta T h^3}{k\nu}, \quad \text{Pr} = \frac{\nu}{k}.$$

The Grashof number is used simultaneously with the Rayleigh number:

$$\text{Gr} = \frac{\alpha g \Delta T h^3}{\nu^2} = \frac{\text{Ra}}{\text{Pr}}.$$

They both can be represented as the ratio of the squares of two times:

$$\tau_\nu = h^2/\nu, \quad \tau_{g'} = \left(h/g'\right)^{1/2},$$

where $g' = \alpha g \Delta T$—gravity reduced in one way or another, $\alpha = 1/T$. For an ideal gas, its compressibility coefficient (Charles' law—I remember from school), for other substances this is a specific thermodynamic quantity. Then:

$$\text{Gr} = \frac{\tau_\nu^2}{\tau_{g'}^2}, \quad \text{Ra} = \frac{\tau_\nu \tau_k}{\tau_{g'}^2}, \quad \tau_k = \frac{h^2}{k}, \quad \tau_\nu = \frac{h^2}{\nu}$$

and at Pr ~ 1 both Rayleigh and Grashof numbers are of the same order. Since convection is excited when these critical numbers reach values of the order of many hundreds, then they must be attributed to $\text{Ra}/\text{Ra}_{\text{cr}}$ for selection last time. The excitation of movements gives the buoyancy flux density: $f = \alpha g f/\rho c_p$.

A non-standard example of the operation of the "fastest reaction rule" is given by the current water in pipes. Consider a pipe with radius α and length l with liquid density ρ and dynamic viscosity μ. The pressure is set at its ends $p_1 - p_2 = \delta p > 0$, i.e. there is a pressure gradient $i = \delta p/l$. From these parameters it is compiled dimensionless quantity—Reynolds number:

$$\text{Re} = \frac{\rho \bar{u} a}{\mu},$$

where u is the flow velocity, which itself must be determined.

Non-standard of this example is that the pressure gradient $i = \delta p/l$ doesn't have dimensions of power, but the term $iS = \pi I\alpha^2$ has the dimension of acceleration of every pillar in the pipe. But representing the Reynolds number as a ratio of two times, viscous and inertial, allows you to reduce the number of dimensional parameters, omitting either density or viscosity. Considerations dimensions are given in the general case:

$$\overline{u} = (ia/\rho)^{1/2} f(\mathrm{Re}).$$

For small numbers Re << 1 the density that determines the forces is insignificant inertia, and then:

$$\overline{u} = c_1 ia^2/\mu,$$

where c_1 is a numerical coefficient. For average pipe flow rate

$$Q = \overline{u}S = c_1 \pi a^4/\mu.$$

The problem of the laminar velocity profile in a pipe was solved analytically in 1840 by Poiseuille. The exact solution gives that $c_1 = 1/8$.

At large Re >> 1 in the zeroth approximation, viscosity can be neglected, and then the remaining parameters will give:

$$\overline{u} = \left(\frac{ia}{\rho}\right)^{1/2}, \quad Q = c_2\pi\left(\frac{i}{\rho}\right)^{1/2} a^{5/2},$$

where the coefficient c_2 is a slow function of the Reynolds number given by experimentally, and the exponent of the radius does not exceed 2.7, which is much weaker α_4 for laminar flows. There is a slight non-self-similarity here. Reynolds number of the form $\mathrm{Re} = \tau_v/\tau_\alpha$, where $\tau_v = \rho\alpha^2/\mu$, acts correctly when Re << 1, and when $\tau_\alpha = < u > /\alpha$, i.e. Re >> 1 acts approximately, maintaining a weak dependence in resistance coefficient:

$$c_2(\mathrm{Re}) = i\left(\rho\overline{u}^2/2a\right)^{-1} = \frac{2a\delta p}{l\rho\overline{u}^2},$$

which at Re << 1 equals:

$$c_D = 2\mu^2/i\rho a^3 = 2\rho v^2/ia^3,$$

and when Re >> 1 the coefficient resistance $c_D = \mathrm{const}\ (\mathrm{Re})$.

This is the situation that Barenblatt has been warning about for about half a century, that the case when the exact value of some similarity parameter can be neglected and some external parameter included in it—rather an exception than the rule, and that the dependence, albeit weak, can persist over the entire interval the existence of similarity criteria (see his book B02). The differences from classical results are small but can be revealed in experiments.

11.2 The Nature of Third Powers of Exponents in Statistical Laws Natural Processes

1. Already in the 1950s, Golitsyn heard that "the laws of all thirds" operate in nature, although at that time the "laws of 2/3 and – 5/3" for structural functions and spectra of velocities and passive impurities, "4/3 laws" for pressure fluctuations and the turbulent mixing coefficient. In the spirit of this book, one might say, that this is the action of the "rule of the fastest reaction": we know the forcing, i.e., the reported power ε into the system, and the only reaction time of such a system of size *r* is:

$$\tau = \left(r^2/\varepsilon\right)^{1/3}. \tag{11.1}$$

The passive scalar θ is described in the same way, for which the average is introduced square of its concentration fluctuations $\Theta = \langle\theta^2\rangle$ and the rate of its generation and dissipation $d\Theta/dt$, and multiplying it by time (11.1) makes it possible to obtain structure function for a scalar, and then its spectrum $k^{-5/3}$, since passive the scalar does not introduce a new time scale.

2. Around the 1940s, the law of earthquake recurrence, EQ, entered science,— Gutenberg Richter law, G–R. For the cumulative number of events with magnitude *m*, it says (see Chap. 3)

$$\lg N(\geq m) = a - b \lg m, \tag{11.2}$$

where α is a constant specific to the region and time of observation. In a sedate form:

$$N(\geq m) = a_1 m^{-b}, \quad a_1 = 10^a \tag{11.3}$$

where $b \approx 2/3$ for $m \leq 7.5$ away from the mid-ocean ridges and turns into $b \approx 1$ V depending on the value of the main similarity parameter for the EQ, Chap. 3:

$$\Pi = \frac{L}{h}\left(\frac{M}{\Delta\sigma}\right)^{1/3} h^{-1}, \tag{11.4}$$

where M is the seismic moment, volumetric characteristic, L is the length of the rupture during the earthquake, h—is the thickness of the cortex at the site of the EQ. When $\Pi < 1$, $b \approx 2/3$, and for $\Pi > 1$, b ≈ 1. Thus, when a crustal rupture of length L exceeds the thickness of the crust, then all the energy in the volume of the EQ leads to G–R in the form $N(\geq m) = -\alpha m^{-1}$, otherwise $N(\geq m) = -\alpha m^{-2/3}$. Here one should remember the concept of the EQ volume, defined as:

$$V = M/\Delta\sigma, \tag{11.5}$$

where $\Delta\sigma$ is the stress in the crust per event, and what is between magnitude and moment there is an approximate statistical relationship:

$$m = \frac{2}{3}\lg M - 6. \tag{11.6}$$

It follows that the magnitude is related to the surface of the crustal rupture, and the moment is related to the volume of the environment in which EQ occurs. There is a magnitude $\log(S/S_o)$, where the S—area ruptures during EQ, $S_o \approx 100$ m^2, reference area at $m = 0$. A similar situation we saw between the volume of precipitation and the area of flooding in Chap. 10.

The G–R law in the form $N\ (\geq m) \sim m^{-b}$ is the cumulative distribution of the number of events and $b = 1$ is the standard view for such events: $N(\geq E) \sim E^{-1}$, when all the energy E is spent on the implementation of an event (see also Chap. 1).

This form of cumulative distribution with the dimension of inverse time follows from relations of Kolmogorov $< u^2 >\sim \varepsilon t$ in the understanding that u^2 there is energy per unit mass, when there are no geometric restrictions. Intra-crustal zones described by area and rock fragmentation near the rupture during the shift of the areas of the cortex adjacent to the rupture and excitation seismic waves, characterized by the similarity parameter Π from (11.4), are spent on the consequences we perceive are the energy of volume (11.5) to a lesser extent than for $\Pi > 1$, and therefore they occur more often for $\Pi < 1$ than in cases for $\Pi > 1$ when the entire volume of EQ is implemented according to (11.5). To put it bluntly and simply, the surface (by size) by its dimension:

$$S \sim V^{2/3}. \tag{11.7}$$

In the case of the spectrum of cosmic rays (Chap. 4), the number of their particles per unit time per unit area, to estimate the latter we use their volumetric density energy, i.e., relations (11.7). As a result, the integral spectrum of cosmic energy rays is proportional to $E^{-5/3}$. Two other examples of "thirds" are in Sect. 11.3.

11.3 Cumulative Area Distributions

11.3.1 Distribution of Lithospheric Plates by Size

In 2003, Bird's article (2003) appeared with the areas of 52 lithospheric plates. If exclude from this list the 6 largest: Pacific, 2.6 steradians in size, African, 1.44, Antarctic—1.43, North American—1.37, Eurasian—1.20, South American—1.03, and 4 smallest and poorly determined plates, then the cumulative number of the remaining 42 plates is approximated depending on the area *S* as:

$$N(\geq S) = 7S^{-0.33}, \tag{11.8}$$

which is shown in Fig. 11.1.

The dimensions of the areas *S* are given in steradians, the area earth's surface 510 million square meters, constituting 4π steradians. With average radius of the Earth $r = 6371$ km area $r^2 = 40.6$ million sq. km. For comparison area of our Russia 17.1×10^6 km^2, which is 1/30 of the total earth's surface.

Recall that the areal scale in the Kolmogorov random distribution (1.33):

$$\langle x^2 \rangle = \varepsilon t^3, \tag{11.9}$$

that in dimension and meaning the left side can be equated to the area *S*. Then cumulative distribution over areas having the dimension of the inverse time, will, according to Chap. 1:

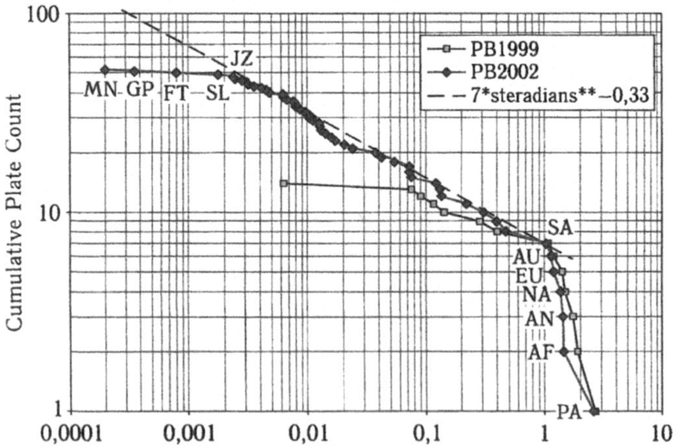

Fig. 11.1 The Cumulative plate count as a function of the area in steradians Gostintsev and Fortov (2007)

$$N(\geq S) = \left(\frac{\varepsilon}{S}\right)^{1/3} C_1, \tag{11.10}$$

where C_1 is a numerical coefficient determined by comparing the last formula with empirical (11.8), which we will do, neglecting the difference between 0.33 and 1/3. This gives:

$$C_1 = 7\varepsilon^{-1/3}. \tag{11.11}$$

The rate of energy generation per unit mass was estimated in Chap. 3 for the quantity convection in the mantle, leading to the generation of elastic energy in the earth's crust, permitted during the EQ process. This value $\varepsilon \approx 110^{-11}\ \mathrm{m}^2/\mathrm{s}^3$ (Golitsyn 2007). However, in approximation (11.8) and Fig. 11.1 the area is in steradians. Therefore, substituting this ε value into (11.11), we get $C_1 = 0.945 \approx 1$, i.e.

$$N(\geq S) = (\varepsilon/S)^{1/3}.$$

In the author's scientific practice an unfortunate mistake was made (corrected in Golitsyn 2017): in the first article (Golitsyn 2008) the rate energy dissipation in the mantle, value ε, was used in units of $\mathrm{cm}^2/\mathrm{s}^3$ a comparison with the empirical formula (11.8), where the areas are in steradians. As a result, the numerical coefficient C_1 in (11.10) meaninglessly turned out to be about 3500. Such large numerical coefficients should always be alarming if you follow Albert Einstein's warning that in dimensionally correct formulas when comparing them with data, the numerical coefficients should not be very small, nor very large (quoted from the book BPW, chapter 8). At the end of the 2000s, the author did not yet remember the results of Kolmogorov in 1934 and their use in 1959 by Obukhov, therefore, based on analogies of the manifestation of traits of small-scale turbulence in flows of crushed ice in glaciers, relative movements in a mass of small balls in closed volumes, in G12 there was formula (11.3) was proposed. However, not being sure of the correctness of such considerations, the author tested experimentally the possibility of implementing similar cumulative distributions of the number of areas proportional to $S^{-1/3}$ by using checkered paper to measure areas.

Three sheets of paper were drawn without looking at them by three persons, and then, in given area intervals, the number of polygons falling into specified intervals. In addition to the two or three largest intervals in terms of area, the number of areas in other intervals are smaller over approximately an order of magnitude values little contradicted the dependence $N(\geq S) = (\varepsilon/S)^{1/3}$. To this, they objected to me, that the paper is flat, and the planet is spherical. The same results were then obtained with shells of randomly crushed eaten eggs (a rare case of using such eggs). These results are shown in Figs. 11.2 and 11.3. A detailed presentation of this point is contained Golitsyn (2008, 2017).

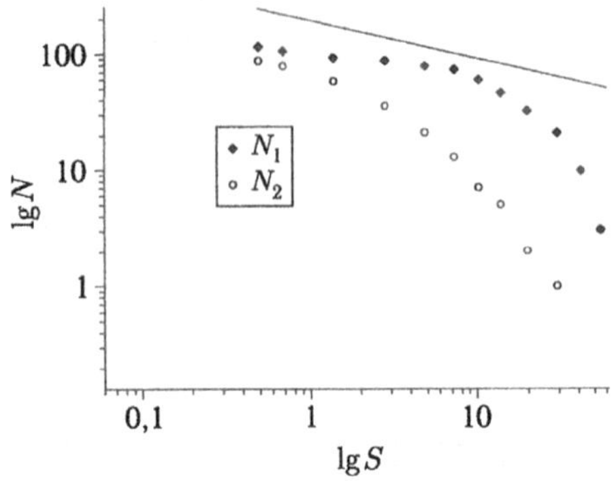

Fig. 11.2 Cumulative distribution by area of the number of polygons during random drawing of G12 paper

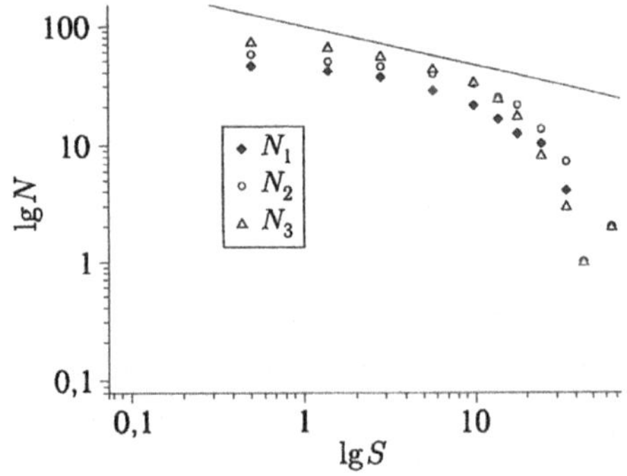

Fig. 11.3 Cumulative distribution of masses of nearby galaxies

11.3.2 Distribution of Galaxy Masses

Figure 11.1 shows the cumulative distribution over the masses of those closest to us galaxies selected within 10 million light years and taken from the book by Surdin (2013). In Fig. 11.4 (not published), the case of 25 objects is shown as their number $N\ (\geq M)$ along with an inclined line proportional to $M^{-1/3}$. This pattern can be understood from the following considerations.

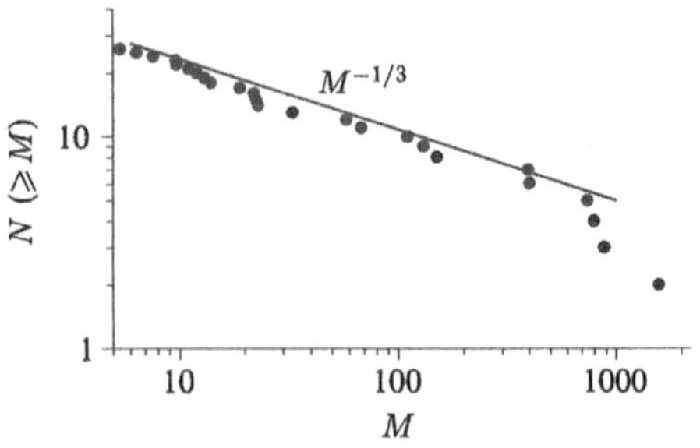

Fig. 11.4 Cumulative area distribution of G12 eggshells

The considered galaxies are disk-shaped with disk area S. For example, the Milky Way has a thickness of 200 parsecs (pc), i.e., about 6×10^{18} m, and a radius of 15,000 pc, 75 times more. Assuming that their masses are uniform along the radius, we can estimate $M = \rho_H S$, where ρ_H—is the average density of galactic matter per unit area. We will consider the time of formation of the object to be the time the substance falls to the center of the disk (Surdin 2013):

$$t = (\rho_H G)^{-1/2},$$

where $G = 0.667210^{-10} = \mathrm{m}^3\,\mathrm{s}^{-2}\,\mathrm{kg}^{-1}$—gravitational constant. It's time to order several hundred million years for large volumes like our Milky Way (Surdin 2013). Then the diffusion coefficient in the Fokker–Planck–Kolmogorov equation in Chap. 1 can be estimated as:

$$\varepsilon_g = \rho_H \varepsilon,$$

which gives a conclusion similar to (11.10):

$$N(\geq M) = C\left(\varepsilon_g / M\right)^{1/3}, \tag{11.12}$$

where C is a dimensionless coefficient to be determined from comparison with observational data, depending, obviously, on the size and homogeneity of the sample. The last statement means that the spread in the number of objects in each of the adjacent logarithmic mass intervals by several orders of magnitude. The common origin of cumulative dependencies is Kolmogorov probability law (4.2) in the form $< x^2 >= \varepsilon t^3 = r^2 \approx S$. To derive a relation (11.12), it is sufficient to assume that the galaxy mass is proportional to its area.

Constructing such cumulative distribution, one should remember that at large N its logarithm increases much slower than $M^{1/3}$ or any power of M. This obstacle could be met in practice when considering 80 swarf galaxies.

11.4 Energy Distribution of Objects Colliding with the Earth

According to the processing of observational data (Werner et al. 2002), the logarithm of the cumulative number of objects colliding with the Earth per year depends on their energy E as:

$$\lg N(\geq E) = a - b \lg E,$$

where $a = 0.568 \pm 0.015$, $b = 0.90 \pm 0.03$, E in kilotons of TNT, 1 kt $= 4.2 \times 10^{12}$ J. Note that this notation is close to the Gutenberg–Richter law for earthquakes, and the difference of b by 0.1 from unity shows that objects with high energy and mass Due to gravity, they are lost more often than small ones. This is a case of incomplete self-similarity according to the terminology of Barenblatt [Btt03, B09], when in the formula for the number of events

$$N(\geq E) = c(\Pi)/E$$

the numerical coefficient c begins to depend on the similarity parameter Π, for which you can choose, for example, $\Pi(G, E, M, p)$, where G is the gravitational constant, S is the area of the planet's disk, M is the mass of the planet. Then to explain the observations there must be $c(\Pi) \sim \Pi^{-0.1}$.

Assuming that the released energy is proportional to the volume of the combustion body, the first relation can be rewritten depending on the diameter of the body (Werner et al. 2002) as:

$$\lg N(\geq d) = c - d \lg d,$$

where d is the diameter of the body in meters, $c = 1.568 \pm 0.030$, $d = 2.70 \pm 0.09$. Note that for lunar craters, recalculated to the diameter of the asteroids that form them (Grant et al. 2006) in the size range from 25 to 200 m, the d index is close to 3, i.e., $b \approx 1$, which is consistent with the scheme of Kolmogorov in Sect. 1.3. This is the same explanation for the cumulative dependence frequency—the volume of volcanic eruptions (Golitsyn 2003), as a cumulative distribution for objects associated with the energy of an individual process (we remind you that in (1.31) the square of the velocity is energy per unit mass).

This also includes data on the distribution of stone masses in the range from 1 to 2000 kg on the surface of Mars for four different locations on its surface (Grant et al. 2006). They also have a power-law appearance $N(\geq m) \sim m^{-n}$, and the n indicators for them are 0.9, 0.9, 1.1, and 1.2. These stones when cosmic bodies hit the surface of the crater Gusev, where the US apparatus landed, and near it they confirm the general formula $N(\geq A) \sim A^{-1}$. Slight differences in indicator values may be due to different composition and properties of rocks and an insufficient sample size. However, ANK34 again acts as (1.31).

11.5 Experimental Test of Kolmogorov's Scales in the Evolution Laws for Spherical Flames

The work by Gostintsev and Fortov (2007) summarizes the theory and experiments to test the laws of propagation of a free turbulent spherical flame in a combustible gas mixture. The experiments were carried out in spherical volumes with a radius from 0.5 to 1 m, limited by a transparent shell through which the photograph was taken on rapidly rotating spherical film. The arson was carried out by burning thin tungsten wire 1 mm long and 0.1 mm in diameter. Used five such mixtures: 30% H_2 + air; $CH_4 + 2O_2$; $C_3H_8 + SO_2$; $H_2 + O_2$; $C_2H_2 + 2SO_2$. High-speed photography gave the positions and velocities of the flame front with accuracy the best 10^{-4} s. All experiments were carried out at normal atmospheric pressure and T = 300 K. For the above mixtures, all the necessary molecular constants: are thermal conductivity coefficient, Prandtl number, and speed of sound. Therefore, we can calculate the rate of generation of kinetic energy ε, per unit representing masses, Kolmogorov microtimes, and velocities in dimensionless coordinates of flame speed and diffusion coefficients at their flame front.

In Fig. 11.5 shows the dimensionless position of the front $R(\tau)$ depending on the time normalized to the Kolmogorov scale given by various symbols average front positions over many experiments. Solid curve—$R \sim \tau^{3/2}$. Figure 11.6 shows the dependence of the Peclet number on the turbulent coefficient stirring:

$$Pe = \frac{D}{\chi} = \frac{dR^2}{2\chi dt} \sim \left(\frac{R}{\lambda}\right)^{4/3},$$

where $\lambda = \left(\chi^3\varepsilon\right)^{1/4}$—Kolmogorov microscale. Along the axis of Fig. 12.4 straight line delayed $\sim R^{4/3}$, the solid bisector on it also demonstrates the Richardson–Obukhov scale (1.32), its existence in such unusual experiments.

Concluding this section, we can say that the Fokker–Planck–Kolmogorov Eq. (1.30) demonstrates the great richness of its content, which proves it by analytical solution at the end of Chap. 1, transforming it by substitutions (1.31) and (1.33) into self-similar form and, finally, by numerical calculation, GIG10, and direct experiments, the results of which are depicted in these figures. All verifiable relationships can be obtained from considerations of similarity and dimension (Gostintsev and Fortov 2007).

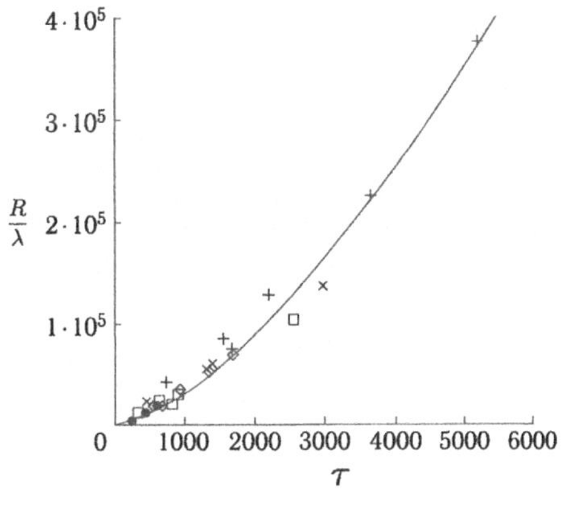

Fig. 11.5 Dependence of R/λ on τ for free turbulent spherical flames (experimental points with different symbols correspond to different combustible mixtures) (Gostintsev and Fortov 2007)

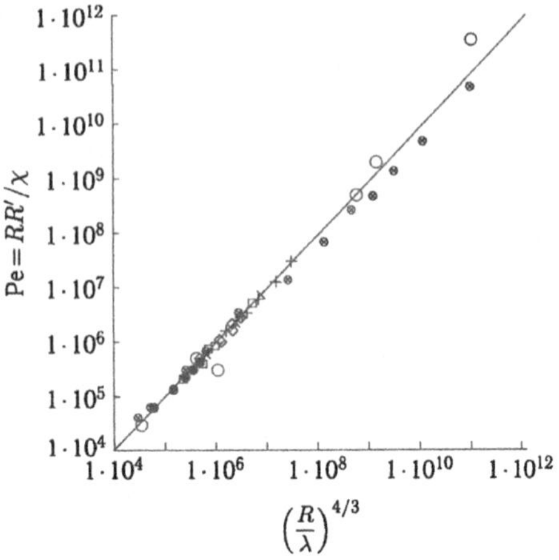

Fig. 11.6 Logarithmic dependence of the Peclet number Pe versus $(R/\lambda)^{4/3}$. Straight line—theoretical dependence, points—experimental values (Gostintsev and Fortov 2007)

11.6 Examples from the Theory of Elasticity

Examples of the appearance of third fractions in exponents in various physical relationships give participation in the processes of physical quantities, related to volumetric and two-dimensional characteristics. Consider the formation of cracks in a solid, leading to its destruction. Macroscopic theory of this process was developed by Barenblatt (1959), a detailed and rigorous description which can be found in book B02. The formation of cracks is determined by the modulus of Young's uniform compression E and surface tension energy T. These two values introduce brittle fracturing:

$$K = \left(\frac{\nu ET}{1-\nu^2}\right)^{1/2},$$

where ν is Poisson's ratio. Young's modulus is a volumetric characteristic with the same dimension as pressure: $\mathrm{ML^{-1}\,T^{-2}}$, surface tension $[\mathrm{T}] = \mathrm{MT^{-2}}$ as energy per unit surface, dimension $[\mathrm{K}] = \mathrm{ML^{-1/2}\,T^{-2}}$. In the engineering system of units, where force F, units are $[\mathrm{E}] = \mathrm{FL^{-2}}$, $\mathrm{T} = \mathrm{FL^{-1}}$, $\mathrm{K} = \mathrm{FL^{-3/2}}$.

In the 1880s, the famous physicist Heinrich Hertz (1837–1894) discovered a new phenomenon in the mechanics of solids. If a stamp is applied to a sample of a solid body with pressure, then a conical crack forms in the body, growing with increasing load, and when under a certain sufficiently large load, the crack grows without increasing the load and the sample falls to pieces.

Analysis of the dimensional force balance for the radius R of a conical crack gives value (Roesler 2013; Benbow 1960):

$$R = C_1 (P/K)^{2/3},$$

where is the numerical coefficient $C_1 \approx 1/\,50$. So, putting into operation a physical unit related to material rupture on some surface leads to the action of volume pressure to the appearance indicator in 2/3.

References

Benbow JJ (1960) Cone cracks in fused silica. Proc Phys Soc B 75:697–699

Bird P (2003) An updated model of plate boundaries. Geochem Geophys Geosyst 4(3):1029/2001GC000252

Brown P, Spalding RE, Revelle DO, Tagliaferri E, Worden SP (2002) The flux of small near-Earth objects colliding with Earth. Nature 420:294–296

Golitsyn GS (2003) Explanation of the frequency-volume relationship of volcanic eruptions Dokl. RAS 390(3):394–396

Golitsyn GS (2007) Energetic cycle of geodynamics and seismic process. Phys Earth No 6:3–6

Golitsyn GS (2008) On the distribution of the number of lithospheric plates by size. Phys Earth No 3:3–8

Golitsyn GS (2017) Distribution of lithospheric plates by areas. Russ J Earth Sci 17:ES5001, https://doi.org/10.2205/2017ES000607

Gostintsev YA, Fortov VE (2007) Laws of Kolmogorov—formation and evolution of a turbulent spherical flame Dokl. RAS 415(5):628–631

Grant JA, Wilson SA, Ruff SW, Golombeck MP, Koestler DL (2006) Distribution of rocks on the Gusev Plains and on Husband Hill. Mars Geophys Res Lett 33:L16202. https://doi.org/10.1029/2006GL026964

Roesler FC (2013) Brittle fractures near equilibrium. Proc Phys Soc 1956(B69):981–992

Surdin VG (ed) (2013) Galaxies.—M.: Fizmatlit.—432 pp

Werner SC, Harris AW, Neukum G, Ivanov BA (2002) The near-Earth asteroid size frequency distribution: a snapshot of the lunar impactor size-frequency distribution. Icarus 156:287–290

Chapter 12
Similarity and Dimension, Rules of Action

12.1 Scaling

This text follows mainly the ideas of Grigory Isaakovich Barenblatt (1927–2018) (Barenblatt 2003). Processes and associated parameters are characterized by certain quantities. These quantities are measured by comparing certain standards of units of measurement with them. These units are currently tied to some natural physical processes, for example, for a unit of length of 1 m, 1,650,736.73 wavelengths of radiation in vacuum are chosen during the transition from the 2P10 level to 5d 5 of the krypton-86 atom. The previously existing standard was equal to one forty-millionth of the length of the Parisian meridian, measured according to the proposal Laplace during the Great French Revolution. The length of such a rod is called a meter, and it is kept in the Chamber of Weights and Measures in Paris.

A second is defined as 1/86,400 of the average solar day. The more accurate and universal modern definition of a second—9,192,621,770 period of radiation of the cesium-133 atom, corresponds to the transition between two hyperfine levels of the ground state of this atom. A sample, also stored in Paris, corresponds to a unit of mass of 1 kg. Initially, it was associated with a volume of (1 dm^2) of water at 0 °C and a pressure of 1 atmosphere.

Units of measurement form systems for measuring (by comparison) certain processes or quantities. One unit of length, for example, 1 m, is sufficient for measuring geometric objects. Two units—length and time—are sufficient for measuring kinematic phenomena, for example, accelerations and velocities. To describe forces, energies, and powers, we need the concept of mass. Systems of three units of measurement LMT, length, mass, and time are sufficient to describe a large number of processes. Russian GOST introduced the meter, kilogram, and second system, following the one introduced in the mid-twentieth century by the International System SI, System International. The scientific community of the world at first took SI with hostility (calling it Satan Invention, the invention of Satan), because they were accustomed to the GGS system, centimeter, gram, second. Currently, the

G. Golitsyn and C. Varotsos, *The Stochastic Nature of Environmental Phenomena and Processes*, https://doi.org/10.1007/978-3-031-77015-9_12

vast majority of scientific disciplines use the SI system. However, both SI and GGS systems are included in the same class of units of measurement LMT.

In technology, the LFT system of units is often used, where the unit of length is 1 m, force is 1 kgf (kilogram-force), and time is 1 s. In electrodynamics, sometimes a convenient system was introduced in 1966 by Kapitsa (1966), where the classical radius of the electron is taken as the unit of length, the mass of the electron is taken as the unit of mass, and time enters through the unit energy of an electron at rest, mc^2. The first author in 2008 (Golitsyn 2008) proposed a system of units based on SI, which uses the 1 J energy unit instead of mass-system LET which is used occasionally in this book. When considering thermal processes, the system of mechanical units of measurement must be supplemented with an independent standard. Convenient to enter the standard temperature in the form of degrees Kelvin 1 K. When considering rather complex processes, you can enter an arbitrary number of units. For example, when considering the speeds of sports academic boats from the number of rowers (Barenblatt 2003; Kapitsa 1966), four units of measurement are introduced: water density, the desired speed, the length of the boat, and N—the number of rowers (see the analysis of the situation in the book (Kapitsa 1966). It turns out that the speed of the boat is proportional to $N^{1/9}$.

It's time to define the dimension. *The dimension is called a function, determining how many times the numerical value of a given physical value changes quantities during the transition from the original system of units to an arbitrary system within this class.* For example, if we move from SI to CGS for speedup, e.g., gravity, then such a transition will be from 9.8 m/sec^2 to 980 cm/sec^2. Dimension acceleration in the general case LT^{-2} in the LMT system of units, and the energy dimension is ML^2T^{-2}. According to Maxwell's proposal, the dimension of a quantity is denoted by square brackets, for example, power dimension $[W] = ML^2T^{-3}$. This unit is in SI called Watt.

Since the time of Buckingham (1914), it has been known that dimension as a physical quantity is monomial (proof in Barenblatt 2003):

$$[y] = AX_1^{\alpha_1}X_2^{\alpha_2}\ldots X_n^{\alpha_n},$$

where X_i—units of measurement, $A, \alpha_1 \ldots \alpha_n$—constants. Such ratios are often found in various branches of science. Power laws never appear by chance, they discover the most important property of the phenomenon under consideration, its self-similarity. This word means that changing in time and space, the phenomenon reproduces itself in changing time and/or spatial scales (in this book, for example, in Sect. 1.3, the main equation by Kolmogorov is reduced by such a replacement to a self-similar form).

In all works (Barenblatt 2003; Kapitsa 1966; Golitsyn 2008), it is noted that if there are n control parameters, physical quantities, or universal constants, which include k units dimensions, and $n - k > 0$, then $m = n - k$ dimensionless quantities. Such quantities are called similarity parameters Π_i. Then the required value can be written as:

$$a = AX_i^{\alpha_1} \dots X_k^{\alpha_k},$$

where the dimensionless quantity A is a function of the dimensionless parameters Π_i, $i = 1 \dots m$. Values of A are found by comparison with measurement data or numerical calculations.

Practical advice for use in real situations is clearly described in Barenblatt (2003) and Barenblatt and Zeldovich (1972). If the problem can be described in mathematical equations, then these equations and initial and boundary conditions and give a set of defining parameters. After this, it is necessary to analyze whether it is possible to form from them dimensionless parameters, similarity criteria. If some of them turn out to be very large or very small, then some of the initial parameters may be omitted in further discussions. Example: if the Reynolds number $\mathrm{Re} = uL/\nu >> 1$, then in the zero approximation we can ignore viscous strength, i.e., such an analysis can simplify the task. In such cases, the resulting results can be called intermediate asymptotics, where power laws. The concept of intermediate asymptotics was introduced by Barenblatt and Zeldovich (1972). This determines the interval action of similar power laws when the dimensionless parameters A in the previous formula are quite constant. Further specific actions consist of writing the dimension of the desired quantity and the dimensions of the defining parameters X_i and compiling a system of linear algebraic equations by equating the exponents on the left to the sum of the exponents on the right. The solution is this system gives the desired result. Sometimes the choice of units of measurement eliminates the need to solve an algebraic system: so, in Chap. 4 in the system units of LET, the energy spectrum of cosmic ray particles is found immediately, it can be said, obvious, although he waited for his approval for over than 40 years.

In real studies, various situations were encountered, in detail described in Barenblatt (2003), concerning the determination of parameter A and exponents, different from standard methods. In this book, which describes the largest natural processes, fortunately, we have not yet encountered any special situations. General rules of action without specific examples do not fit into the framework for long the minds of researchers. Therefore, we will illustrate with two instructive, from our points of view, research. The first, relatively recent one, belongs to Barenblatt and Monteiro (2010) and refers to crystalline bodies (Barenblatt and Golitsyn 2017; Landau and Lifshitz 1976). Grigory Isaakovich told the first author, quanta has never been used in his practice actions $h = 6.626\ 10^{-27} = \mathrm{g\ cm^2\ s^{-1}}$. Solids are characterized by their elastic modulus μ and, of course, density ρ. These three quantities are characterized by three units measurements, so you can immediately obtain the length scale from them (Barenblatt and Golitsyn 2017):

$$\lambda = \alpha\left(\frac{h}{(\rho\mu)^{1/2}}\right)^{1/4} = \alpha\left(\frac{h}{c\rho}\right)^{1/4}, \tag{12.1}$$

where $c = (\mu/\rho)^{1/2}$—speed of propagation of longitudinal elastic waves. Accounting Poisson's ratio is needed only in the formula for transverse velocity, but considering the small degree of 1/4, the result will change insignificantly (Barenblatt and Monteiro 2010). Estimates are given for scale λ values of the order of 1 nm $= 10^{-9}$ m.

In the same conversation, I told Grigory Isaakovich that the discovery of such a nanoscale like a diamond, but it still needs to be processed to turn it into a brilliant. This is how work (Barenblatt et al. 2014) with professional mineralogists arose. We tried to give a physical meaning and origin to the scale λ, but G.I. said no all at once, which was done in Golitsyn (1970), which was supplemented in comparison with (Barenblatt and Golitsyn 2017) one metal and six simple molecules. As a result, for 33 substances comparison scale λ with real size data $\alpha\lambda_r$ crystal lattice, i.e., interatomic distances, given for the coefficient $\alpha = 1.986 \approx 2$. This is once again illustrating Einstein's statement that in the correct dimensional formulas, numerical factors obtained by comparing theoretical predictions with real data must be neither very small nor very large, i.e. be O(1). Here, in contrast to Barenblatt et al. (2014) and Barenblatt and Golitsyn (2017), the calculations used the value $h = h/2\pi$.

Along the way, in Barenblatt and Golitsyn (2017) a formula was derived for estimating the density of crystalline bodies by formula $\rho_T = m\lambda^{-3}$, where m is the mass number of an atom or molecule in a crystalline grate. Density calculations using this formula for 23 crystals gave $\rho_o/\rho_T \approx 4$, i.e., again this is several of the order of one. Of course, in reality, the density is found simply by weighing certain volumes, but this formula of ours only illustrates possibilities of the dimension theory. Physical origin of scale λ such that it is increased by two or three order of magnitude Heisenberg uncertainty relation $\Delta r \cdot \Delta p \geq= h/4\pi$, What is obtained by using the simplest Debye theory of condensed matter (Landau and Lifshitz 1976) and using the relationship between the energy of a vibrating atom ε and him impulse

$$\varepsilon = p^2/2m.$$

The second example is the theory of similarity for planetary atmospheres (Golitsyn 1970, 1973) and its additions 40 years later in monograph G12. Its peculiarity is that all the equations are there: continuity, movement, state, energy. To identify similarities in this the system needs to non-dimensionalize these equations by choosing the appropriate scales: for lengths, this is the radius of the planet α, the speed scale is the adiabatic speed of sound:

$$c_a = (\gamma RT_e/\mu)^{1/2}, \quad \gamma = c_p/c_v, c_p - c_v = R/\mu$$

- constant for gas with molecular weight μ,

$$T_e = [q(1 - A)/4\sigma]^{1/4}$$

- effective temperature of the exhaust heat radiation, q—solar radiation arriving at the disk of the planet. $\sigma = 5.67\,10^{-8}\,\mathrm{W/m^2K^4}$ is a constant in the Stefan–Boltzmann radiation law. For the Earth on distance 1 au. (astronomical unit)

$q = 1365 \pm 2\,\mathrm{W/m^2}$, i.e., per unit the earth's surface, considering albedo $A = 0.3$, accounts for 239 $\mathrm{W/m^2}$. For scale time taken $\tau_e = \alpha/c_e$, which, as is known, determines the establishment local thermodynamic equilibrium on scales on the order of the radius of the planet α.

After reducing the equations to dimensionless form, 3 parameters appear in the similarities:

$$\Pi_\omega = \frac{\omega_a a}{c_e}$$

with a term c of the Coriolis acceleration called the rotational Mach number, at gravity acceleration term:

$$\Pi_g = \frac{RT_e}{\mu g a} = \frac{H}{a},$$

where H is the atmospheric scale height. The energy equation:

$$\frac{dT}{dt} = \frac{Q}{c_p},$$

where $Q = Q(z)$—heat influx per unit mass, associated mainly with the balance of solar and thermal radiation. When averaging it over height as:

$$\frac{1}{M}\int\limits_0^\infty \rho(z)\frac{Q(z)}{c_p}dz$$

an important similarity parameter appears, the criterion:

$$\Pi_M = \frac{q\tau_e}{Mc_pT_e}.$$

Note that in Gierasch et al. (1970) and Golitsyn (1970, 1973) these criteria appeared by simple search defining external parameters included in the system equation. Viscous dick is neglected, considering that the Reynolds number is very high and its exact value immaterial. Similarity parameter Π_M is related to the mass of the atmosphere, and it can be interpreted as a relation between two times τ_e/τ_o, where $\tau_o = Mc_pT/q = w_o/q = Mc_pT_e/q$—heat content (enthalpy) of an atmospheric column of mass M. In Gierasch et al. (1970) this parameter was interpreted as a measure of the nonadiabaticity of the movements.

Similarity parameters Πg and Π_M for the planets of the solar system all are small 10^{-2}–10^{-5}. Therefore, we can accept the hypothesis that those included only in parameters g and M are insignificant for estimating the total kinetic energy atmospheric movements. Rotational number Π_ω order 10^{-2} for Venus and Titan, a satellite of Saturn, about 1 for the Earth and Mars, and noticeably more than one—from 4 to 15 for giant planets. Four dimensional values are remaining: the density of the incoming radiation surface unit $[q_A] = \mathrm{MT}^{-3}$; radius $[\alpha] = \mathrm{L}$, heat capacity $\left[c_p\right] = \mathrm{L}^2\mathrm{T}^{-2}\mathrm{K}^{-1}$, constant $[\sigma] = \mathrm{MT}^{-3}\mathrm{K}^{-4}$, have 4 independent dimensions, and they can form a quantity with the dimension of energy:

$$E_k = 2\pi B\sigma^{1/8} c_p^{-1/2} q_A^{7/8} a^3.$$

This quantity does not contain mass, it cannot be potential energy, no heat content, and only kinetic energy remains; the 2π factor is introduced to simplify further formulas. Dimensionless numerical coefficient B, according to Barenblat (2003), may depend on all three similarity parameters. However, the parameters Π_g and Π_M are very small and can be neglected. Addiction remains $B = B(\Pi_\omega)$. Let us assume that $B = 1$ for planets with $\Pi_\omega \leq 1$, i.e., for Venus, Mars, Earth, and Saturn's moon Titan, slowly rotating with a dense atmosphere. For the Earth, the average speed is 17 m/s, and our theory gives 11 m/s (Gierasch et al. 1970; Golitsyn 1973); for Venus at the surface 0.4–0.7 m/s, and in its main atmosphere ~ 2 m/s, near Mars—according to observations from 5 to 20 m/s, theory—30 m/s, for Titan ~ 0.5 m/s, theory 0.4 m/s. Let us present here for Jupiter—100–200 m/s according to observations and 120–140 m/s—theory (G12).

For giant planets, Taylor series expansion:

$$B(\Pi_a) = 1 + a\Pi_\omega^2$$

gives at $\alpha = 0.9 \pm 0.2$ [see (G12)] velocity values for Jupiter, also consistent with measurements. A similar consideration of movements in the solar atmosphere also leads to reasonable results (G12). These books provide useful formulas for estimates of average wind speeds:

$$U = \left(\frac{2E_k}{M_0}\right)^{1/2} \approx \Pi_M^{1/2} c_e,$$

where $M_0 = 4\pi\alpha^2 M$—full mass of the atmosphere and $\delta\mathrm{T} \approx \Pi_M^{1/2} T_e$—an estimate of the temperature difference on the surface of the planets, setting the atmosphere in motion, i.e., exciting the winds. The formula for the kinetic energy of winds can be presented as a short formula:

$$E_k \simeq \pi r^2 q_A \cdot \tau_e,$$

which states that the energy of atmospheric movements is of the order of solar power energy incident on the disk $Q_A = \pi r^2 q_A$, multiplied by thermodynamic time relaxation on a planetary scale, which is the minimum temporary scale in the system—a good illustration of Sect. 12.1.

12.2 Astrophysical Applications

This author has worked considerable time on stellar, galactic and galactic cluster astrophysics (Golitsyn, 2015, 2016, 2017) and has explained a number of empirical observations.

Measured are up to five characteristics and there is the universal gravity constant G. Beside a non-dimensional similarity parameter Π_v, virial, other non-dimensional parameters may exist. The luminosity W with G form a velocity scale $U_d = (WG)^{1/5}$, with temperature velocity scale $U_T = (3kT/\mu m)^{1/2}$ it forms the new similarity value $\Pi_2 = U_d/U_T$. For 30 clusters of quite different ages $\Pi_2 = 0.26 \pm 0.02$. Due to excess number of the measured quantities each of which is determined by two units of mass, time and length many possibilities exist to find connections between these quantities which are described by Golitsyn (2017), which may be used to oversee the general quality of observations.

References

Barenblatt GI (2003) Scaling. Cambridge University Press, 171 pp

Barenblatt GI, Golitsyn GS (2017) Similarity criteria and scales for crystals. Phys Mesomech 20(1):116–119

Barenblatt GI, Monteiro P (2010) Scaling laws in nanomechanics. Phys Mesomech 13(5):41–45

Barenblatt GI, Zeldovich YB (1972) Self-similar solutions as intermediate asymptotics. Ann Rev Fluid Mech 4:285–312

Barenblatt GI, Golitsyn GS, Eremin NN, Urusov VS (2014) Universality of the linear nanoscale Dokl. RAS 458(5):528–530

Buckingham E (1914) On physically similar systems. Phys Rev 4:354–376

Gierasch P, Goody R, Stone P (1970) The energy balance of planetary atmospheres. Geophys Fluid Dyn 1(1):1–13

Golitsyn GS (1970) A similarity approach for the general circulation of planetary atmospheres. Icarus 13(1):1024

Golitsyn GS (2008) Visualization for a number of problems of choosing energy as a unit of measurement instead of mass. Physics 178(7):753–755

Golitsyn GS (1973) Introduction to the dynamics of planetary atmospheres.—L.: Gidrometeoizdat—104 p (NASA Transl. No. TTF-15, 627, Dec. 1974)

Golitsyn GS (2015) Galaxy clusters, similarity parameters, and ratios between measurable characteristics. Usp Fiz Nauk 185(12):1323–1331. https://doi.org/10.3367/UFNr.0185.201512c.1323

Golitsyn GS (2016) Results of galaxy cluster analysis by the theory of similarity and dimension Dokl. RAS 466(2)

Golitsyn GS (2017) Theory of similarity and dimension for galaxies: explanation of observational results Dokl. RAS 475(4):495–499

Kapitsa SP (1966) Natural system of units in classical electrodynamics and electronics. Physics 136(2):188–190

Landau LD, Lifshitz EM (1976) Statistical physics, §§62–65.—M. Fizmatlit

Chapter 13
Convection

13.1 Introduction

Convection is the most common type of fluid movement in nature, in technology, and even in everyday life. His research dates back just over a century, and during this time they acquired a solid basis, which we will try to present here. Thousands of articles and dozens of books are devoted to the description of convection, but the physical side of the process is rarely given enough attention. This will be done using the methods of mathematical physics, similarity and dimension theory, and connections with the scale of the theory of random motions of A. N. Kolmogorov 1934 and their development in the works of his students will be traced. They will also be brief on the main manifestations of convection in nature are discussed: movements from instantaneous and constant heat sources, convection in a rotating fluid, deep convection in the ocean, in the earth's mantle, etc. Many of these aspects are presented in the book G12 and are found in the previous paragraphs of this book. Quite systematic the presentation of the subject seems useful, but there are clarifications and their further development.

Convection is a manifestation of Archimedes' law: heavy bodies sink, light bodies float. Due to the equation of state, density is mainly a function of temperature, and for low speeds compared to the speed of sound pressure can be neglected, but in a gravity field the pressure is close to hydrostatic. The general condition for the occurrence of convective instability of a horizontal layer of liquid is the requirement for a vertical density gradient—its value (in absolute value) must be greater than adiabatic:

$$\gamma_{pa} = -\frac{d\rho}{dz} = -\frac{dp}{d\rho}\frac{dp}{dz} = -\frac{\rho g}{c^2} \tag{13.1}$$

where c^2 is the square of the adiabatic speed of sound, equal to $RTc_p/c_v\mu$ due to the equation of state. In Landau and Lifshits (1986) the derivation of the vertical temperature gradient with the condition that the entropy of a substance does not

G. Golitsyn and C. Varotsos, *The Stochastic Nature of Environmental Phenomena and Processes*, https://doi.org/10.1007/978-3-031-77015-9_13

increase with height, from which:

$$\gamma_{pa} = -\frac{\alpha g T(z)}{c_p} = -\frac{T(z)}{H}, \quad H = \frac{c_p}{\alpha g}, \tag{13.1'}$$

where α is the coefficient of thermal expansion of the substance under consideration, and H is the internal height parameter: for air at T = 300 K the value is $H = 30$ km, for water at $T = 20$°C, $\alpha = 2 \times 10^{-4}$ and $H = 2000$ km, for the mantle $H = 6000$ km.

When density inhomogeneity occurs, a buoyancy parameter appears:

$$b = -\frac{g}{\rho}\langle w'\rho'\rangle \tag{13.2}$$

where the primes mean fluctuations of vertical velocity and density, and angular brackets indicate averaging. Boussinesq's approach $\rho'/\rho = -T'/T$ operates in the case when the height of the layer under consideration $h \ll H = c_p/\alpha g$. Otherwise, it is necessary to use the so-called "deep convection approximation" (anelastic approximation). This incident (and other moments) are described in detail in Kerry's book Emanuel (1994). For example, for convection in the earth's mantle, which extends over 3000 km, and the magnitude $H \approx 6000$ km, however, the refined equations do not give significantly new effects (Golitsyn and Vulfson 2007).

13.2 Basic Equations

To describe convection, the Navier–Stokes equations are used, as the equation heat diffusion and equation of state:

$$\rho\frac{\partial u_i}{\partial t} + \rho u_k\frac{\partial u_i}{\partial x_k} + 2\rho\varepsilon_{ijk}\omega_j u_k = -\frac{\partial p}{\partial x_i} + \rho F_i + \rho\frac{\partial}{\partial x_k}\nu\frac{\partial u_i}{\partial x_k}, \quad F_i = g_i, \tag{13.3}$$

$$\frac{\partial T}{\partial x_i} = k\frac{\partial^2 T}{\partial x_i^2}, \quad p = p(\rho, T) \tag{13.4}$$

where ω_i—vortex vector, k—thermal diffusivity coefficient, density ρ considered to be a function of temperature only. Multiply scalarly (13.3) by u_i and let's integrate vertically over the layer thickness from 0 to h. Then we get the equation energy balance, G12, and Golitsyn (1979, 1980):

$$\frac{dK}{dt} = G - D, \quad K = \frac{1}{2}\rho\int_0^h \langle u^2 + v^2 + w^2\rangle dz \tag{13.5}$$

kinetic energy of a unit column of liquid,

$$G = \alpha\rho g \int_0^h \langle w'T' \rangle dz \tag{13.6}$$

the rate of generation of kinetic energy from potential energy,

$$D = \rho\nu \int_0^h \left\langle u_i \frac{\partial^2 u_i}{\partial x_k^2} \right\rangle dz \tag{13.7}$$

rate of kinetic energy dissipation due to viscosity. *B* statistically inpatient case $dK/dt/ = 0$ and G = D, as well as $\varepsilon = b$.

During convection, the heat flux consists of molecular transport and heat, carried by movements. At an arbitrary level within the layer under consideration full flow:

$$f_0(z) = \rho c_p \left(-k \frac{dT}{dz} + \langle w'T' \rangle \right)$$

and, integrating it from 0 to *h*, we get:

$$f_0 h = \rho c_p k \Delta T + f_e h, \tag{13.8}$$

where $\Delta T = T_1 - T_2$—temperature difference at the boundaries of the layer, $f_0 = \rho c_P \langle w'T' \rangle$—heat flow, carried by movements. Divide (13.8) by $f_0 h$ and we get:

$$\frac{f_e}{f_0} = 1 - \mathrm{Nu}^{-1} \equiv 1 - \alpha_N, \tag{13.9}$$

where the most important sought-for quantity appears in the convection process, the Nusselt number:

$$\mathrm{Nu} = \frac{f_0 h}{\rho c_p k \Delta T}, \tag{13.10}$$

showing how many times the amount of heat flux imparted to the layer is superior to conductive molecular flux. Note that from (13.9) it is clear that flow f_e, carried by movements in $(1 - \alpha_N)$ times less than the total heat flux supplied layer. If the horizontal and vertical scales are noticeably different, then their ratio will appear in the continuity equation $\Pi = h/L$, the so-called aspect ratio. A common and little-known result is, so to speak, Convection efficiency, i.e., what fraction of the heat supplied to the layer is used for the generation of kinetic energy (Golitsyn 1980), which gives the ratio (13.6) to (13.8). Taking into account (13.9):

$$\frac{G}{f} = \frac{\alpha g h}{c_p} \frac{f_e}{f_0} = \frac{h}{H}(1 - \alpha_N) \cong \frac{h}{H}, \tag{13.11}$$

i.e., in this sense, the efficiency can be very small, and the condition h << H justifies the approach of Boussinesq.

Next is mathematical physics, begun by Lord Rayleigh (Emanuel 1994; Rayleigh 1916). He considered the problem of starting motion in a flat layer of thickness h with a lower boundary $T(0) - T(h) > 0$. Introduced scales of speed k/h, time h^2 /k and temperature ΔT, and then in equation movement, two dimensionless parameters appear, subsequently called the number Rayleigh:

$$\text{Ra} = \alpha g \Delta T h^3 / k\nu \tag{13.12}$$

and the Prandtl number:

$$\text{Pr} = \nu / k. \tag{13.13}$$

The presence of two similarity numbers in the equations of the process indicates that the number Nusselt is their function Nu $= f$ (Ra, Pr), a problem found in (Emanuel 1994). Helpful is the so-called Rayleigh flux number considering (13.2)

$$\text{Ra}_f = \text{Ra} \cdot \text{Nu} = \frac{bh^4}{k^2 \nu}. \tag{13.14}$$

From the equality of generation and dissipation in a steady state, we can estimate for convection buoyancy flux through velocities, to estimate which we use the Stokes formula (Landau and Lifshits 1986)

$$\varepsilon = \nu \frac{\partial u_i}{\partial x_k}\left(\frac{\partial u_i}{\partial x_k} + \frac{\partial u_k}{\partial x_k}\right) = b = \frac{\alpha g f}{\rho c_p}, \tag{13.15}$$

where the buoyancy flux is expressed using formulas (13.2) and (13.15). Let's evaluate velocity derivatives:

$$\frac{\partial u_i}{\partial x_k} \sim \frac{U}{h/2},$$

through the desired total speed U and half layer height h for both solid walls. In our flat case in (13.15) there will be 8 terms, there are 18 of them in three dimensions. Inserting these expressions into (13.15) and resolving it relative to U^2, we obtain, taking into account the fact that $\varepsilon = b$ in the quasi-stationary state:

$$U^2 = \left(\langle u^2 \rangle + \langle w^2 \rangle\right) \approx 2\langle w^2 \rangle$$

and onwards

$$\overline{u} \approx \overline{w} = \left(\frac{\varepsilon}{av}\right)^{1/2} h = \left[\frac{\alpha g f}{a \rho c_p v}(1 - \alpha_N)\right]^{1/2} \approx \frac{k}{h}\left(\frac{\text{Ra}_f}{a}\right)^{1/2}\left(1 - \frac{\alpha_N}{2}\right)$$
$$= \frac{k}{ah}\text{Ra}_f^{1/2}\left(1 - \frac{\alpha_N}{2}\right), \tag{13.16}$$

where $\alpha = 8$ or 18. Later in Golitsyn and Grachev (1981) the exact coefficients a were calculated and it was it was found that their value weakly depends on the type of boundary conditions and coincides within 20% with our elementary estimates. Note that the expression for:

$$U^2 \approx \varepsilon h^2 / v$$

coincides with the structural one up to a numerical coefficient velocity function for viscous scales in the theory of local homogeneous and isotropic turbulence Kolmogorov–Obukhov (Monin and Yaglom 1975). From comparison (13.16) and (13.14) it follows that:

$$\overline{u} \sim \text{Ra}_f^{1/4}.$$

13.3 Convective Instabilities

Convection arises from rest when the Rayleigh number reaches some critical value, it is obvious that at this point the Nusselt number Nu = 1, so it starts to make sense from one. This problem was solved to the end in 1916 by Lord Rayleigh (Rayleigh 1916; Emanuel 1994), who obtained for two free boundaries of the layer the final value of the critical number $\text{Ra}_{\text{cr}} = 27\,\pi^4/4 = 657.5$. Two other cases of boundary conditions were considered by Chandrasekar (1961) and Golitsyn and Vulfson (2007), who found that at the upper free boundary, and the lower solid boundary $\text{Ra}_{\text{cr}} = 1100.7$, and with both solid boundaries $\text{Ra}_{\text{cr}} = 1707.8$, for the determination of which the transcendental algebraic equations. It is interesting to note that if the boundary from above is free and the wind blows over it wind, the value of Ra_{cr} increases with the Reynolds number for the flow (Gray 1968). One of the many conditions for hurricane formation, according to long-term observations, is the absence of appreciable wind shear in height. In the heat diffusion equation before the second derivative of the temperature at disintegration, the Peclet number appears:

$$\text{Pe} = Uh/k = \text{RePr}, \tag{13.17}$$

which is the product of Reynolds and Prandtl numbers. At Pe >> 1 it means that near the wall there appears a thermal boundary layer with a thickness of

$$\delta_T \approx h\mathrm{Pe}^{-1/2}, \tag{13.18}$$

and the main vertical temperature changes will be in such layers. From the continuous heat flux during the transition from the boundary layer to the bulk, close to the isothermal fluid mass, it follows that $f' = f/\rho c_p = \mathrm{k}\Delta T/2$, where $\Delta T/2$—is the temperature drop at one wall. Then it follows from (13.18) that Nusselt number:

$$\mathrm{Nu} = \frac{1}{2}\mathrm{Pe}^{1/2} \tag{13.19}$$

When studying the convection mode of a viscous fluid, it is natural to take (13.16) as the scale of the velocity to take (13.16), and then with the help of (13.19) and (13.16)

$$\mathrm{Nu} = \frac{1}{2}\left[\frac{\mathrm{Ra}_f}{a}(1 - \alpha_N)\right]^{1/4} \tag{13.20}$$

and, remembering that $\mathrm{Ra_f} = \mathrm{RaNu}$, we obtain from here after simple manipulations

$$\mathrm{Nu} = \mathrm{Ra}^{1/3}\frac{(1 - \alpha_N)^{1/3}}{2(2a)^{1/3}} \cong \frac{1}{2}\left(\frac{\mathrm{Ra}}{2a}\right)^{1/3}\left(1 - \frac{\alpha_N}{3}\right) \sim \mathrm{Ra}^{1/3}. \tag{13.21}$$

This is the most important relation in the theory of convection, long-established from experience and justified by considerations of similarity, since at $\mathrm{Nu} \sim \mathrm{Ra}^{1/3}$ the heat transfer does not depend on the height of the layer h, if there are no heat sources or sinks in it. At $\alpha_N/3 = (3Nu)^{-1} << 1$ the dependence on h remains in this term, but already at $\mathrm{Nu} = 7$ this correction reduces the empirical relation (13.21) by about 5%. The dimensionless proportionality factor in (13.21) is $\left[2(2\alpha)^{1/3}\right]^{-1} = 0.15$ at $\alpha = 18$. Extracting this value from data from numerous experiments in Golitsyn (1979, 1980) yielded a value of $\beta \approx 0.14 \pm 0.015$ Thus, simple theoretical estimates with the involvement of experimental data justify the law:

$$\mathrm{Nu} = \beta\mathrm{Ra}^{1/3}, \tag{13.22}$$

where the value β may depend on the Prandtl number Pr, which is not considered here, *h/L* ratio, characterize boundary conditions, etc.

At the interface between water and air, only molecular mechanisms of heat-mass transfer (vaporization), as well as the thermal radiation balance. If we know the total heat flux (enthalpy) in water, then $\Delta T \sim f^{3/4}$. With moderate winds and $f = 100$ W/m^2, $a = 14 \pm 1$, $\rho = 10^3$ kg/m^3, $\nu = 10^{-1}$ m^2/s, Pr ≈ 7, $\alpha = 2 \times 10^{-4}$ K (T = 20 °C), hence we get $\Delta T \approx 0.3$ K—is the cold film. In tropical storms $f \approx 700$ W/m^2 and more, $\Delta T \sim 1$ K, for polar hurricanes $f \approx 2$ KW/m^2 and then $\Delta T \approx 2$ K. In hurricanes, there are strong winds and waves. Therefore, the cold film is breaking down all the time. Its lifetime is estimated by the formula (h^2/k), independent of the film thickness. In these three cases, the times will be 190, 72, and 48 s. However, strong and gusty winds, the collapse of the waves, and the limited periods of less than 10 s do not make it possible to the possibility of any more or less universal analytics, although conversations with scientists using advanced experimental techniques show that water gives hurricanes their total enthalpy ≥ 1 kW/m^2, G12.

A few words about convection in the mantle. It's caused by a geothermal flux of heat, not uniform over the surface, but equal to 86 mW/m^2 on average. The mantle is about 3000 km thick, with $c_p = 10^7$ J/kg K. For scale $H = 6000$ km at kinematic viscosity $\nu \sim 10^{17}$ m^2/s and using formula (13.16) we obtained convective velocities of the order of 5 cm/year, p. 12.2. Such velocities are confirmed by GPS for individual plates moving from 1 to 18 cm/yr.

13.4 Time Criteria and Heat Transfer

Time scales have long remained without much attention. They can be found by knowing the velocity scale and the dimensions. For 3 N >> 1, it follows from (13.16) that:

$$\tau = \frac{h}{u} = a\left(\frac{\mu c_p}{\alpha g f_e}\right)^{1/2} = \frac{ah^2}{k}\mathrm{Ra}_f^{-1/2}, \tag{13.23}$$

whence it can be seen, taking into account (13.14), that the time scale does not depend on the thickness of the layer. This is confirmed by Fig. 13.1, which shows the maxima of the temperature spectra in depending on Ra_f, taken from Boubnov and Golitsyn (1990), where the same figure is available, and the following are given (in Fig. 13.2) also time spectra of temperature fluctuations in a rotating vessel and at different speeds of its rotation, differing by a factor of 30.

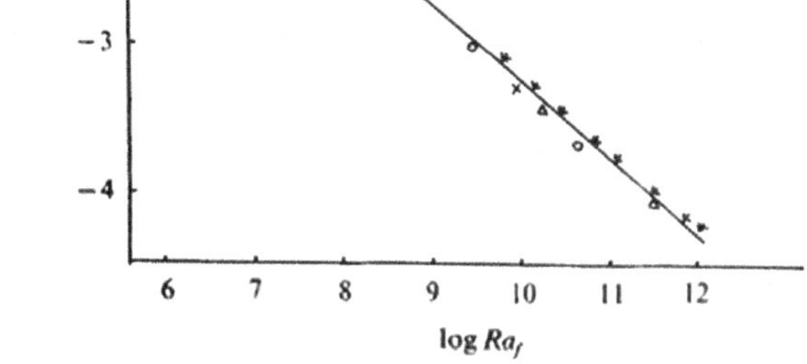

Fig. 13.1 Dependence of the most probable period of temperature fluctuations fluctuations τ_0 on the Rayleigh flux number Ra_f, the straight line corresponds to dependence $\tau_o \sim \mathrm{Ra}_f^{-1/2}$ (Boubnov and Golitsyn 1990; Obukhov 1949)

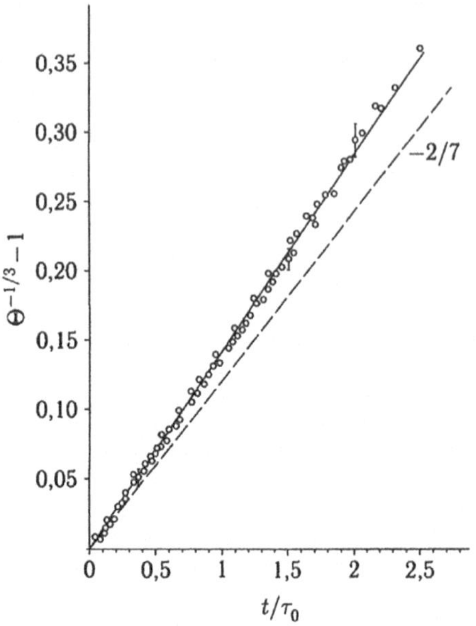

Fig. 13.2 Solution of the energy balance Eq. (13.30) by a cooled liquid with time. The dotted line is the solution of the same problem with the law of heat transfer law Nu ~ Ra with $n = 2/7$ instead of 1/3. The straight line is drawn with slope β = 0.144 (Golitsyn and Grachev 1980)

Normalized by the square of the variance, they all lie well on the curve ω^{-2}. This is interpreted in the spirit of Obukhov (1949) about fluctuations of the passive scalar when it has the same spectral characteristics as the velocity field (see also Sect. 11.2). According to Obukhov (1949), both temperature and other passive scalar impurities, in turbulent flow have an inertial interval with the same spectra as the velocity. According to ANK34, Sect. 1.3, in Lagrangian variables, the mean square of the velocity fluctuations is $< u^2(t) > = \varepsilon t$, and its frequency spectrum is ~ ω^{-2} (Golitsyn 2018).

Considerations of dimensionality allow us to estimate fluctuations of temperature, which turn out to be equal to:

$$\sigma_1 = \delta T = \frac{\langle W'T' \rangle}{\overline{w}} = \frac{f_e'}{\overline{w}} = \frac{a}{h}\left(\frac{f\nu}{\rho c_p \alpha g}\right)^{1/2}. \tag{13.24}$$

Outside the thermal boundary layer, the formula for the temperature fluctuations was obtained by Prandtl (1932) and Obukhov (1960) from the considerations of dimensionality:

$$\sigma_T = C_\sigma f_e'^{2/3} (\alpha g z)^{-2/3}, \tag{13.25}$$

where z is the height. The value of C_σ varies from 1 to 2.5 depending on the conditions of the experiment, which are not always identical (Foster and Waller 1985). It is convenient to have this formula also in the form:

$$\sigma_T/\Delta T = C_\sigma \mathrm{Nu}\left(\mathrm{Ra}_f \mathrm{Pr}\right)^{-1/3}. \tag{13.25'}$$

In the last 15 years of the last century there appeared works on verification of the relation (13.22)—Nu ~ Ra^n, in which at numbers Ra > 10^7 the values of n equal to 2/7, 0.309, etc. (cf. Kadanoff 2001), where it is stated about impossibility of a more or less unified theory of convection at arbitrary Prandtl and Rayleigh numbers, at different boundary conditions and station times of processes inside the layer, etc.). For such studies, one should always take into account the temperature dependence of the medium parameters. Thus, when T changes from 20° to 60°, the Prandtl number for water changes from 7 to 3. Our understanding is that the law (13.22) works well in the case of convection with a free surface and large Rayleigh numbers, i.e. (Ra/Ra_{cr}) >> 1.

To verify this, we set up two varieties of special experiments: (1) on cooling with time of a layer of water with a known heat input and a layer of aqueous alcohol solutions of different concentration with full control of the environment—temperature, humidity, etc. (Golitsyn and Grachev 1980, 1984). Considering the enthalpy balance in the liquid then:

$$\frac{dW}{dt} = -f, \tag{13.26}$$

where W is the total enthalpy of its components. The right part is the sum of direct heat removal from the surface and evaporation from there and radiation balance on it. The rate of heat removal by convection into the air is determined by the reduction in it gravity, for the moment we will consider this only for water (see Chap. 9)

$$g' = g\frac{\Delta T}{T}\left(1 + \mathrm{Bo}^{-1}\right), \quad \mathrm{Bo}^{-1} = \frac{\mu_w}{\mu}\frac{L\Delta e}{\rho c_p \Delta T} = 0.614 L_1 \frac{\Delta e}{T}, \tag{13.27}$$

where Bowen number is the ratio of apparent and latent heat fluxes, L—is the heat of vaporization (weak function of temperature according to the Clapeyron-Clausius law), Δe—relative humidity of air, related to the mixture ratio $q = 0.622\,(e/p)$, p—air pressure, equal to 1.013×10^5 Pa at mean sea level. It is further assumed that. at the surface $e = e_s = 100\%$, i.e. the vapour is saturated, and away from it the measured humidity, is determined by the air temperature and then $\Delta e = e_s - re(T)$, where r—relative humidity. The moisture flux from the surface is assumed to be governed by a law of the type (13.22), i.e.

$$\mathrm{Nu}_v = \mathrm{Ra}_w^{1/3}, \text{ where } \mathrm{Ra}_w = \frac{\alpha g \Delta e h^3}{v k_\alpha}, \tag{13.28}$$

where index w refers to water vapour, $k\alpha$—its diffusion coefficient in air, $\alpha\Delta e$—dimensionless value. Similar expressions could be written for other vapors, see Fig. 13.3.

When working with alcohol solutions, the same relationships are written for ethyl alcohol vapors (Golitsyn et al. 1984). With appropriate nondimensionalization, the enthalpy balance equation in a liquid (Golitsyn and Grachev 1986) is reduced to the form:

$$\frac{d\theta}{d\tau} = -\frac{1}{3}\theta^{4/3}, \quad \theta = \frac{\Delta T(\tau = 0) - \Delta T(\tau)}{\Delta T_0}, \tag{13.29}$$

whose solution at the initial condition $\theta(\tau = 0) = 1$ looks like:

$$\begin{aligned} &\theta(\tau) = (1 + \tau)^{-3} \\ &\theta^{-1/3}(\tau) - 1 = \tau \end{aligned} \tag{13.30}$$

This solution is shown in Fig. 13.2, where the experimental points fit well on the theoretical curve (13.30) as a function of time. The dotted lines correspond to the curve with $n = 2/7$ instead of 1/3 in the law (13.27). The same hypotheses, with $n = 1/3$, were used when processing data with alcohol solutions assuming that there are no ethyl alcohol vapors in the laboratory room.

Figure 13.3 presents data on mass flows of matter from the surface of alcohol solution, controlled using analytical balances, in comparison with the theoretical solution mass loss curve, calculated using relationships like (13.28) for the mass of vapors in a specialized Institute of State Standards of the USSR as the third co-author of the work Golitsyn et al. (1984). So, these two figures experimentally confirm the operation of law (13.22), at least for convection in a layer with an open surface. A complex combination at the horizontal axis gives an idea of the combined Rayleigh number with n = 1/3, the numerical coefficients α, β, and γ all correspond to the general relationship $\mathrm{Nu} = 0.147\ \mathrm{Ra}_f^{1/3}$. The alcohol concentration in solutions varied from 0 and reached 69.5%.

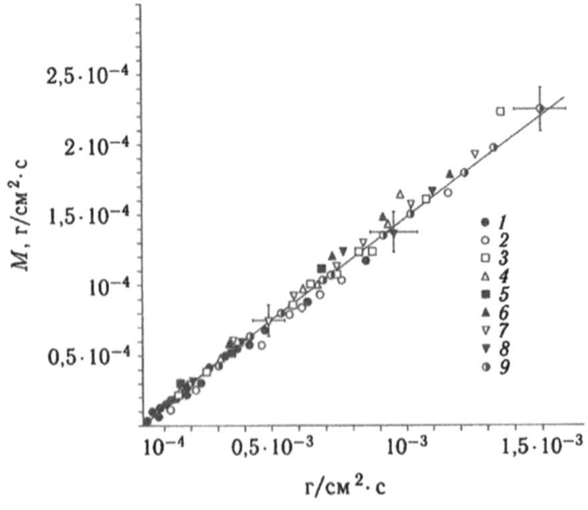

Fig. 13.3 Data on mass flow from the surface of alcohol solutions in motion with a theoretical curve obtained at $n = 1/3$ and $\beta = 0.147$ (Golitsyn et al. 1984)

On this basis, a formula was proposed for heat and mass transfer between the atmosphere and the ocean (Golitsyn and Grachev 1986), considering real measurements of heat and moisture flows in nature and in the laboratory. It was seen at the European Medium Term Center weather forecasts. Its use has significantly improved global weather forecasting, which the director of this center told me about in 1992. True, later they slightly improved our parameterization with Grachev, including weak horizontal winds (Grachev 1990), associated with the suction of mass into rising thermals, and did not refer again.

In 1985, a young and highly active employee of the nuclear center in Aldermaston, Great Britain, showed keen interest in the second experiment with alcohol, but I limited myself to referring to Grachev (1990). One can imagine that these laws (Golitsyn 1983) are needed for some technological processes. In the 1980s, I, as the "discoverer" of the "nuclear winter" effect (the results of the first author were published three months earlier than in the USA), having an analogy with global dust storms on Mars, quickly realized in April 1983 that clouds of dust and smoke as a consequence of a large-scale nuclear war will have the same anti-greenhouse effect as Martian dust storms and the same meteorological consequences—increasing the static stability of the atmosphere, absorbing solar radiation with dust, but a surface transparent to thermal radiation, suppression the formation of cyclones, a decrease in evaporation, all due to an increase in the stability of the atmosphere (see Budyko et al. 1986).

13.5 Convection in Rotating Fluids

The role of the rotation is described in Chaps. 5 and 9. It adds for consideration the Coriolis parameter $l_c = 2\Omega$, multiplied on a sphere by $\sin\theta$, θ is addition to latitude. The Nusselt number here is the function of the three non-dimensional parameters $Nu = Nu(Ra, Ta, \mathrm{Pr})$. The problem of fluid instability at rest was solved by Chandrasekhar (1953, 1961) and was laterly experimentally confirmed.

Golitsyn (1980) and Golitsyn and Grachev (1981) has dimensionally obtained the velocity scale $u = (b/\Omega)^{1/2}$, where b is the buoyancy flux and tested by home experiments with $\Omega = 78$, 45 and 33 revolutions per minute.

By stopwatch measured the time of travels of a fixed distance by dry tea particles in a warmed fluid evaporation of which supplied the necessary flux b. For each set we performed 30 measurements and found that at 33 rpm velocities of particles were about 1.5 times faster than at 78 rpm: $(78/33)^{1/2} = 1.54$.

At that time, end of 1980s, there was no laser velocity measurements and no routines to get results. But after these measurements the work Golitsyn and Grachev (1981) was send in press. In the paper of 1980 there was an estimate of the magnetic Reynolds number $\mathrm{Re}_m = UL/\nu_m \approx 100$, and at that time such a value was considered to be enough for the geomagnetic dynamo; the velocities of order 5 km/year *was* close to the drift velocities of the geomagnetic field.

In 1983 I had a graduate student Boris Boubnov (1953–1999), a great master in all hands: experimenter, for routines, analytical calculations and we started studies of rotating fluid convection for more than 10 years. The results are described in the book by Boubnov and Golitsyn (1990). In the difficult for Russia 1990s Boris worked in the prestigious hydrodynamical laboratories in England, France, USA, Germany, quickly organized simple but meaningful experiments, analyzed results and wrote papers. He was quite popular abroad, but died at 46. In 2001 in USA was an International conference for geophysical hydrodynamics and its participants dedicate this conference to his memory.

The work with Boris started with the test of the appearance of instabilities at changes of both similarity parameters Ra and Ta. At Fig. 13.4 the classification of varior convective regimes is presented in logarithmic coordinates Ra_f and Ta at $\mathrm{Pr} = \mathrm{O}(1)$ for a rotating plane layer. The region between lines *a* and *b* contains the hexagonal structure of Chandrasekhar, which at the laboratory experiments is seen at Fig. 13.6 as a triangular one. At Fig. 13.4 the roman numbers note the regions of the basic convective regimes: I—the rest, *a*—the line instability, II is the region of regular geostrophic convection as at Figs. 13.5 and 13.6, where the velocity scale is Eq. (5.3) with numerical coefficient $k = 1.7$ (see Boubnov and Golitsyn 1990, 1995). The region III is the irregular geostrophic convection as is Fig. 13.7 in the lab and Fig. 13.8 on the sea surface. Figure 13.6 shows the process of transformation of hexagonal vortex grid by appearance of vortices dancing around one another, formation of the double helix and then its coalescence in one larger vortex.

At Fig. 13.4 the arabic numbers denote various geophysical and astrophysical objects (Golitsyn 1991) at an assumption, that some phenomenological determination of Ra and Ta one could determine sufficiently stable configurations similar to those at Fig. 13.7. The comparison of Figs. 13.7 and 13.8 gives a hope that it is sensible, though the sizes differ by 10^6 times. But it would require to change the molecular transport coefficients on the macroscopic ones. In other words, those are hydrodynamical structures of the same nature, at Fig. 13.8 the vortices are miniharricanes (see Sect. 9. 3) and at Fig. 13.7 those are microharricanes in its structure. Tropical hurricanes sometime also form in pairs and once even 5 hurricanes have been observed over Atlantics.

Atmosphere of Jupiter presents a large variety of motions. At the temperate latitudes it has a system of bounds of few latitudes wide were winds are opposite directions at neighbouring bounds. In the South hemisphere about 350 years ago Robert Hook found an oval Large Great Spot of about 20 thousand kilometers wide. Later many smaller spots were discovered of a few thousand kilometers wide living for several years.

Due to the small angle of the own rotation, $3°7'$ high polar latitudes have not been observed. Rotational Mach number $M_\omega = \omega r/c$, r—the planetary radius, c—adiabatic sound velocity, $M_\omega = 15$.

Hough (1898) has proposed the tidal equation to describe linear perturbations on a rotating planet. The parameter $\gamma = 4M_\omega^2 = 1200$ for Jupiter. The eigen functions of the Laplace tidal equations are found within ± 30 around equator (Golitsyn and Diky

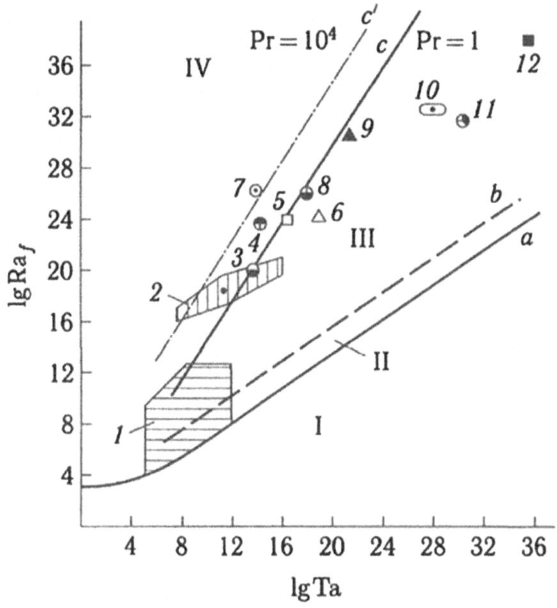

Fig. 13.4 Convective regimes for a rotating liquid layer: I at rest; II region of regular motions; III region of irregular geostrophic convection; IV region of thermal turbulent convection; (a) critical curve of the onset of convection; (b) boundary between regular and irregular convection for Pr = 0.7; (c) boundary between geostrophic and normal thermal convection for Pr = 1; (c′) same for Pr = 104; (1) parameter range for our experiments; (2) range of values for parameters of basaltic magma; (3) ocean, $h = 200$ m; (4) atmospheric boundary layer; (5) atmosphere of Jupiter; (6) hot neutron star; (7) solar granulation; (8) ocean, $h = 2$ km; (9) cold pulsar; (10) accretion disk; (11) Earth's liquid core; (12) interior of Jupiter. G12

1966; Longuet-Higgins 1968) and further on decrease exponentially towards poles, and we see the bands at the Jupiter and Saturn only there, for them both $\gamma = O(10^3)$.

In 1960s for all larger planets internal sources of heat were found because their outgoing thermal radiation was higher than the one if only solar radiation was acting. These internal sources had been clearly forgotten toward 2016 when US station Juno started to have meridional orbit around Jupiter and found at latitudes ±82° eight cyclones near North pole and five cyclones near South one (Adriani et al. 2018). During 2 or 3 years the author of papers on the subject hopelessly said about mystery, nature of the cyclones before Ingersole et al. (2021) after thorough analyses have concluded that the nature of the flour there is convective.

Golitsyn (2021) not at once paid an attention to the problem and has estimated the buoyancy flux due to the internal heat to be $b = 0.1\ \mathrm{m^2/s^3}$. The formulas (1.31) and (1.33) for $t = 35,424$ s, the period of planetary rotation. Using this time and $b = 0.1\ \mathrm{m^2/s^3}$ we obtain the mean velocity of 60 m/s and the size of about 2850 km. The first observations presented tangential velocities of about 80 m/s with the scatter of about 20%. At the near North pole diameters of 5 cyclones 5600–7000 km and

Fig. 13.5 Horizontal current lines in a rotating layer during convection in a dense layer (Chandrasekhar 1961)

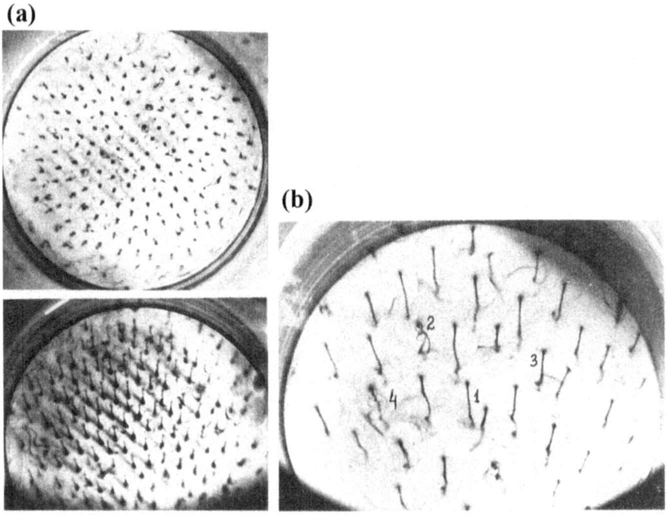

Fig. 13.6 **a** Regular vortex lattice in region II. **b** The regular regime can be transformed into an irregular one by rotating vortices around each other (1), forming double spirals (2) and (4), and merging two vortices (3)

at the South pole for 8 cyclones 4000–4600 km. Remembering that at the Eq. (1.31) should be a numerical factor about 1.6–2 (and $\sqrt{1.8} = 1.3$) and at (1.33) this factor of order 10 (and $10^{1/2} = 3$) we may say that ANK34 Eqs. (1.31) and (1.33) works satisfactory with orders of magnitude for both sizes and velocities of Jupiter.

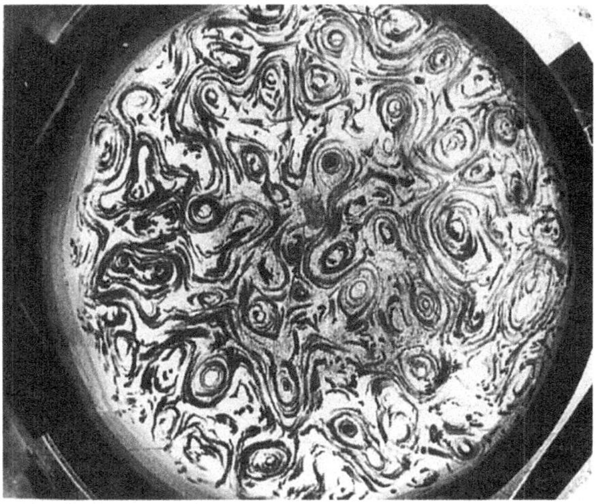

Fig. 13.7 Irregular vortex grid in sector 3

Fig. 13.8 Spiral vortices on the surface of the Baltic Sea (ESA)

The next planet Saturn is also big and for it the value $\gamma = 4M_{\omega}^2$ is also of order 10^3. Therefore, it has also belts in its temperature latitudes but the one cyclone only at a pole. The Jupiter's appearance has a quantity number of irregular feature in a number cyclones, a single Great Red Spot, but none at Saturn, also with an internal heat source Fig. 13.9. But both planets have different environment: between orbits of Jupiter and Mars there is an asteroid belt with occasional very large objects, but nothing is between Jupiter and Saturn. The encounter of Jupiter with multiple-body of the comet Levi–Schumacher in August 1994 left long-lived traces in its atmosphere.

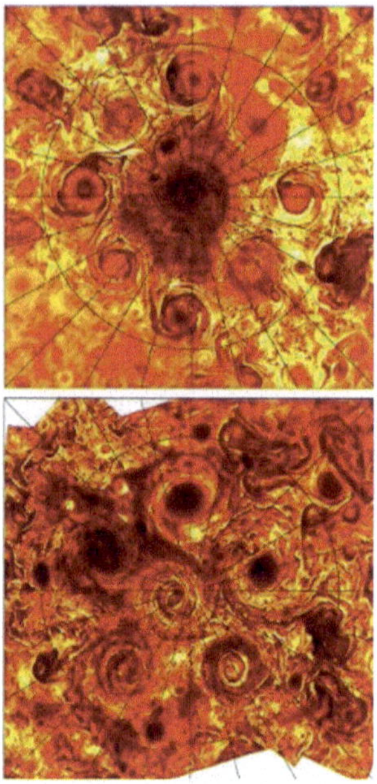

Fig. 13.9 Large-scale vortices near both poles of Jupiter

So one may say that meteorites could leave their own traces in the atmosphere of the planet.

The first author has been told not once by his colleagues that in the atmosphere over sand deserts, and within oceans at the intrusion on them large very cold air masses forests of convective columns are formed. Our laboratory measurement is presented at Figs. 13.4, 13.5, 13.6 and 13.7 showing various patterns of convective regimes give a hope that redetermination of molecular transport coefficients into large-scale ones may explain as those "forests" but approaching at his age 90 the first author leaves this problem to the much younger generations.

References

Adriani A et al (2018) Clusters of cyclones encircling Jupiter's Poles. Nature 555:216–219

Boubnov BM, Golitsyn GS (1990) Temperature and velocity field regimes of convective motions in a rotating fluid layer. J Fluid Mech 290:215–239

Boubnov BM, Golitsyn GS (1995) Convection in rotating fluids. Dordrecht: kluwer Acad Publ 224 p

Budyko MI, Golitsyn GS, Israel Yu A (1986) Global climatic disasters.—M.: Gidrometeoizdat,—159 po. Springer Verlag

Chandrasekhar S (1953) The instability of a layer heated below and subject to Coriolis forces. Proc R Soc Lond A217:306–327

Chandrasekhar S (1961) Hydrodynamic and hydromagnetic stability. Oxford University Press, Clarendon Press, 652 pp

Emanuel KA (1994) Atmospheric convection. Oxford University Press, 580 pp
Foster TD, Waller S (1985) Experiments on convection at very high Rayleigh numbers. Phys Fluids 28:455–462
Golitsyn GS (1979) Simple theoretical and experimental study of convection with geophysical applications and analogues. J Fluid Mech 95(3):568–608
Golitsyn GS (1983) Speech at the all-union conference of scientists for ridding humanity of the threat of nuclear war, for disarmament and peace. Moscow, May 17–19, 1983. Bull USSR Acad Sci 9:57–60
Golitsyn GS (1991) Convective regimes on various rotating geophysical and astrophysical objects. Izv Atmos Ocean Phys 27(1):13–21
Golitsyn GS (2007) Energy cycle of geodynamics and seismic process. Izv RAS Phys Earth 6:3–6
Golitsyn GS (2009) Tropical cyclones and polar lows: velocity, size, energy scales and relation to 26°C origin criteria. Adv Atmos Sci 26(3):585–590
Golitsyn GS (2018) Laws of random walks by A.N. Kolmogorov 1934. Meteorol Hydrol 3:5–15
Golitsyn GS (2021) Simple arguments for jupiter's circumpolar cyclones. Izv RAS Phys Atmos Ocean 57(6):627–631
Golitsyn GS, Grachev AA (1981) Average speeds of Benard convection and heat transfer. Izv Acad Sci USSR Mech Liquid Gas 5:11–18
Golitsyn GS, Grachev AA (1986) Free convection of multicomponent media and parametrization of air-sea interaction at light winds. Ocean Air Interact 1(1):57–78
Golitsyn GS, Vulfson AN (2007) On the conservative form of deep equations convection and thermal energy conversion efficiency for polytropic models of the Earth's mantle. Izv RAS Phys Earth 8:15–23
Golitsyn GS, Grachev AA (1980) Speed and heat and mass transfer during convection in two-component medium. Dokl Acad Sci USSR 255(3):548–552
Golitsyn GS, Grachev AA, Lapshin AI (1984) Heat transfer with free convection in a multicomponent gas environment. Dokl Acad Sci USSR 277(4):858–862
Golitsyn GS (1980) Study of convection with geophysical applications and analogies.—L.: Gidrometeoizdat—56 pp
Grachev AA (1990) Laws of friction in the approximation of free convection. Izv RAS FAO 26(11):850–859
Gray WM (1968) Global view of the origin of tropical disturbances and storms. Mon Weather Rev 96:669–700
Hough SS (1898). On the application of harmonic analysis to the dynamical theory of the tides. Part II. Philosophical Transactions of the Royal Society of London. Series A, Containing Papers of a Mathematical or Physical Character, vol. 191, 139–185
Kadanoff LP (2001) Turbulent heat flow: structures and scaling. Phys Today 8:34–39
Landau LD, Lifshits EM (1986) Hydromechanics.—M.Fizmatlit,—736 pp
Longuet-Higgins MS (1968) The eigenfunctions of Laplace's tidal equation over a sphere. Phil Trans Roy Soc A262(1132):511–607
Monin AS, Yaglom AM (1975) Statistical hydromechanics. T. 2.—M.: Fizmatlit,—720 pp (V. 2, MIT Press, 1975)
Obukhov AM (1960) On the structure of temperature and velocity fields under conditions free convection. Izv Acad Sci USSR Ser Geophys 9:1392–1396
Obukhov AM (1949) Structure of the temperature field in a turbulent flow. Izv USSR Ser Geographer Geophys 13(1):58–69
Prandtl L (1932) Meteorologische Anwendungen der Strömungslehre. Z Phys Für Atmosphäre 19(3):188–202
Rayleigh L (John William Strutt) (1916) On convection currents in a horizontal layer of fluid when the higher temperature is underside. Philos Mag 529–546

Chapter 14
Clouds and Turbulence, Self-similarity, and Peculiar Invariants

In the mid-1970s, Mandelbrot (1977) introduced the concept of a fractal, a power-law statistical pattern between two random variables, and the last quarter of the last century was filled with articles on the results of searches for such patterns, the values of exponents were given, often without estimating them uncertainties and discussion of their nature and connection with the general laws of probability theory. To characterize random figures, it was proposed to use the relationship between the area of such figures A and the length of their perimeter P. He proposed to calculate the average radius $R = A^{1/2}$ of the figure perimeter

$$P \equiv CR^{\beta}. \tag{14.1}$$

Lovejoy (1982) (LS) took advantage of this suggestion and analyzed satellite cloud measurements and radar data on rain clouds and found that $B = 1.35$. He noticed that this value is very close to 4/3, i.e. only 1/60 less than 1.35. The analysis (Lovejoy 1982) turned out to be not very complete; no estimates of the pre-fractal factor B in the formula were not presented by:

$$A = BP^{\alpha} = B\left(CR^{\beta}\right)^{\alpha} \tag{14.2}$$

and no 95% confidence intervals in determining the value of B. The authors of Golitsyn et al. (2022) had only Fig. 1 from LS in hand, in which the values of the areas A were compared to the lengths of the perimeters of the corresponding cloud fields. Therefore, the experimental data of Fig. 1 from Lovejoy (1982) were renumbered and calculated a new fractal $A = BP^{\alpha}$. It was found that $\alpha = 1.50 \pm 0.03$ with a 95% confidence level, and $B = 0.15\ \text{km}^{1/2}$. The sizes of the analyzed clouds ranged from 1 to $1.2 \cdot 10^6\ \text{km}^2$. From relationship (14.2) under the condition $R = A^{1/2}$ it follows that

$$\alpha\beta/2 = 1, \tag{14.3}$$

G. Golitsyn and C. Varotsos, *The Stochastic Nature of Environmental Phenomena and Processes*, https://doi.org/10.1007/978-3-031-77015-9_14

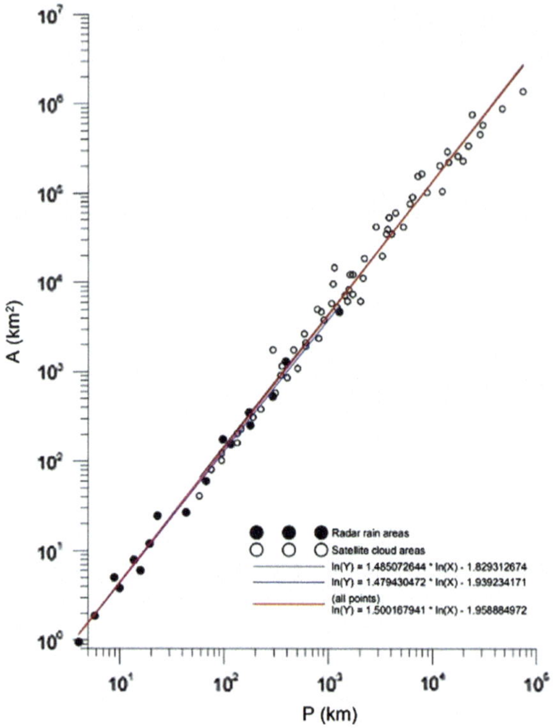

Fig. 14.1 Area of rain clouds (●) and satellite clouds (○) as a function of cloud perimeter. as a function of cloud perimeter

$$BC^{\alpha} = 1, \tag{14.4}$$

i.e. $C = B^{-1/\alpha}$, and if $\pm\alpha = 3/2$, then $\beta = 4/3$ and $\mathrm{C} = 0.15^{-2/3} = 3.7\mathrm{km}^{-1/3}$, all with 95% accuracy and from experimental data. Figure 14.1 shows the recalculated data of Fig. 1 from LS in Golitsyn et al. (2022) for the fractal (14.2). The last three digits given on it gave an arithmetic mean of – 1.91, whence $B = \mathrm{e}^{-1.91} = 0.148 \approx 0.15\,\mathrm{km}^{1/2}$. In Lovejoy (1982), the fractal (Mandelbrot 1977) $A = BP^{\alpha}$ was not noted in any way, although relation (14.2) was clearly known and used by the author to estimate the value of the index β. Later, modelling numerical calculations of the cloud shapes (Luo and Liu 2007), which gave $\beta = 1.35 \pm 0.1$. The very proximity of the index β to 4/3 clearly worried the scientific community, and in Luo and Liu (2007) the result of a statistical analysis of the fields of already noctilucent clouds (von Savigny et al. 2011) was published, which gave the value of β again 1.35. Therefore, we can think that the proximity of β to 4/3 is a law of nature for random fields.

The launch in the USA of a special satellite for measuring the geometrical characteristics of clouds (Guillaume et al. 2018) allowed us to establish that the spectrum of horizontal dimensions of clouds and cloud-free areas in the interval from 1 to 1000 km has a power character with the exponent of degree -1.66, which differs

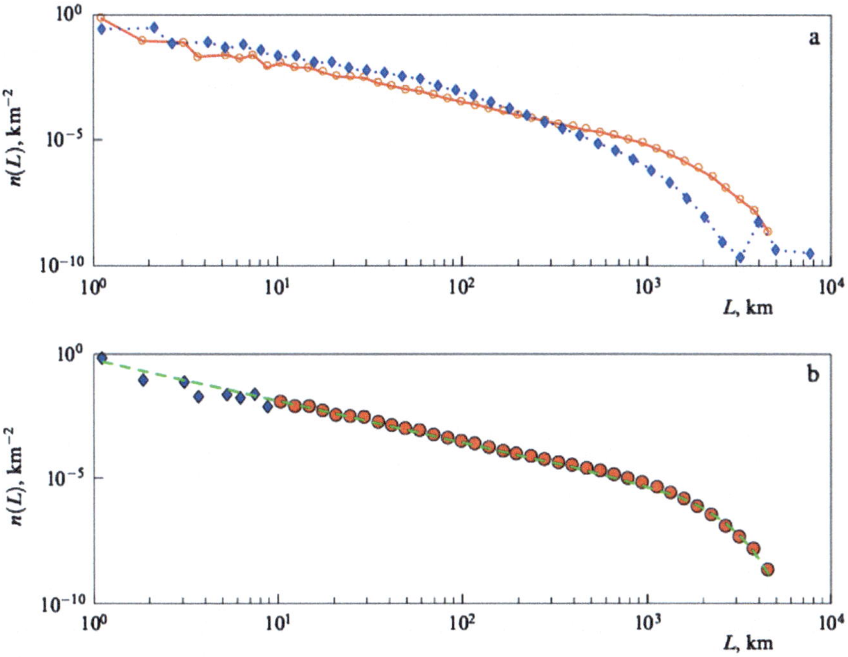

Fig. 14.2 Distributions of horizontal lines $n(L)$ for cloud and cloud-free segments, open circles and gray traces, respectively, top; gray line at the bottom $-n(L) = L^{-\beta}\exp\left[-(L/L_x)^2\right]$, where $B = 1.66 \pm 0.00$ and $L_x = 1850\,\text{km}$

only by 1/150 from – 5/3. Figure 14.2 from Guillaume et al. (2018) is probably the clearest demonstration of the operation in nature of ideas about the local structure of turbulence not only for the clouds themselves as a passive scalar but also for the cloud-free dimensions of a clear sky.

To see this, we should use the results of A. N. Kolmogorov and A. M. Obukhov, stated in Sect. 1.3 (Kolmogorov 1934; Obukhov 1959), practically unknown or unused in the West, and at us, in particular, for the second moments of the distribution function of the 6-dimensional random distribution vector time scale $p(t, u_i, x_i)$ having the form (1.31)–(1.33), whence follows

$$\tau = \left(R^2/\varepsilon\right)^{1/3}, \tag{14.5}$$

where ε is the rate of generation/dissipation of turbulence kinetic energy, and R^2- is the mean square of the distance between the observation points.

All the previous formulas (14.1)–(14.5) include only the length dimension in some degree α or β. Therefore, the dimension formulas include dimensional numerical multipliers arising due to the fractal nature of the fields under study. In such a situation, we need to define the length scale. In the general case, it can be written as:

Table. 14.1 Data on cloud fields from Fig. 14.1 of the article by Lovejoy (1982) and calculated according to formula (14.2) for the perimeter and radius and formula (14.9) for the similarity parameter (14.8)

A (km^2)	P (km)	R (km)	Π
10	16.4	3.16	5.4
10^2	76.3	10	8.0
10^3	354	31.6	11.7
10^4	1640	100	17.2
10^5	7630	316	25.2
10^6	35400	1000	37.2

$$\mathrm{L} = (Kt)^{1/2}, \tag{14.6}$$

where K is the turbulent mixing coefficient determined by formula (1.32). At Mandelbrot, the mean cloud size is defined as the root of the area, and the perimeter, obviously larger than $A^{1/2}$, is determined by turbulence with time scale τ by (14.5). The area, on the other hand, grows due to slower processes determined by diffusion, i.e., the value of A is proportional to time. As a result, we can form a dimensionless similarity parameter:

$$\Pi = \frac{P}{R} = \frac{B^{-1/\alpha}A^{1/\alpha}}{\sqrt{A}} = \left[\frac{K_1\tau}{K_2t_2}\right]^{1/2} = \left(\frac{a\left(R^2/\varepsilon\right)^{1/3}}{t}\right)^{1/2}, \tag{14.7}$$

where $\alpha = K_1/K_2$—is the ratio of the diffusion coefficients for perimeter and area, and both of these quantities are estimated below from measured cloud field data. The resulting similarity number is:

$$\Pi = P/R = \left(B^{-1/\alpha}A^{1/\alpha}\right)/A^{1/2} = B^{-1/a}A^n, n = 1/a - 1/2,$$

and if $\alpha = 3/2$, then n $= 1/6$. When B $= 0.148 \approx 1.50\,\mathrm{km}^{1/2}$ and C $= 3.54\mathrm{km}^{-1/3}$ we get with 95% confidence

$$\Pi = 3.54\,\mathrm{A}^{1/6}. \tag{14.8}$$

The similarity parameter increases with the size of the cloud $\mathrm{A}^{1/6} \sim R^{1/3}$ as illustrated by the proximity of β to 4/3.

The main hypothesis for obtaining all the results, formulated by Kolmogorov, is the Markov property of the forces (accelerations) acting in the system, and the entry for the probability distributions of the 6-dimensional vector $p(t, u_i, x_i)$ in the form of the Fokker–Planck–Kolmogorov equation in the form (1.29). A. M. Obukhov found the second moments of the distributions of the vector p(t, u_i, x_i) and determined the time scale (14.5), which describes the structure of turbulence in the inertial interval and the Richardson–Obukhov law for vortex mixing. Further, here, using similarity considerations, the length scale is determined, and processing with its help the fractal patterns of the analyzed fields for tropospheric and noctilucent clouds leads to the same fractal indicator in the power-law dependence of the cloud perimeter on its average radius. Thus, it is the result of random movements of the 6-dimensional vector *ui, xi*. This movement is characterized by two interconnected fractal relations (14.3) and (14.4). In this case, not only the indicators of these fractals are connected, but also pre-fractal factors with the corresponding dimensions, fractional dimensions of length L. In the notation introduced by Maxwell, $[B] = L^{1/2}$, and $[C] = L^{-2/3}$. Meaning that $B = 4/3$, and $\alpha = 3/2$. Unfortunately, in Lovejoy (1982) and subsequent articles the second fractal between area and perimeter is not even mentioned, although the relation $\beta = 2/\alpha$ is used in von Savigny et al. (2011). Apparently, 4/3 is more attractive to those familiar with small-scale turbulence than 3/2.

Thus, we can accept that, within 95% accuracy, all the above numerical relationships are observed, in which fractal exponents and pre-fractal factors participate, and that these are characteristics of random movements in the spaces of velocities and coordinates, obtained from the moments of the probability distribution function of time evolution 6 dimensional vector x_i, u_i. All this opens up a wide field for research with the introduction, for example, of internal time correlation for acceleration fields, as was done in Golitsyn to explain the features of the relief spectrum of the surfaces of the Moon and planets, as at the end of Chap. 8 of this book.

Equations (1.31) and (1.33) can be used to derive the probability distributions for energy and areas (lengths $l^2 = S$). The integral distribution, or histogram, has the dimension of frequency $f = 1/t$, and the differential distribution for the parameter A have dimension 1/At. Therefore from (1.31)

$$N(\leq E) = \frac{\varepsilon}{E}, \quad N(E) = \frac{\varepsilon}{E^2}, \tag{14.9}$$

and from (1.33) we will have

$$N(S) = \left(\frac{\varepsilon}{S}\right)^{1/3}, \quad N(< S) = \frac{\varepsilon^{1/3}}{S^{4/3}}. \tag{14.10}$$

These are distributions for areas. In Guillaume et al. (2018) the differential distributions for lengths $l = S^{1/2}$ of clouds are presented $N(l) = l^{-n} \exp\left(-l^2/L_0^2\right), n = 1.66 \pm 0.00...$, $L_0 = 1850\text{km}$. Therefore

$$N(l) = \frac{1}{l}\left(\frac{\varepsilon}{l^2}\right)^{1/3} \sim l^{-5/3},$$

with excellent correspondence to Guillaume et al. (2018). This 5/3 is not for the energy spectrum, but for the differential presentation; the histogram for horizontal lengths of clouds will be $\sim l^{-2/3}$. We see that both observational results of Lovejoy and Guillaume et al. correspond well to theories of turbulence of 1941 and 1934.

References

Golitsyn GS, Chkhetiani OG, Vazaeva NV (2022) Clouds and turbulence theory: peculiar self-similarity, 4/3 fractal exponent and invariants. Izv Atmos Oceanic Phys 58(6):645–648

Guillaume A et al (2018) Horizontal and vertical scaling of cloud geometry inferred from CloudSat data. J Atmos Sci 75(7):2187–2197. https://doi.org/10.1175/JAS-D-17-0111.1

Kolmogorov AN (1934) Zufallige Bewegungen. Ann Math 35:116–117. https://doi.org/10.2307/196812

Lovejoy S (1982) Area-perimeter relation for rain and cloud areas. Science 216(4542):185–187. https://doi.org/10.1126/science2164542.185

Luo Z, Liu C (2007) A validation of the fractal dimension of cloud boundaries. Geophys Res Lett 34(3). https://doi.org/10.1029/2006GL028472

Mandelbrot B (1977) Fractals, form, chance and dimension. W.H. Freeman and Co., San Francisco

Obukhov AM (1959) Description of turbulence in terms of Lagrangian variables. Adv Geophys 6:113–116. https://doi.org/10.1016/S0065-2687(08)60098-9

Oort AH (1964) On estimates of the atmospheric energy cycle. Mon Weather Rev 92(11):483–493. https://doi.org/10.1175/1520-0493(1964)092%3c0483:OEOTAE%3e2.3.CO;2

von Savigny C et al (2011) First determination of the fractal perimeter dimension of the noctilucent clouds. Geophys Res Lett 38(2). https://doi.org/10.1029/2010GL045834

Chapter 15
The Global Sea Level Dynamics

15.1 Detrended Fluctuation Analysis (DFA) and Multifractal DFA (MF-DFA)

The temporal evolution of the global mean sea level (GMSL) has been recently studied (Varotsos et al. 2024). This study was performed by using monthly mean averages derived from two distinct datasets: a reconstructed timeseries and a satellite altimeter timeseries. To examine the scaling characteristics of the dataset, Varotsos et al. (2024) employed two established methodologies: detrended fluctuation analysis (DFA) and multifractal DFA (MF-DFA). The key finding of this study was that both timeseries of the global average sea level exhibit power-law long-range correlations as well as multifractality. Interestingly, the scaling features of the 134-year and the most recent 28-year GMSL-dataset were remarkably found similar, indicating that the long-range correlations may primarily arise from factors of nature. This highlighted the need for further extensive research on the association among the fluctuations of climate change and sea-level, as their indirect processes have significant implications for ecology and conservation.

Recently several investigations have aimed to explore the sea-level scaling evolution observed. These studies have focused on exploring the potential for long-range dependence, self-similarity, and fractal behavior within sea-level dataset. The concept of "long-range dependence" refers to a quantity's property where certain values remain correlated with each other even after a significant period of time has passed. On the other hand, "self-similarity" and "fractal" imply that a dataset exhibits similarities or patterns that resemble a part of itself (Varotsos et al. 2013). About this, Fraedrich and Blender (2003) conducted a study on temperature temporal correlations over the oceans, which is considered a fundamental thermodynamic quantity influencing dynamic sea-level changes. The findings revealed a scaling exponent that was nearly equal to unity, indicating a long-term dependence.

G. Golitsyn and C. Varotsos, *The Stochastic Nature of Environmental Phenomena and Processes*, https://doi.org/10.1007/978-3-031-77015-9_15

The investigation conducted by Varotsos et al. (2024) was largely centered on analyzing the long-range correlation (LRC) and multifractal characteristics of two-time datasets, namely the satellite altimeter dataset (SAD) and the reconstructed dataset (RD). To achieve this, they utilized the DFA and MF-DFA techniques, as introduced by Peng et al. (1994).

To examine the multifractal characteristics of the dataset, the MF-DFA method was employed. The steps involved in this process are outlined below:

(1) The differences between the N measurements of $y(i)$ and their mean were first used to integrate the time series $y(i)$ over time.
(2) Subsequently, the integrated dataset $x(i)$, was split into distinct intervals with a length τ. After that, the algorithm was run again, this time beginning at the end of the profile, resulting in $2N_\tau$ intervals (where N_τ represents the integer portion of N/τ).
(3) Next, each interval's polynomial least-square fit (of order l) had to be found, and the associated variance had to be computed using a set of formulas (Kantelhardt et al. 2002):

(a) for each interval $j = 1, \ldots, N_\tau$:

$$F^2(j,\tau) = \frac{1}{\tau}\sum_{t=1}^{\tau}\left[x((j-1)\tau + i) - t(i)\right]^2 \tag{15.1}$$

(b) for each interval $j = N_\tau + 1, \ldots, 2N_\tau$:

$$F^2(j,\tau) = \frac{1}{\tau}\sum_{i=1}^{\tau}\left[x((N - j - N_\tau)\tau + i) - t(i)\right]^2 \tag{15.2}$$

where $t(i)$ is a locally best polynomial fitted trend (of second degree) to the τ data.

(4) By averaging the variances in each interval, the q-th order fluctuation function was calculated, where q is the variable moment:

$$F_q(\tau) = \left[\frac{1}{2N_\tau}\sum_{j=1}^{2N_\tau}\left[F^2(j,\tau)\right]^{q/2}\right]^{1/q} \tag{15.3}$$

When $q \to 0$, Eq. (15.3) takes on the following form:

$$F_0(\tau) = \exp\left[\frac{1}{2N_\tau}\sum_{j=1}^{2N_\tau}\ln\left[F^2(j,\tau)\right]\right] \tag{15.4}$$

(5) Finally, $F_q(\tau)$ was plotted against vs. τ (in log–log graph) for various q values. When there is multi-scaling pattern, $F_q(\tau)$ exhibits a power-law, where $h(q)$ stands for the generalized Hurst exponent.

$$F_q(\tau) \sim \tau^{h(q)} \tag{15.5}$$

Furthermore, a multifractal series was characterized by deriving the singularity spectrum $f(n)$ from $h(q)$ via the modified Legendre transform, where n represents the singularity strength or Hölder exponent (Kantelhardt et al. 1999).

It is important to remember that the DFA tool, which is mainly used to analyze mono-fractal features, is extended by the MF-DFA technique (Peng et al. 1994).

The fluctuation function is computed by the original DFA tool. Equation (15.3) is used to calculate $F_d(\tau)$ for $q = 2$ and a first-degree fitted trend. This computation is carried out starting at the end of the profile and without having to repeat the methodology, that is to say,

$$F_d(\tau) = \left[\frac{1}{N_\tau} \sum_{j=1}^{N_T} \left[F^2(j, \tau) \right] \right]^{1/2} \tag{15.6}$$

In the case of a fractal series, it is expected that $F_d(\tau)$ exhibits a power-law pattern (i.e., $F_d(\tau) \sim \tau^a$) where a represents the monofractal exponent. If the a-exponent falls within the region $(0, 0.5)$, power-law anticorrelations (antipersistence) are found. On the other hand, an a-exponent value in the region (0.5, 1.5) indicates long-range power-law correlations (persistence). The series is classified as white noise when $a = 0.5$ and as $1/f$ noise when $a = 1$. A process with a frequency spectrum in which the power spectral density is inversely proportional to the signal frequency is known as $1/f$ noise, also known as pink noise.

In real systems, the scaling results of the DFA method can be distorted by the existence of trends in noisy signals. Hu et al. (2001) carried out a thorough investigation into how trends affect DFA outcomes. Both the GMSL dataset (obtained from the reconstructed timeseries and satellite altimeter data) were first detrended using a sixth-order polynomial best fit in order to avoid such interference. The Wiener filter, which offers an estimate of a random pattern through linear time-invariant filtering, was then used to perform deseasonalization (Wiener 1950). It is noteworthy to emphasize that the time series under examination retained the pertinent long-term oscillations and the sixth-order polynomial fitting produced the most significant results (at a 95% confidence level).

15.2 The Multifractality of the Global Mean Sea Level from Satellite Altimeter Data

For various moments q, the fluctuation function $F_q(\tau)$ can be estimated using the MF-DFA technique. This method demonstrated the expected power-law scaling behavior (i.e., $F_d(\tau) \sim \tau^a$), for all selected positive (negative) moments q (see Fig. 15.1a) on large scales $\tau > 8$ months ($\tau > 12$ months) when applied to the D&D GMSL time series derived from SAD (Varotsos et al. 2024; Ihlen 2012).

The generalized Hurst exponent $h(q)$ was plotted as a function of q-values in order to verify the multifractality of the analyzed time series (see Fig. 15.1b). Exponent $h(q)$ values for the GMSL time series exhibit multifractal behavior and persistent LRC, as evidenced by their variation with q and higher-than-0.5 values. Additionally, it was noted that lower $h(q)$-values were correlated with larger fluctuations (positive q-values), which is consistent with the usual features of multifractal time series.

The multifractality found in GMSL has the potential to greatly advance our knowledge of science today. We have additionally plotted the singularity spectrum $f(n)$ against the singularity strength n to offer a more thorough explanation of this feature.

The maximum value of $f(n)$ is associated with $q = 0$, and the values of $f(n)$ that lie either side of the maximum signify positive or negative moments (Fig. 15.1c). It is crucial to underline and make clear that the right mathematical strategy for revealing the fractal nature of geophysical parameter evolution, like GMSL, is to combine the MF-DFA tool with the two Maraun criteria (Varotsos et al. 2024).

Fig. 15.1 **a** An examination of the generalized Hurst exponent h(q) and the MF-DFA fluctuation factor $F_q(\tau)$ for the D&D GMSL time series (SAD) spanning 1993 to 2020. For all chosen positive (negative) moments q, the log–log plot in **a** illustrates the power-law scaling behavior of $F_d(\tau) \sim \tau^a$ at large scales $\tau > 8$ months ($\tau > 12$ months). On to **b**, where scales ranging from 6 months to 6.3 years are used to illustrate how $h(q)$ depends on q-values. Lastly, the relationship between the singularity strength n and the singularity spectrum $f(n)$ is demonstrated in **c** (Varotsos et al. 2024)

15.3 Conclusions on the Multifractality of the Global Mean Sea Level

The analysis by Varotos et al. (2024) yielded the following primary conclusions:

A scaling exponent $a = 0.77 \pm 0.02$ ($a = 0.76 \pm 0.02$), is obtained by applying the DFA technique to the D&D GMSL datasets from the satellite altimeter time-series (reconstructed timeseries) for the years 1993–2020 (1880–2013), indicating a positive correlation between the variations in average sea-level values over shorter and longer time periods.

$F_d(\tau)$ on large scales τ for all the selected positive and negative moments showed a power-law scaling behavior, according to the MF-DFA technique applied to both GMSL datasets used. Furthermore, there was multifractality and persistent long-range correlations revealed by the higher-than-0.5 h(q) values, suggesting that the generalized Hurst exponent $h(q)$ depends on q.

The trends and scaling characteristics of the two GMSL datasets were compared for the shared period, which runs from January 1993 to December 2013. The two-datasets' scaling characteristics were similar, indicating that the historical data set could be utilized in any how to verify the dataset from satellite altimeters.

The multifractality characteristics mentioned above that were found in the GMSL may add to the current discussion surrounding estimates of both global and regional mean sea-level rise and aid in integrating a comprehensive understanding for creating temporary solutions and lessening unavoidable effects (Frederikse et al. 2020; Vahsen et al. 2023). Accordingly, given the rapid rise in sea level, it is currently impossible to predict how the climate, mangroves, and marshes will evolve in the future. For instance, salt marshes can react to sea level rise by vertically accumulating, contracting, and expanding laterally (Roman et al. 2024).

References

Fraedrich K, Blender R (2003) Scaling of atmosphere and ocean temperature correlations in observations and climate models. Phys Rev Let 90:108501

Frederikse T, Landerer F, Caron L, Adhikari S, Parkes D, Humphrey VW, Dangendorf S, Hogarth P, Zanna L, Cheng L et al (2020) The causes of sea-level rise since 1900. Nature 584:393–397

Ihlen EA (2012) Introduction to multifractal detrended fluctuation analysis in Matlab. Frontiers in physiology 3:141

Hu K, Ivanov PC, Chen Z, Carpena P, Stanley HE (2001) Effect of trends on detrended fluctuation analysis. Phys Rev E 64:011114

Kantelhardt JW, Berkovits R, Havlin S, Bunde A (1999) Are the phases in the Anderson model long-range correlated? Phys A Stat Mech Appl 266:461–464

Kantelhardt JW, Zschiegner A, Koscielny-Bunde E, Havlin S, Bunde A, Stanley HE (2002) Multifractal detrended fluctuation analysis of nonstationary time series. Phys A Stat Mech Appl 316:87–114

Peng CK, Buldyrev SV, Havlin S, Simons M, Stanley HE, Goldberger AL (1994) Mosaic organization of DNA nucleotides. Phys Rev E 49:1685–1689

Roman CT, Lynch JC, & Cahoon DR (2024) Twenty-year record of salt marsh elevation dynamics in response to sea-level rise and storm-driven barrier island geomorphic processes: fire island, NY, USA. Estuaries Coasts 47(7):1903–1917

Vahsen ML, Blum MJ, Megonigal JP, Emrich SJ, Holmquist JR, Stiller B, Todd-Brown KEO, McLachlan JS (2023) Rapid plant trait evolution can alter coastal wetland resilience to sea level rise. Science 379:393–398

Varotsos CA, Melnikova I, Efstathiou MN, & Tzanis C (2013) 1/f noise in the UV solar spectral irradiance. Theor Appl Climatol 111:641–648

Varotsos C, Mazei Y, Sarlis NV, Saldaev D, Efstathiou M (2024) On the impacts of the global sea level dynamics. Fract Fraction 8(1):39

Wiener N (1950) Extrapolation, interpolation and smoothing of stationary time series. MIT Technology Press, Wiley, New York, NY

Chapter 16
The Intrinsic Properties of Precipitation and Rainfall

16.1 Introduction

Efstathiou and Varotsos (2012) based on the fact that the assessment of prospective expected rainfall and precipitation, both in the short and long term, is essential, due to the significant impact their changes can have on socioeconomic and ecological aspects investigated their intrinsic properties.

Therefore, detrended fluctuation analysis has been utilized to study the temporal variability of normalized rainfall and precipitation fluctuations in the Sahel. The findings revealed that Sahel precipitation anomalies from 1900 to 2010 display persistent long-range correlations over time lags varying from 4 months to 28 years. This indicates that there is a power law relationship between the variations in Sahelian precipitation anomalies over short and long periods. In contrast, Sahelian standardized rainfall anomalies from 1948 to 2001 exhibited behavior akin to an almost random walk.

These findings may significantly improve both the sophisticated modeling of the variability of the global climate system and precipitation forecasting. The long-range correlations seen in historical precipitation data must be reflected in precipitation forecast models. A forthcoming publication will provide more analysis on this topic.

16.2 Description of the Problem

Over the past ten years, there has been a significant interest in studying the patterns, periodicities, and similarities of various atmospheric factors. These factors include ozone content, solar ultraviolet radiation, and properties of solid particles. Researchers have focused on understanding the behavior of these quantities and their impact on the environment.

G. Golitsyn and C. Varotsos, *The Stochastic Nature of Environmental Phenomena and Processes*, https://doi.org/10.1007/978-3-031-77015-9_16

The study of precipitation has received special attention because there are differing opinions on how to quantify the impacts of climate change on hydrology.

Several studies have presented analytical rainfall patterns over different time scales and regions, such as the whole basin of the Amazon and all of its subbasins. Negative rainfall trends have been observed in the whole Amazon basin, but the northern and southern sub-basins show contrasting trends. Other studies have employed spectral analyses to identify variations in different time scales. Southern Amazonia exhibits decadal variations, while both interannual and decadal scale variations have been observed in northern Amazonia. These findings contribute to our understanding of the complex dynamics of the atmosphere and its impact on the hydrological cycle.

Multichannel singular spectrum analysis was used by Krishnamurthy and Shukla (2007) to investigate the lagged components of rainfall anomalies in India. Three enduring components in seasonal monsoon rainfall were found, along with 45 and 20-day oscillations, after they examined gridded daily rainfall data spanning 70 years. It's interesting to note that the seasonal mean rainfall was not significantly affected by the dominant intra-seasonal oscillations that were observed during active and break periods. On the other hand, the interannual variability of monsoon rainfall was significantly influenced by the enduring seasonal components.

Ip et al. (2011) looked into the multi-scale variability and trends of North China's precipitation data. They examined a long-term historical flood/dryness grade dataset from 1470 to 2000 as well as a historical series of precipitation data spanning more than a century. In the precipitation dataset, the study found clear seasonal variations, a quasi-biennial oscillation, an interannual 4–7-year parameter, and an interdecadal 19-year pattern. Furthermore, the historical flood/dryness grade time series revealed a quasi-10-year interdecadal fluctuation, a quasi-24-year portion, a 50–80-year centurial recurrence, and a 4-to-5-year ENSO interannual transient response.

Andrade et al. (1998) conducted an earlier investigation on the rainfall scaling pattern analyzing a long precipitation dataset provided by various weather stations worldwide. The research indicated that in semiarid regions, droughts follow a power law fashion. This result implied a relationship between rainfall and the idea of self-organized criticality (SOC), where small-time interval fluctuations are linked to larger time interval fluctuations in a power law fashion (as seen at the critical point during phase transitions) (Turcotte 1999). In non-equilibrium systems with a high degree of nonlinearity, like the complex climate system, SOC is usually observed.

The primary objective of the research implemented by Efstathiou and Varotsos (2012) was to analyze the variations in Sahel precipitation fluctuations (Sahel normalized rainfall) within the geographical coordinates of 20° N–10° N and 20° W–10° E, spanning from January 1900 to October 2010 (specifically over 20° N–8° N and 20° W–10° E, from January 1948 to June 2001). Briefly, the study aimed to determine if these fluctuations demonstrate persistent long-term correlations. Additionally, the study seeks to explore how rainfall and anomalies in precipitation are distributed throughout the Sahel.

16.3 Methodology and Analysis

The monthly average values of the Sahel precipitation anomaly (SPA) above the area 20° N–10° N, 20° W–10° E, were utilized in this study by Efstathiou and Varotsos (2012). The period covered by the data was January 1900–October 2010.

To assess the data homogeneity in both 1900 and 2010, we employed the nonparametric Mann–Whitney test. This test compared different segments of the time series to determine if they followed the same distribution structure. The normalized Mann–Whitney parameter V in each of these individual tests remained below 1.96, the 95% confidence interval's lower bound.

This result indicates that the data is homogeneous. The Global Historical Climatology Network (GHCN) V2 monthly timeseries from the National Oceanic and Atmospheric Administration (NOAA) provided the data for this study. This timeseries includes gridded anomalies of precipitation and is based on data from 20,590 raw precipitation stations (without homogeneity adjustments) worldwide and 2,064 precipitation stations (from the USA, Canada, and the former Soviet Union). All the gridded fields underwent homogeneity testing, and in cases where no homogeneity-adjusted data was available, the GHCN raw data with the greatest possible global coverage was utilized.

Every month, 2,592 gridded data points (72 longitude $\times$ 36 latitude grid boxes) are generated for the entire world on a 5×5 degree scale. The information covers the period from January 1900 to this month. Using the classic anomaly method, gridded precipitation anomalies were obtained from averaged station anomalies within each 5×5 degree grid box based on the 1950–1979 period. The same procedure was used to compute anomalies from raw station data for grid boxes without adjusted data. Without any homogeneity adjustments, the SPA measurements used in this survey were obtained from raw precipitation stations.

The average Sahel region was calculated using a rotated principal component analysis of African precipitation, as suggested by Janowiak (1988). Our inquiry was restricted to this particular region because historical records were available here. Additionally, the Sahel standardized rainfall (SSR) average monthly values for the regions 20° N–8° N and 20° W–10° E from January 1948 to June 2001 were used. The NOAA Earth System Research Laboratory-Physical Science Division website provided the source of this timeseries.

In their time series, the average monthly SPA and SSR values showed a slight long-term trend and high seasonality (Fig. 16.1). Using particular statistical techniques, these features were removed by detrending and deseasonalization. While deseasonalization was accomplished by using the classical Wiener method (1958) to eliminate seasonal changes such as 6-month and 12-month oscillations, a quasi 38-year portion, and a 50–55-year periodicity, detrending involved subtracting the average monthly SPA values from their linear best fit.

Then, taking into account the non-stationarities in the data, a novel statistical method called detrending fluctuation analysis (DFA) was applied to the detrended

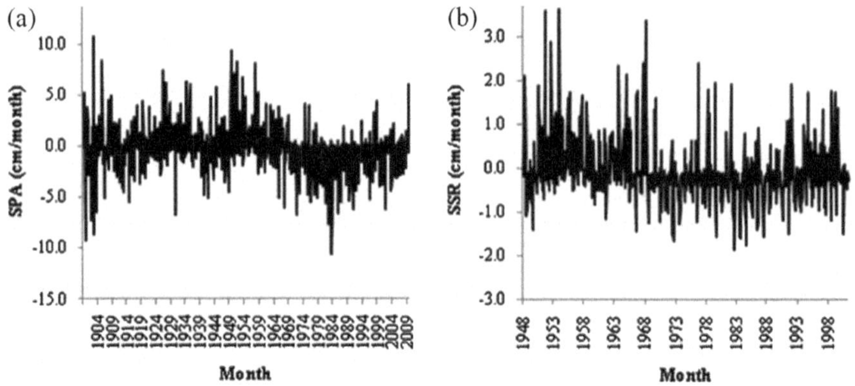

Fig. 16.1 Temporal sequence of the average monthly **a** SPA values from January 1900 to October 2010 and **b** SSR values from January 1948 to June 2001

and deseasonalized average monthly SPA and SSR time series to examine their intrinsic properties.

However, a challenge with traditional time series analysis techniques is that noise in the data fails to differentiate attenuation shapes at long lags. DFA provides an efficient way to observe the autocorrelation decay pattern over time, which was the primary driving force behind this research. The DFA method, rooted in random walk theory, enables the identification of intrinsic self-similarity in non-stationary time series commonly found in various research fields. The sequential steps of DFA, proven effective in analyzing complex systems with self-organizing property, are detailed in Varotsos et al. (2007).

16.4 Interpretation of the Results

To examine their distribution, the average monthly values of SPA and SSR were divided into equal-length classes. The smoothed line of the Gaussian distribution and the percentage relative frequency histogram of the mean monthly SPA and SSR values x (centimeters per month) are shown in Fig. 16.2. However, after performing statistical best-fit tests like Chi-square, Anderson–Darling, and Kolmogorov–Smirnov, it was found that, at a 95% confidence level, the hypothesis that the SPA or SSR values follow a normal distribution was rejected.

Looking at Fig. 16.2, it is clear that the main cause of the poor fit of the normal distribution on SPA and SSR appears to be the "mode" of the values or the value that occurs the most frequently in a dataset (− 0.061 and − 0.13 for SPA and SSR, respectively). Furthermore, an evaluation was conducted on the SPA and SSR timeseries to determine if they conformed to an exponential, geometric, simple power law, generalized power law (also known as the Zipf-Mandelbrot), or lognormal distribution.

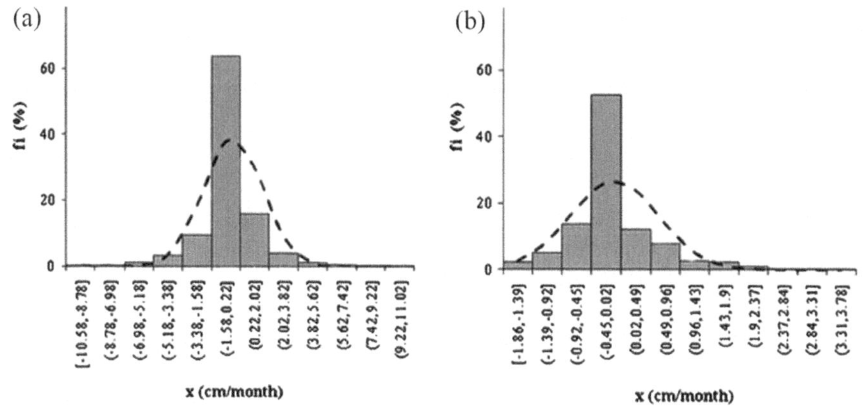

Fig. 16.2 The average monthly percentage relative frequency histogram for **a** SPA values x from January 1900 to October 2010 and **b** SSR values x from January 1948 to June 2001. The normal distribution is shown by a smooth line

Nevertheless, there was no statistically significant conformance to the previously mentioned distributions in the observed data.

Plotting a semilogarithmic graph of the probability $P(X > x)$ of exceeding a given value x allowed for the determination of the distribution of Sahel precipitation anomalies. By applying linear regression analysis, a statistically significant association was discovered among the probability $P(X > x)$ and the precipitation values. This relationship follows the Gutenberg Richter law for values higher than the mode of the data. The correlation between the empirical and semi-log distribution was confirmed using statistical tests at a 95% confidence level according to:

$$P(X > x) \sim 10^{\alpha x}. \tag{16.1}$$

Equation (16.1) states that SPA (SSR) values that are greater than the data's mode fall into the Gutenberg-Richter law's distribution (Goldstein et al. 2004; Rundle 1989). This finding Fig. 16.3 was validated through the use of the Kolmogorov–Smirnov statistical best-fit test at a 95% confidence level. Additionally, Fig. 16.4 demonstrates a clear correlation ($r = 0.99$) among the cumulative function of the empirical and semi-log shape for both SPA and SSR tineseries.

As was previously mentioned, there is currently a lot of interest in the rainfall index's scaling behavior. Numerous regions' worth of extensive data analyses have revealed the existence of self-similarity in the precipitation and other hydrologic variables' time series and spatial fields (Zhu and Liu 2003). The DFA method was applied to the detrended and deseasonalized mean monthly SPA values from 1900 to 2010 to examine the intrinsic properties of precipitation. With a scaling exponent of $\alpha = 0.59 \pm 0.01$ for all time lags from 4 months to 28 years, persistent long-range power-law correlations were found (Fig. 16.5).

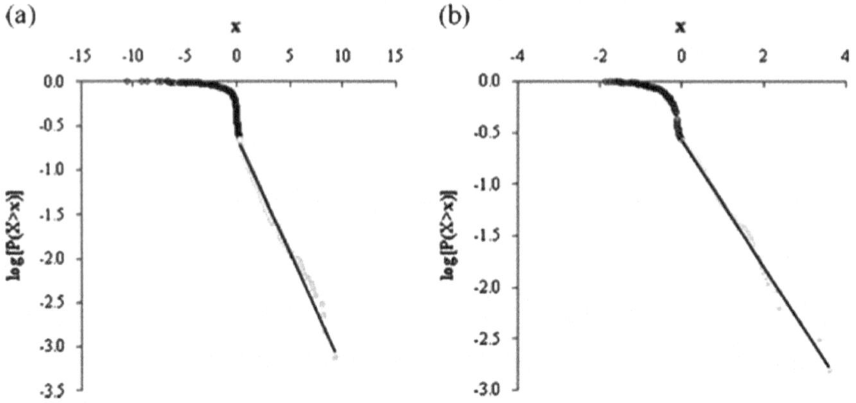

Fig. 16.3 Plotting a fixed SPA and SSR value x on a semilogarithmic graph, with the empirical probability $P(X > x)$ of exceeding the value. The lines show the SSR values over the mode of each dataset **a** $y = -0.26x - 0.66$ with $R^2 = 0.99$ and **b** $y = -0.62x - 0.56$ with $R^2 = 0.99$, respectively and the least-square fit for the SPA

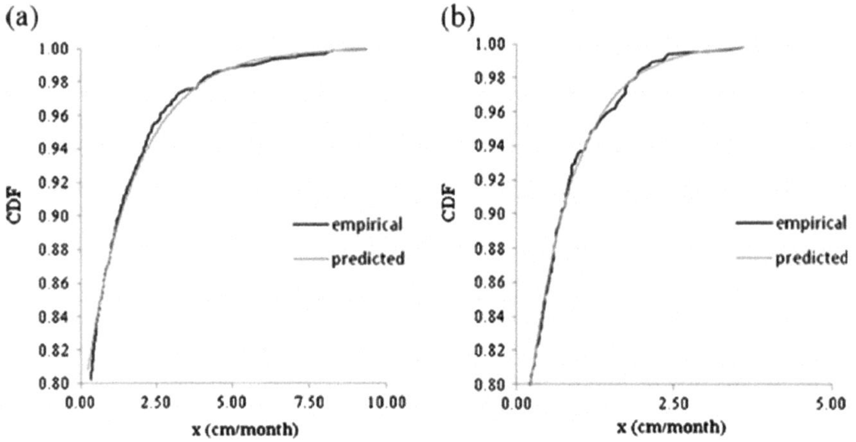

Fig. 16.4 The cumulative function consists of the actual and anticipated distributions of the average monthly values of **a** SPA values (x) from January 1900 to October 2010, and **b** SSR values (x) from January 1948 to June 2001

In plainer language, it was discovered that there was a power law pattern of positive correlation between the variations in SPA over shorter time intervals and those over longer time intervals. Compared to the spectrum characteristics of other meteorological data, this behavior is very different. The data suggests that the SPA time series exhibits long memory and is related with fractal behavior. It is crucial to note that long-range dependence and "long memory" are equivalent ideas (e.g., Varotsos 2003, 2005; Varotsos and Kirk Davidoff 2006; Varotsos et al. 2006). More

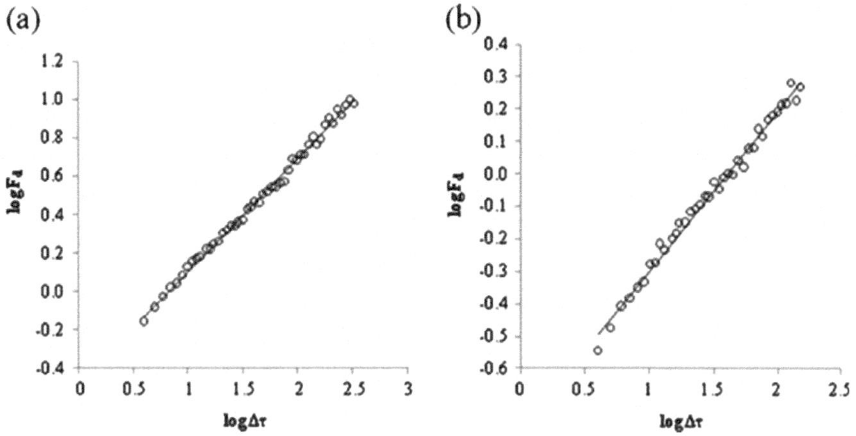

Fig. 16.5 The log–log fit of the root-mean-square fluctuation function ($F_d(\tau)$) against the temporal interval τ (measured in months) is presented for the detrended and deseasonalized dataset of the average monthly SPA values (**a**) and SSR values (**b**) along with the corresponding best-fit equations ($y = 0.59x - 0.48$ with $R^2 = 1$ and $y = 0.49x - 0.79$ with $R^2 = 0.99$, respectively)

precisely, a power law relationship seems to be followed by the averaged square of the detrended fluctuation function $F(\tau)$ spanning N/τ intervals with length τ.

$$\langle F^2(\tau)\rangle \sim \tau^{2\alpha} \tag{16.2}$$

According to Kantelhardt et al. (2002), the power spectrum function fluctuates with $S(f) = 1/f^{\beta}$, where $\beta = 2a - 1$. It is noteworthy to emphasize that the SSR's α exponent is roughly 0.5 ($\alpha = 0.49 \pm 0.01$), suggesting a random walk-like pattern.

References

Andrade RFS, Schellnhuber HJ, Claussen M (1998) Analysis of rainfall records: possible relation to self-organized criticality. Phys A 254:557–568

Efstathiou MN, Varotsos CA (2012) Intrinsic properties of Sahel precipitation anomalies and rainfall. Theor Appl Climatol 109:627–633. https://doi.org/10.1007/s00704-012-0605-2

Goldstein ML, Morris SA, Yen GG (2004) Problems with fitting to the power-law distribution. Eur Phys J B 41:255–258

Ip WCh, Zhang L, Wong H, Xia J (2011) Multi-scale variability and trends of precipitation in North China. Water Resour 38 (1):18–28

Janowiak JE (1988) An investigation of inter-annual rainfall variability in Africa. J Clim 1(3):240–255

Kantelhardt JW, Zschiegner SA, Koscielny-Bunde E, Havlin S, Bunde A, Stanley HE (2002) Multifractal detrended fluctuation analysis of nonstationary time series. Phys A 316(1–4):87–114

Krishnamurthy V, Shukla J (2007) Intraseasonal and seasonally persisting patterns of Indian monsoon rainfall. J Climate 20:3–20

Rundle JB (1989) Derivation of the complete Gutenberg-Richter magnitude-frequency relation using the principle of scale invariance. J Geophys Res 94:12337–12342
Varotsos C (2003) What is the lesson from the unprecedented event over Antarctica in 2002? Environ Sci Pollut R 10(2):80–81
Varotsos C (2005) Power-law correlations in column ozone over Antarctica. Int J Remote Sens 26(16):3333–3342
Varotsos C, Kirk-Davidoff D (2006) Long-memory processes in ozone and temperature variations at the region 60 degrees S-60 degrees N. Atmos Chem Phys 6:4093–4100
Varotsos C, Assimakopoulos MN, Efstathiou M (2007) Technical note: long-term memory effect in the atmospheric CO_2 concentration at Mauna Loa. Atmos Chem Phys 7:629–634
Wiener N (1958) Nonlinear problems in random theory. Technology Press of Massachusetts Institute of Technology, Cambridge; Wiley, New York, pp 28, 41, 88
Zhu J, Liu Z (2003) Power-law distribution of acid deposition. Int J Environ Pollut 19:11–21

Chapter 17
The Global Vertical Atmospheric Ozone Long-Memory

17.1 Column Ozone Variability and Power-Law

The first use of detrended fluctuation analysis was carried out by Varotsos (2005) on daily column ozone (also known as total ozone, or TOZ) in the springtime along the border and inside the Antarctic ozone hole. This was accomplished by combining data from satellite-borne instruments from 1979 to 2003 with ground-based observations from 1972 to 2003.

The results can be summed up like this. Extreme column ozone fluctuations were first seen to follow a power-law distribution with exponents, suggesting that substantial fluctuations are more likely to occur inside the ozone hole than on its periphery. Moreover, on time scales longer than a year, there were more pronounced long-range power-law correlations in column ozone fluctuations from 1979 to 1992. Nevertheless, antipersistence (persistence) was found for time lags more than (less than) ten days across the board in the dataset after the long-term trend was eliminated. This transition demonstrates how planetary waves affect the scaling characteristics of the Antarctic ozone hole's spatiotemporal variability. Lastly, a shift in the intrinsic dynamics of column ozone along the border of Antarctica has been noted since 1996, which is not observed in the Antarctic ozone hole.

17.2 How Likely Are the Antarctic Column Ozone Extreme Values to Occur?

The ozone hole over Antarctica is a phenomenon that has captured global attention. Nowadays it is considered among the serious environmental problems. In September 2002 the ozone hole displayed a transient phenomenon breaking up into two holes as a result of an extraordinary major sudden stratospheric warming over that region (e.g., Varotsos 2002, 2003, 2004).

G. Golitsyn and C. Varotsos, *The Stochastic Nature of Environmental Phenomena and Processes*, https://doi.org/10.1007/978-3-031-77015-9_17

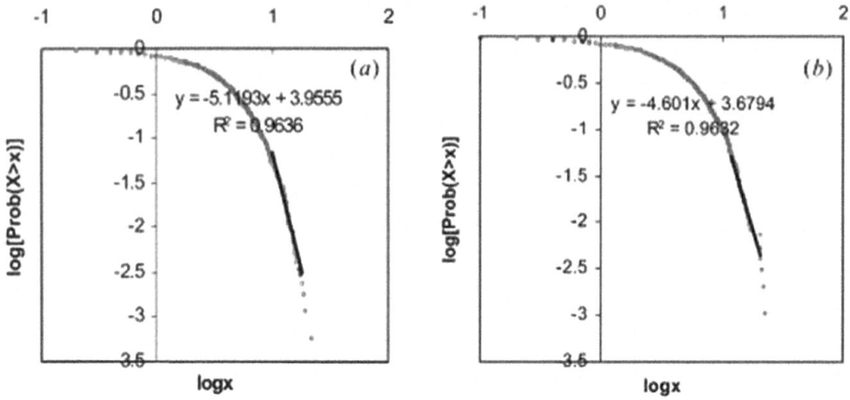

Fig. 17.1 The empirical probability to observe TOZ fluctuations with amplitude larger than some value $x = |y_{i+1} - y_i|$. An asymptotic power law scaling is found for fluctuation amplitudes between 10 and 15 DU, during **a** 1979–1992 and **b** 1996–2003 (Varotsos 2005)

Varotsos (2005) calculated the empirical probability Prob (X > x) to witness TOZ fluctuations with an amplitude greater than a certain value $x = |y_{i+1} - y_i|$ in order to address these extreme phenomena. Next, Fig. 17.1a (1979–1992) and 17.1b (1996–2003) plotted the acquired result. These distributions' tails show coherence with a power law, Prob (X > x)$-x^{-\mu}$, suggesting that probability decreases as TOZ fluctuations become more intense.

We estimated $\mu = 5.12$ for Fig. 17.1a and $\mu = 4.60$ for Fig. 17.1b using a linear least-squares fit, taking into account the amplitudes of oscillations between 10 and 15 DU (2σ), or about (e.g., 100 DU—Dobson Units) = 1 mm thickness of pure ozone on the Earth's surface.

Both the TOZ fluctuations' distributions are non-Gaussian and non-symmetrical (small positive skewness). A linear least-squares fit yields an estimate of $\mu = 3.81$ for the period 1979–1992 and $\mu = 2.77$ for 1996–2003, where TOZ fluctuations vary from 20 to 40 DU. The tails of these distributions follow a power law Prob $(X > x) - x^{-\mu}$. Large TOZ changes are more likely to occur at 87.5°S (20–40 DU) than at 62.5°S (10–15 DU), according to a study of the μ-values at 62.5°S and 87.5°S. It is important to highlight that the μ-values for both latitudes are outside the range for stable Levy distributions (0 < μ < 2) or Gaussian distribution ($\mu = 2$) (Ausloos and Ivanova 2001). Consequently, large TOZ fluctuations are more probable than what a Gaussian distribution would anticipate in both time periods.

17.3 Scaling in Column Ozone Fluctuations

In Fig. 17.2a, the function below is displayed when using the above-mentioned two TOZ time-series:

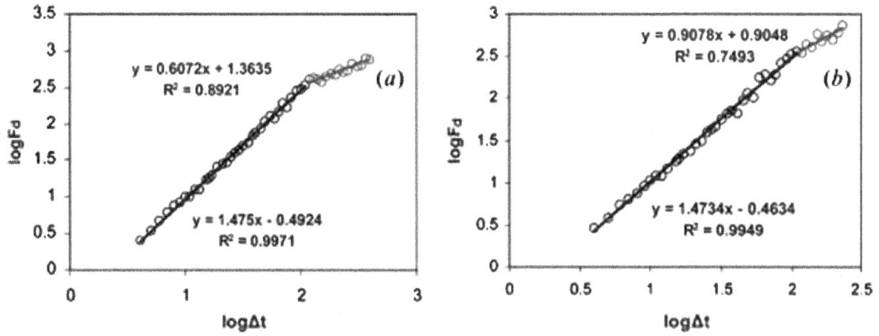

Fig. 17.2 DFA-function for the TOZ data from **a** 1979–1992 and **b** 1996–2003 in a log–log plot (Varotsos 2005)

$$F_d = \sqrt{\langle F^2(\tau) \rangle}$$

The function roughly mimics Brownian noise at a time interval of around 4 days to 4 months (one year in this case), according to the value of $a_1 = 1.475$ (⊣1.5). Moreover, the oscillations of the function F_d exhibit a distinct crossover at a year.

A crossover to $a_2 = 0.6$ is seen when looking at time-scales longer than a year, particularly those around four years. This implies that with time lags longer than a year, there may be long-lasting long-range power-law correlations in the variations of TOZ.

The DFA results for the second period are shown in Fig. 17.2b. The value of a_1 in this instance is roughly equal to that of the first dataset (1.473), demonstrating once more the existence of Brownian noise in the TOZ fluctuations for time lags shorter than a year. On the other hand, $a_2 = 0.91$ ($\sim$ 1) is discovered for time lags of more than one year, indicating that the TOZ variations for time lags longer than one year most likely correlate to $1/f$ noise shape. Thus, the primary inference made from the TOZ data at 62.5°S is that, while there are fluctuations that presumably fluctuate between white and Brownian noise during the ozone recovery era of 1996–2003, stable long-term correlations in the TOZ variations predominate during the period of 1979–1992.

The DFA approach was applied to the available daily zonal mean TOZ measurements for the months of October through November during 1979–1992 and 1996–2003, in order to investigate the temporal correlations of TOZ fluctuations for the inner half of the Antarctic ozone hole at 87.5°S. The function Fd once again closely resembles Brownian noise ($a_1 = 1.53$) during the years 1979–1992, and this is true for a time interval of roughly 4 days to 2 months (one year in this example). A crossover to $a_2 = 0.6$ is shown at time-scales longer than a year, i.e., approximately four years, suggesting the existence of long-term correlations that are consistent in the TOZ variations over time delays greater than a year.

A value of $a_1 = 1.50$ was noted for the period 1996–2003, suggesting the existence of Brownian noise in the TOZ fluctuations for time lags shorter than a year. On the other hand, a value of 0.93 ($\sim$ 1) was found for time lapses longer than a year, which most likely corresponds to $1/f$ noise.

Similar outcomes were achieved when employing the DFA technique on the daily average TOZ data for the Faraday (now Vernadsky) and Halley ground-based stations situated on the periphery of Antarctica (65°S and 75.5°S, respectively) within the O_3 hole region. The results derived by both a_1 and a_2 values of the DFA analysis performed atthe aforementioned datasets are consolidated in Table 17.1, indicating that in both time periods the a-values are:

$$1.0 < \alpha_1 \leq 1.5 \text{ and } 0.5 < \alpha_2 \leq 1.0.$$

The initial relationship implies the existence of enduring long-range correlations; specifically, $a_1 = 1.5$ represents Brownian noise. On the other hand, the subsequent relationship (for a_2) suggests the presence of enduring long-range power-law correlations from 1979 to 1992 for time lags exceeding one year; $a_2 = 1.0$ (predominantly observed from 1996 to 2003) corresponds to 1/f noise.

To summarize, fluctuations in springtime TOZ at the periphery and within the Antarctic ozone hole display persistent scaling behavior for time lags surpassing one year. Notably, since 1996, TOZ fluctuations exhibit correlations primarily characterized by noise. This highlights the challenge in resolving the issue of the recovery of the Antarctic ozone using raw TOZ observations.

Now let us examine whether the dynamics of the Antarctic O_3 hole have changed since 1972. A topic of great interest is the varying times at which the "ozone recovery" emerged. Within the available dataset, Varotsos (2005) also sought to ascertain whether the scaling patterns of the most current Antarctic TOZ data are different from those of earlier years. Varotsos (2005) used the multifractal DFA (MF-DFA) (see Chap. 15) to accomplish this. Compared to traditional DFA, this approach involves two extra phases, which are briefly explained below:

(i) computing the qth-order fluctuation function for multiple scales l by averaging across all intervals l:

$$F_q(l) = \left\{ \frac{1}{N} \sum_{v=1}^{N_l} \left[F^2(l, v) \right]^{q/2} \right\}^{1/q}$$

Table 17.1 The power law of the DFA function for the TOZ fluctuations at the edge and in the Antarctic ozone hole, with a crossover at one year, and its two exponents, a_1, a_2

Latitude (°S)	1979–1992		1996–2003	
	α_1	α_2	α_1	α_2
87.5	L5	0.6	1.5	0.9
75.5	1.2	0.8	1.4	1.0
65.0	1.3	0.8	1.3	1.0
62.5	1.5	0.6	1.5	0.9

(ii) for each q, plot log $F_q(l)$ vs log$l$ to determine the relationship:

$$F_q(l) \sim l^{h(q)}$$

We were able to draw conclusions that were in line with earlier research by examining the preliminary springtime daily average TOZ readings from the Faraday station at the border of Antarctica (65°S) between 1972 and 2003. For scales longer than ten days, the MF-DFA analysis of the TOZ changes from 1996 to 2003 showed statistically significant differences from the previous sub-series. This suggests that since 1996, the intrinsic dynamics of TOZ at Antarctica's edge have changed.

When the MF-DFA method is applied to the interim springtime daily mean TOZ values for the Halley station (75.5°S) in the interior region of Antarctica between 1972 and 2003, the same results for a-values are found. However, over time scales longer than ten days, there is no appreciable variation in the MF-DFA functions of TOZ fluctuations between the three time periods. This suggests that, at time periods longer than ten days, the intrinsic dynamics of the TOZ dataset in Antarctica's interior have not changed since 1972.

17.4 Scaling Effect in Planetary Waves Over Antarctica: Impact on Ozone

Planetary wave-breaking can be compared to the breaking of ocean waves on a shore. Charney and Eliassen (1949) utilized a basic linear barotropic beta plane model to simulate stationary planetary waves in the troposphere. They hypothesized that these waves were induced by a consistent eastward wind blowing over the surface topography. Additionally, Smagorinsky (1953) considered the impact of thermal sources on wave generation. Subsequently, Charney and Drazin (1961) analyzed stratospheric planetary waves, including stationary waves, using the quasi-geostrophic theory on a beta-plane. Matsuno (1970) conducted a quantitative analysis of vertically propagating stationary planetary waves in the stratosphere (waves 1 and 2) by employing a linearized quasi-geostrophic potential vorticity equation. Notably, linear models of planetary disturbances in the middle stratosphere suggest that the disturbance amplitudes do not increase indefinitely over time (stable disturbances). However, the linear theory predicts a range of disturbances with amplitudes that grow without bounds (unstable disturbances). According to Charney and Stern's theorem (1962), a crucial condition for the instability (barotropic or baroclinic) of a basic zonal flow on a beta plane to conservative quasi-geostrophic disturbances is a change in the basic northward quasi-geostrophic potential vorticity gradient within the flow domain.

The DFA tool was utilized by Varotsos et al., (2008) to analyze the total ozone planetary waves 1 and 2 in the middle and high latitudes of the southern hemisphere. These waves were obtained from daily observations of total ozone using the TOMS instrument since 1979. The primary objective of this analysis was to investigate the ozone layer as a system that remains unchanged across a wide range of scales characterized by a fractal structure. Additionally, the study aimed to determine the presence of long-range correlations.

The findings indicate that fluctuations in the amplitude of wave 1 exhibit persistent long-range power-law correlations for time scales exceeding 4 days but less than 3 months. This suggests a close relationship between fluctuations in shorter and longer time intervals, following a power-law pattern within the range of approximately 4 days to 3 months. Conversely, the analysis of wave 2 revealed the absence of long-range correlations.

17.5 Scaling in Column Ozone at the Region 60°S–60°N

In order to detect long-memory processes, Varotsos and Kirk-Davidoff (2006) analyzed global column O_3 and tropospheric temperature measurements gathered from ground-based (1964–2004) and satellite-borne (1978–2004) instrumentation.

It was found that there was a power-law positive correlation between temperature and ozone fluctuations throughout a range of time periods. Across longer time periods, the exponent of this link was larger in the mid-latitudes for temperature than in the tropics, and higher in the tropics for ozone. Either stronger positive feedbacks or more inertia could be the cause of the enhanced persistence that was seen. As a result, the poleward growth in climate sensitivity predicted by global climate models may be related to the steeper slope of the temperature power distribution in mid-latitudes at longer time scales, as opposed to the tropics. Long-range correlation errors can be found using the detrended fluctuation analysis of both the model and the observed time series. Accurately representing these flaws would greatly increase reliability in the modeling of the long-term climate and atmospheric chemistry patterns.

17.5.1 The Temporal Scaling of the Total Ozone Variations

The mid-latitude zone and the extra-tropics were studied by Varotsos and Kirk-Davidoff (2006), who focused on the TOZ observations for the latitude zones 25°N–60°N and 25°S–60°S between 1964 and 2004. The analysis of the deseasonalized TOZ (D-TOZ) observations of the zone 25°N–60°N using DFA-1 demonstrates the persistence of long-range correlations again.

As Fig. 17.2a illustrates, for time scales less than or equal to two years, the correlations in TOZ variability show a "stronger memory" at $\alpha_1 = 1.22 \pm 0.04$ compared to $\alpha_2 = 0.63 \pm 0.04$ (for time scales between approximately two and eleven years).

Using the first order DFA (DFA-1) on the D-TOZ data from WDN, a log–log plot of the root-mean-square fluctuation function $F_d(\Delta t) = F(n)$ is shown across the zone 25°S–25°N in Fig. 17.2b. Based on $\alpha = 1.1\ (\pm 0.04)$, it is inferred that TOZ

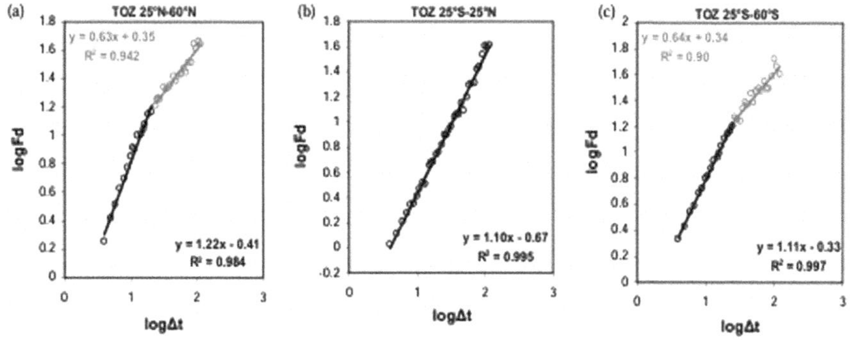

Fig. 17.3 The graph displays a log–log plot of the root-mean-square fluctuation function (F_d) for total O_3 (TOZ) as a function of the temporal box Δt (measured in months) for deseasonalized TOZ observations. The data was collected between 1964 and 2004 by the WMO Dobson Network, with a crossover occurring at $\Delta t \approx 28$ months over the mid-latitudes at both hemispheres (**a**, **c**) and over the tropics (**b**) (Kirk-Davidoff and Varotos 2006)

variations in the tropics exhibit enduring long-range correlations ($1/f$ noise-like) over a period of about 4 months to 11 years. The detected long-range correlations show that there are dynamic links between long- and short-term behaviors rather than the existence of cycles with distinct periodicities (Fig. 17.3).

Figure 17.2c shows the results obtained from applying the DFA-1 approach to the D-TOZ values in the latitude range of 25°S–60°S. The endurance of TOZ fluctuations is shown once more. To be more precise, $\alpha_1 = 1.11 \pm 0.02$ applies to time spans shorter than roughly two years, whereas $\alpha_2 = 0.64 \pm 0.06$ applies to longer time scales. As a result, the tropics show almost similar persistence at shorter time scales but a more marked persistence at longer time scales than the extratropics.

Varotsos and Kirk-Davidoff (2006) applied the DFA-l approach to the identical D-TOZ time series to confirm the previously described results. The results collected showed no discernible departures from DFA-1. They looked into whether the persistence in TOZ time series is due to the time development of the TOZ values or the values themselves to learn more about it. They achieved this by randomly shuffling TOZ data throughout the tropics using DFA-1. The result was $\alpha = 0.51 \pm 0.01$. Therefore, the distribution of the TOZ values has no bearing on the persistence in the TOZ time series, which results from the sequential organization of the TOZ values. For the TOZ time series over the extratropics and mid-latitudes in both hemispheres, comparable outcomes were seen.

Furthermore, the log–log plot obtained by applying DFA-1 to the global D-TOZ data shows that $\alpha = 1.1 \pm 0.02$, suggesting that the global ozone layer's variability is dominated by the strong persistence noted previously in the tropics and mid-latitudes. To summarize, over all-time lags ranging from around 4 months to 11 years, the TOZ oscillations throughout the tropics, extra-tropics, and mid-latitudes in both hemispheres, as well as globally, show persistent long-range relationships. But with time lapses of roughly 2–11 years, its persistence diminishes over the extratropics. These results are consistent with the first findings reported by Varotsos (2005).

The retrospective time-series of TOZ at Arosa (46.47°N, 9.40°E) and Poluy River (66.1°N, 68.3°E), which have been recently constructed, have been utilized by Efstathiou et al. (2009) to identify the presence of long-range correlations. By employing multifractal detrended fluctuation analysis on the average yearly values of the reconstructed total ozone content at both locations, it has been observed that the fluctuations in TOZ display enduring long-range power-law correlations across all time lags ranging from 4 to 90 years. Furthermore, when the same analysis is applied to the yearly increments of the reconstructed TOZ values, it reveals a crossover point at approximately 11 years. Before the crossover point, the correlations exhibit persistent long-range power-law correlations, while after the crossover point, they demonstrate persistent (antipersistent) long-range power-law correlations.

17.5.2 Conclusions on the Column Ozone Scaling

Varotsos and Kirk-Davidoff (2006) have found different results for temperature and ozone variations in the mid-troposphere. Temperature fluctuations show considerable persistence in mid-latitudes and random noise in the tropics, whereas ozone fluctuations show the strongest persistence over long time scales in the tropics but are weaker there.

There are several possible explanations for the variations in temperature and ozone persistence. Stronger positive feedbacks or more inertia typically lead to greater persistence. Global climate models indicate a poleward increase in climate sensitivity, which may be related to the lower slope of the temperature power distribution in the tropics over extended periods of time when compared to the mid-latitudes. On the other hand, while mid-tropospheric temperatures are anticipated to rise consistently with latitude, this prediction only pertains to surface temperatures.

The fact that jet stream changes have a significant impact on zonal average TOZ fluctuations in the mid-latitudes explains the latitude dependence of the persistence in ozone oscillations. The border between higher tropopause heights on the tropical side (less TOZ) and lower tropopause heights on the poleward side (more TOZ) is marked by these changes. Since seasonal weather forecast would be simpler in the absence of such variations, it is anticipated that they will last for only a few months at most. The TOZ distribution is more strongly correlated with temperature gradients related to the jet position than it is with temperature, which could account for the disparity in persistence patterns between ozone and temperature.

Even though there are a number of established linked mechanisms between temperature and O_3, our understanding of the overall effects of these interactions and feedbacks is still lacking. A way for visually representing the long-range correlations within data sets is provided by DFA analysis. We have shown that temperature and total ozone, when averaged over large regions, show remarkably different long-range correlation patterns over a range of time durations.

As a result, any model that attempts to predict the overall amount of ozone or global temperature over a long period of time needs to be able to reproduce the long-range correlations shown in DFA analysis. The accuracy of model projections about climate and ozone levels would be more confidently predicted if DFA curves based on observable data could be successfully simulated.

17.6 Scaling in Tropical Stratospheric Ozone Fluctuations

Varotsos et al., (2017) explored the distinctive characteristics of the intrinsic dynamics of ozone vertical distribution in the tropical stratosphere using modeled data. Their investigation involved analyzing the intrinsic scaling properties within the time series of the tropical monthly mean ozone vertical mixing ratio (VMR) from 1980 to 2014 at altitudes of 35, 45, and 50 km. The focus of their research was on identifying potential long-range correlations in the ozone VMR of the tropical middle and upper stratosphere. The study deliberately refrained from employing a non-linear regression model on satellite data sets (such as Atmospheric Chemistry Experiment: 2004–2013, Halogen Occultation Experiment: 1991–2005, and SAGE II: 1985–2005) due to the significant temporal variability present in these datasets.

The study conducted by Varotsos et al. (2017) utilized the DFA technique to analyze the tropical monthly mean ozone VMRs from 1980 to 2014. Three model simulations were employed at altitudes of 35, 45, and 50 km to investigate potential power-law scaling features. To validate the findings obtained through DFA, the researchers utilized the autocorrelation function and the method of local slopes of the fluctuation functions. The results from the analysis of the three models at different altitudes did not show evidence of exponential decay in the autocorrelation function for large scales or constancy in the local slopes within a significant range. Consequently, the establishment of long-range dependence for ozone VMRs was not possible.

In summary, the modeled ozone VMRs do not exhibit the same long-range correlations observed in the data from other parameters, such as temperature, at similar stratospheric heights. This suggests that the observational ozone VMRs may display a power law scaling effect (long-range dependence) that is not present in the modeled data. Therefore, accurately quantifying the scaling effect on upper stratospheric ozone remains a topic of ongoing scientific inquiry.

17.7 Scaling in Surface Ozone Fluctuations

Varotsos et al. (2012) investigated whether surface air pollution entirely vanishes within a range and looked at the relative predictive power of the pollution at various time frames. To this goal, the deseasonalized and detrended mean monthly values of SOC (Surface Ozone Concentration) during the daytime and nighttime periods of 1901–1940 and 1987–2007 were analyzed using the DFA approach. Using this approach, it was possible to find long-term power-law correlations that remained constant throughout a range of time lags, from four months to ten years ($a = 0.76$). It also showed how a fractal structure, which promotes predictability, is linked to long memory. It was discovered that rather than the distribution of its variations, the long memory effect in SOC results from its temporal evolution.

Additionally, the characteristics of SOC's persistence in the beginning of the twentieth century and the beginning of the twenty-first century were comparable. This implies that the Athens basin's industrialization and improvement of in situ photochemistry had no effect on SOC's fractal behavior. Furthermore, for both time periods, there was very little difference between the SOC changes during the day and the night. These results may aid in the creation of more precise simulation models that forecast differences in the predictability of surface air pollution and fluctuations in its levels across various time frames.

Thus, despite its current doubling, Varotsos et al. (2012) proposed that the SOC variations at Athens since 1900 show long-range dependence (long-memory). Using new analytical tools, Varotsos et al. (2015) later confirmed the long-memory in SOC that was postulated by Varotsos et al. (2012).

References

Ausloos M, Ivanova K (2001) Power-law correlations in the southern-oscillation index fluctuations characterizing El Nino. Phys Rev E 63:047201

Charney JG, Eliassen A (1949) A numerical method for predicting the perturbations of the middle latitude westerlies. Tellus 1:38–54

Charney JG, Drazin PG (1961) Propagation of planetary-scale disturbances from the lower into the upper atmosphere. J Geophys Res 66:83–109

Charney JG, Stern ME (1962) On the stability of internal baroclinic jets in a rotating atmosphere. J Atmos Sci 19:159–172

Efstathiou MN, Tzanis C, Varotsos CA (2009) Long-term memory dynamics of total ozone content. Int J Remote Sens 30(15–16):3897–3905. https://doi.org/10.1080/01431160902821817

Matsuno T (1970) Vertical propagation of stationary planetary waves in the winter northern hemisphere. J Atmos Sci 27:871–883

Smagorinsky J (1953) The dynamical influence of large-scale heat sources and sinks on the quasi-stationary mean motions of the atmosphere. Q J R Meteorol Soc 79:342–366

Varotsos C (2002) The Southern Hemisphere Ozone Hole Split in 2002. Environ Sci Pollut Res 9:375–376

Varotsos C (2003) Why did a 'no-ozone-hole' episode occur in Antarctica. Eos 84:183

Varotsos C (2004) The extraordinary events of the major, sudden stratospheric warming, the diminutive Antarctic ozone hole, and its split in 2002. Environ Sci Pollut Res. https://doi.org/10.1065/espr2004.05.205

Varotsos C (2005) Power-law correlations in column ozone over Antarctica. Int J Remote Sens 26(16):3333–3342. https://doi.org/10.1080/01431160500076111

Varotsos C, Kirk-Davidoff D (2006) Long-memory processes in ozone and temperature variations at the region 60° S–60° N. Atmos Chem Phys 6:4093–4100. https://doi.org/10.5194/acp-6-4093-2006

Varotsos CA, Efstathiou MN, Cracknell AP (2017) On the temporal evolution of the tropical stratospheric ozone. J Atmos Solar Terr Phys 157:1–5

Varotsos C, Efstathiou M, Tzanis C et al (2012) On the limits of the air pollution predictability: the case of the surface ozone at Athens, Greece. Environ Sci Pollut Res 19:295–300. https://doi.org/10.1007/s11356-011-0555-8

Varotsos C, Tzanis C, Efstathiou M, Deligiorgi D (2015) Tempting long-memory in the historic surface ozone concentrations at Athens, Greece. Atmos Pollut Res 6(6):1055–1057

Varotsos CA, Milinevsky G, Grytsai A, Efstathiou M, Tzanis C (2008) Scaling effect in planetary waves over Antarctica. Int J Remote Sens 29(9):2697–2704

Chapter 18
The Air Temperature Scaling Effect

18.1 The Long-Range Correlations in the Tropopause

Varotsos et al. (2009) conducted a study on the fluctuations of tropopause height across different latitudinal zones in both hemispheres. Between 1980 and 2004, they looked at the monthly tropopause height anomalies around the world and in seven latitude zones. These anomalies were found to be non-stationary and comprised both periodic and aperiodic waves, according to the research. It was found that the underlying correlations in the observations were frequently hidden by these non-stationarities. To meet this difficulty, the researchers stressed the necessity of using a specialist analytical tool.

Time records of monthly anomalies of tropopause height (Z) were investigated by Varotsos et al. (2009) to explore the existence of long-range relationships in the tropics. Their goal was to find out if there was any relationship at all between the Z value at one point in time and its value at another. The discovery that many environmental variables show long-range dependence—that is, that their values continue to correlate even after a considerable amount of time has passed—led to the question.

To ensure accurate insights into this problem, Varotsos et al. (2009) utilized the monthly anomalies of tropopause height after removing trends and seasonal variations. This approach helped avoid obscuring any potential scaling behavior due to long-term trends or well-known cycles. The researchers employed DFA-1 (Detrended Fluctuation Analysis) to analyze the data of tropopause height anomalies in the tropics. The results, presented in Fig. 18.1 (left), displayed a log–log plot of the root-mean-square fluctuation function $F_d(\Delta t) = F(n)$. They came to the conclusion that Z variations over the equator showed persistent long-range correlations resembling $1/f$ noise, with an α value of 0.97 ($\pm$0.04) for latitude 0°. This link, which lasted from roughly 4 months to 6 years, was statistically significant.

It should be mentioned that because most water vapor enters the stratosphere in this region, the $1/f$ noise fascinating behavior of the tropical tropopause is very significant. It's also critical to emphasize that the long-range correlations found show

G. Golitsyn and C. Varotsos, *The Stochastic Nature of Environmental Phenomena and Processes*, https://doi.org/10.1007/978-3-031-77015-9_18

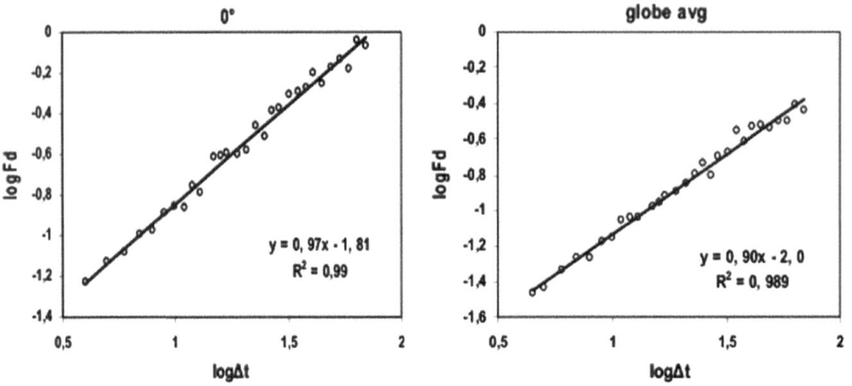

Fig. 18.1 Log–log plot of the tropopause height (Z) root-mean-square fluctuation function ($F_d(\Delta t)$) for detrended and deseasonalized monthly Z values, recorded by the radiosonde network over the equator and the world between 1980 and 2004 (Varotsos et al. 2009) versus temporal interval Δt (in months)

that there are dynamic links between actions on both longer and shorter time scales rather than pointing to the existence of cycles with distinct periodicities.

Subsequently, Varotsos et al. (2009) applied the aforementioned methodology to analyze the dataset of monthly fluctuations of tropopause height (Z) on a global scale. The findings, as depicted in Fig. 18.1 (right), reveal a strong persistence of fluctuations in the average Z of the globe, with a time period ranging around 4 months–6 years (ENSO), approximately, given that $\alpha = 0.90\ (\pm 0.04)$.

A study by Varotsos et al. (2009) looked at a number of latitude zones, such as the high latitudes, mid-latitudes, and subtropics. They specifically examined the Z data from 1980 to 2004 for six latitude bands in both hemispheres. They discovered that, as seen in the deseasonalized and detrended Z data, consistent long-range correlations are present in all latitude bands of both hemispheres using the DFA-1 approach. As Fig. 18.2(right) illustrates, the correlations in Z fluctuations revealed that memory increased from high to low latitudes, with the Northern Hemisphere showing higher memory than the Southern Hemisphere.

It is noteworthy that the DFA-l approach can be used as a reasonable means of verifying the previously reported findings. The DFA-l analysis's findings did not substantially differ from Figs. 18.1 and 18.2's results. However, the researchers looked at whether this persistence comes from the values of Z itself or from their time evolution to identify the variables leading to the long-range persistence in tropopause height. They used DFA-1 on randomly shuffled Z data spanning the tropics to investigate this. With $\alpha = 0.51 \pm 0.02$ as the analysis's outcome, it can be concluded that the distribution of the Z values has no bearing on this persistence, which only results from the Z values' sequential ordering.

To summarize, the Z fluctuations show consistent long-range correlations with time lags ranging from around 4 months to 6 years throughout multiple latitude zones, including the tropics, subtropics, middle, and polar latitudes of both hemispheres, as

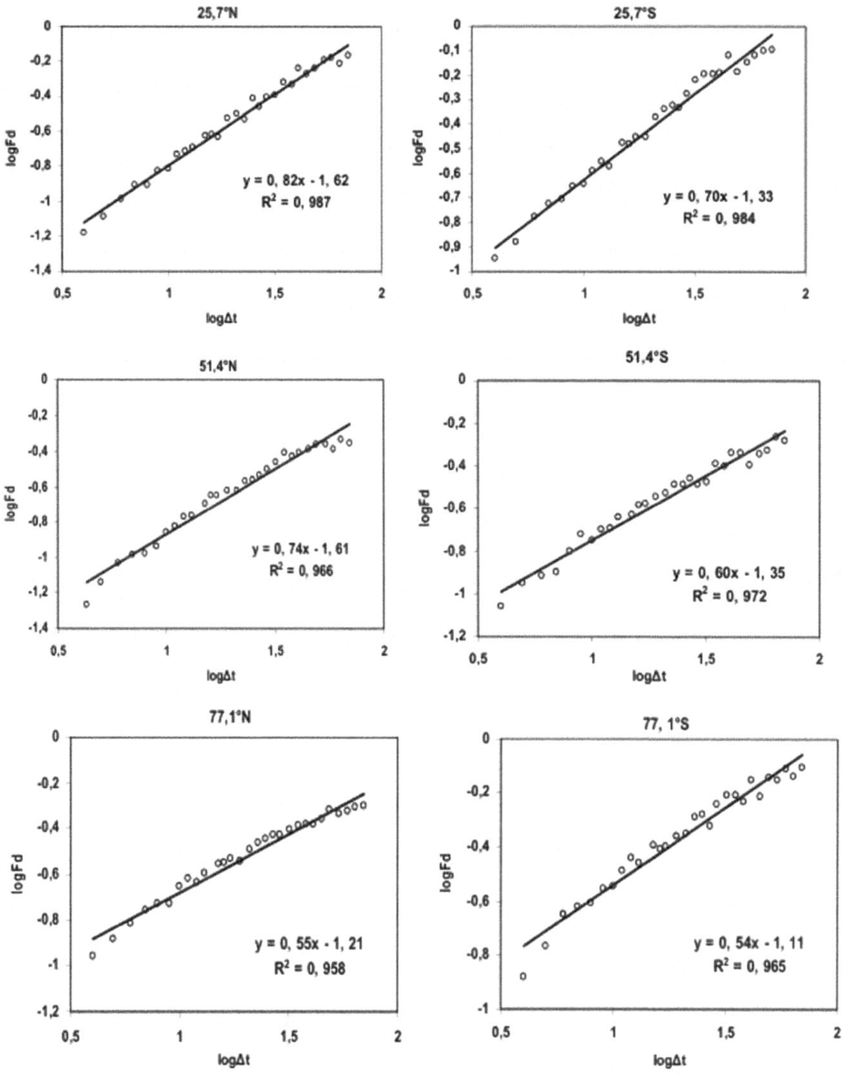

Fig. 18.2 Tropopause height (Z) root-mean square fluctuation function ($F_d(\Delta t)$) for detrended and deseasonalized monthly Z values: a log–log plot of the tropopause height (Z) observed by the radiosonde network over the tropics, middle latitudes, and high latitudes of the Northern and Southern Hemispheres between 1980 and 2004 versus (Δt)) (Varotsos et al. 2009)

well as globally. Over the tropics, this persistence gets greater as it gets closer to the $1/f$ mode.

18.2 The Temporal Scaling of Tropospheric Temperature Variations

Varotsos and Kirk-Davidoff (2006) also investigated the presence of time scaling in the fluctuations of tropospheric temperature (TRT), a parameter commonly used to measure global warming. The researchers utilized data from passive microwave temperature soundings obtained from various satellites, including TIROS-N, NOAA-6 to NOAA-12, NOAA 14, NOAA-15 to NOAA-17, and AQUA, covering the time period from 1978 to 2004.

In Fig. 18.3b, a log–log plot of the function $F_d(\Delta t)$ is presented, which was obtained by applying DFA-1 to the averaged D-TRT dataset in pentads of days within the latitude zone of 25°S–25°N.

The research showed that there were long-range correlations between the TRT fluctuations for time periods smaller than about two years ($\alpha_1 = 1.13 \pm 0.04$). But at time scales like the El Nino-Southern Oscillation (ENSO), which lasts for around two to seven years, the fluctuations have a random walk pattern ($\alpha_2 = 0.50 \pm 0.04$). The crossover point was identified as the moment at which the errors in the two linear best fits were minimized. This occurred at around two years.

The extratropics and the mid-latitude zone—that is, the areas between 25°N and 60°N and 25°S and 60°S—came into sharper focus after that. DFA-1 was used to analyze the D-TRT data in both latitude zones (Fig. 18.3a and c). The results indicated the presence of long-range power-law correlations with $\alpha = 0.80 \pm 0.01$ at time periods spanning from around 20 days to 7 years in both hemispheres.

This suggests that long-range persistence is exhibited by global TRT variations. These results were validated by additional study using DFA-2 to DFA-7, which produced α-values varying between 0.78 to 0.86. DFA-1 was also applied on the

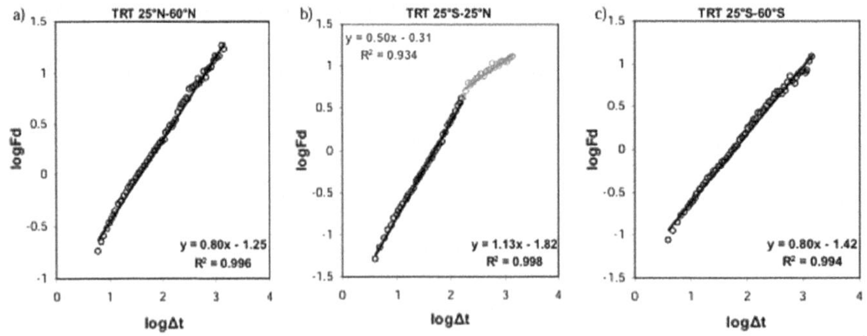

Fig. 18.3 The tropospheric brightness temperature (TRT) root-mean-square fluctuation function (F_d) is plotted log–log against the temporal interval Δt (averaged in pentads of days) for deseasonalized TRT (mid-tropospheric temperature) values as observed by the multi-satellite instrumentation between 1978 and 2004 (crossover at $\Delta t \approx$ 28 months over the tropics **b**) and mid-latitudes of both hemispheres **a**, **c**. Kirk-Davidoff and Varotos (2006)

shuffled TRT anomalies to confirm the persistence seen earlier, and the results showed that there were no persistent fluctuations present once more.

18.3 The Long-Range Correlations in the Land-Sea Surface Temperature

The mean monthly land and sea surface temperature (LSST) anomalies from January 1850 to August 2008 were used by Varotsos and colleagues (2009). These datasets were sourced from the Climatic Research Unit (http://www.cru.uea.ac.uk/cru/data/temperature/). The terrestrial and marine data were integrated into a single average to produce a worldwide coverage dataset that included coastal and island gridboxes.

Strongly persistent long-range power-law correlations were found when the DFA approach was used to the global mean monthly LSST anomalies time series (Fig. 18.4a); the scaling exponent varied between 0.86 for all time lags between (4 months–39 years) (Fig. 18.4b).

The substantial persistence observed indicates a power-law pattern of positive correlation between the LSST anomaly variations at small time intervals and bigger ones (up to 39 years). This finding implies that rather than following the traditional Markov-type stochastic behavior, which exhibits an exponential reduction with time, the correlations between the fluctuations in LSST anomalies show more slowly fading correlations. The latter is in line with recent climate model projections, which state that the higher troposphere should warm more quickly than the surface. Tropospheric convection, which in turn depends in part on SST, dynamically results in the cooling of the tropical tropopause, especially in the tropical zone.

Finally, Varotsos et al. (2009) looked at whether the persistence mentioned above is caused by the LSST anomaly levels alone, as opposed to their temporal evolution.

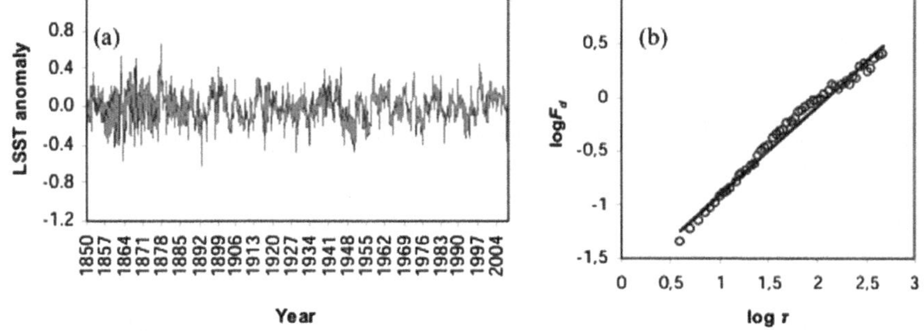

Fig. 18.4 **a** The dataset of the worldwide average monthly LSST anomalies that were detrended and deseasonalized from January 1850 to August 2008. **b** The log–log plot of the DFA-function for the dataset is shown in **a** with the best linear fit equation $y = 0.86 \times -1.77(R^2 = 0.98)$ (Varotsos et al. 2009)

For example, they used DFA on shuffled data of the LSST anomalies at random, and the exponent that they got was $\alpha = 0.53 \pm 0.01$. Consequently, the "long memory" observed in the time series of the LSST anomalies originates from their temporal evolution rather than their value distribution.

Furthermore, for all time lags between 4 months and 39 years, Efstathiou et al. (2011) discovered a "long memory" in the global mean LSST anomalies time-series for the months of January 1850–August 2008 in both the northern and southern hemispheres. Rather than originating from the values distribution of the LSST anomalies, these long-range correlations are a result of their temporal history.

18.4 The Scaling Effect in Global Land Surface Air Temperature

Detrended fluctuation analysis was performed by Varotsos et al. (2013) on the mean values of LSAT anomalies on an annual and monthly basis in both hemispheres and the world from 1880 to 2011. Examining the intrinsic dynamical properties was the goal. The following conclusions were reached after the above-mentioned analysis and discussion:

1. Globally and in both hemispheres, the time series of mean LSAT anomalies, both annually and monthly, show a constant power-law scaling. We used the autocorrelation function and the local slopes of the fluctuation function approach to validate this result. The obtained results show that, at large scales, the autocorrelation function rejects exponential decay, and the local slopes remain constant within a sufficient range. Both results support the presence of long-range dependence (LRD) in the LSAT anomalies, as do the computed error estimates.
2. The $a = 0.65$ (0.73–0.75) scaling exponents of the annual (monthly) mean LSAT anomalies are roughly identical in both hemispheres and approach the $a = 0.68$ (0.80) scaling exponent of the worldwide yearly (monthly) mean LSAT anomalies. By contrasting this finding with our earlier finding (Efstathiou et al. 2011) that the LSAT anomalies show a larger scaling exponent ($a = 0.89$) in the Southern Hemisphere (SH) as opposed to the Northern Hemisphere (NH) ($\alpha = 0.78$), we can deduce that the sea surface temperature (SST) is primarily responsible for the difference in scaling exponents between sea and land surface air temperature. In the SH compared to the NH, there is more scaling of the SST. This result is consistent with the basic knowledge that seas have a larger capacity to retain heat, which causes them to control land temperature less persistently.
3. The latitude zones 24°N – 44°N, 24°S – 44°S, 44°N – 64°N, 44°S – 64°S, 24°N – 90°N, and $24 \circ S{-}90 \circ S$ show consistent power-law scaling in the yearly average values of LSAT anomalies. With increasing latitude, the scaling exponents progressively grow and become more noticeable.

18.5 Symmetric Scaling in Global Surface Air Temperature Anomalies

The dataset of mean monthly land surface air temperature (LSAT) anomalies, obtained from the National Aeronautics and Space Administration Goddard Institute for Space Studies (in 0.01 °C), covering the years 1880–2013, was used by Varotsos and Efstathiou (2015). The land–ocean temperature index (LOTI), which they derived from the same source and used over a shared timeframe, was the combined land-surface air and sea-surface water temperature anomalies. It is commonly known that when the signal under study incorporates scaling, such as in the following cases, the power-law behavior for the root-mean-square fluctuation function $F_d(\tau)$ is observed:

$$F_d(\tau) \sim \tau^a$$

with a self-affinity parameter signifying the long-range power-law correlation and a as the scaling exponent.

First, the upward fluctuation function $F_d^+(\tau) \sim \tau^{a+}$, and the downward fluctuation function $F_d^-(\tau) \sim \tau^{a-}$, where a^+ and a^- are the upward and downward scaling exponents, respectively, were estimated using the A-DFA technique. Analyzing the sign of the slope b_k of the linear local trend fitted in each box of τ values allows for trend discrimination. Within the k box of length τ, a positive (resp. negative) trend is shown when $b_k > 0$ (resp. $b_k > 0$). As a result, the root-mean-square fluctuations are calculated using:

$$F_d^+(\tau) = \sqrt{\frac{1}{M^+\tau}\sum_{i=k\tau+1}^{(k+1)\tau}\frac{\left[\text{sign}(b_k)+1\right]}{2}[y(i)-z(i)]^2} \text{ and}$$

$$F_d^-(\tau) = \sqrt{\frac{1}{M^-\tau}\sum_{i=k\tau+1}^{(k+1)\tau}\frac{-\left[\text{sign}(b_k)-1\right]}{2}[y(i)-z(i)]^2},$$

$$k = 0, 1, 2, \ldots, (M-1),$$

where, given that $b_k \neq 0$ for all $k = 0, 1, 2, \ldots, (M-1), M = M^+ + M^- = N/\tau$, denotes the number of boxes with positive (resp. negative) trends.

The autocorrelation function and the technique of the local slopes of the fluctuation functions (i.e., the two criteria suggested by Maraun et al. 2004) were also employed to verify the existence of long-range correlations in the time series of LSAT anomalies and LOTI.

First, the A-DFA method was applied to the LSAT anomaly time series. The worldwide root-mean-square fluctuation function $F_{\text{d}}^+(\tau)$, $F_d(\tau)$, $\text{F}_{\text{d}}^+(\tau)$ and $F_d^-(\tau)$ for the detrended and deseasonalized mean monthly LSAT anomalies dataset is presented against the time scale τ (measured in months) in Fig. 18.5a, a double logarithmic graph. The persistent dynamics are indicated by the DFA scaling exponent $a = 0.75$.

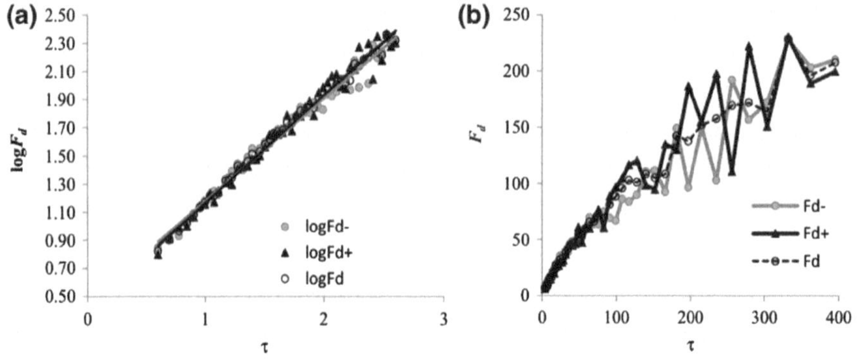

Fig. 18.5 **a** The root-mean-square fluctuation function $F_d(\tau)$, $F_d^+(\tau)$, and $F_d^-(\tau)$ are plotted twice logarithmically against the time scale τ (measured in months) for the detrended and deseasonalized mean monthly LSAT anomalies set (global) using the best-fit equations ($logF_d(\tau) = 0.75\log\tau + 0.42$ with $R^2 = 0.994$, $F_d^+(\tau) = 0.77\log\tau + 0.4$ with $R^2 = 0.98$, $F_d^-(\tau) = 0.73\log\tau + 0.45$ with $R^2 = 0.98$). **b** For the detrended and deseasonalized mean monthly LSAT anomalies set (globally), the root-mean-square fluctuation function $F_d(\tau)$, $F_d^+(\tau)$, and $F_d^-(\tau)$ versus time scale τ (in months) (Varotsos and Efstathiou 2015)

Furthermore, concerning Fig. 18.5a, the calculated value of $a- = 0.73$ appears slightly lower than the value of $a+ = 0.77$, suggesting that the decreasing dynamics of LSAT anomalies are just as persistent as the rising dynamics (a t-test was used to confirm this similarity, and at a 95% confidence level, the hypothesis $a- = a+$ was not rejected). To elaborate, there are no noticeable differences in fluctuation between $F^+(\tau)$ and $F^-(\tau)$ across all time scales, indicating a symmetric persistence. The root-mean-square fluctuation function $F_d(\tau)$, $F_d^+(\tau)$, and $F_d^-(\tau)$ displayed against the time scale τ (in months) for the detrended and deseasonalized mean monthly LSAT anomalies dataset (globally) provides a clearer picture of this symmetry.

The aforementioned findings were also analyzed for the Northern NH and Southern SH. It was found that there was no notable distinction in terms of the persistence of increases and decreases in LSAT anomalies. This is due to the fact that in both hemispheres, the scaling exponents (NH $a^+ = 0.7$, $a^- = 0.69$, and SH $a^+ = 0.68$, $a^- = 0.66$) were extremely close.

Similar results were obtained by Varotsos and Efstathiou (2015) when applying the same technique to the LOTI dataset.

To summarize, the examination of the upward and downward scaling patterns of LSAT anomalies indicates that there were no notable variations between the increases and decreases in LSAT anomalies across the entire globe and within each hemisphere. Conversely, the merging of land-surface air and sea surface water temperature anomalies appeared to disrupt symmetry, with the increases in temperature anomalies of the land and sea surface showing greater persistence compared to the decreases.

18.6 The Temperature Scaling Altitude Dependence at the Global Troposphere

The mean monthly and latitudinal averaged temperature values from different height layers of the global troposphere from 1980 to 2004 were used by Efstathiou and Varotsos (2010). The National Oceanic and Atmospheric Administration's (NOAA) National Climatic Data Center (NCDC) Integrated Global Radiosonde Archive provided the temperature readings for these 100 stations. These time series were chosen with an emphasis on covering various tropospheric height layers, including 850 hPa, 500 hPa, 300 hPa, 200 hPa, 150 hPa, and 100 hPa, depending on the availability and completeness of the daily data record. For the tropopause height in the tropics, the 100 hPa level is thought to be a dependable approximation; in the mid-latitudes, the 200 hPa level is appropriate; and in both polar regions, the 300 hPa level is a good approximation. The global tropopause temperature can be roughly estimated from the temperature at 100 hPa.

The link between the DFA-function and the temporal interval (t in months) for detrended and de-seasonalized temperature readings at two distinct pressure levels—(a) 850 hPa and (b) 100 hPa—for the years 1980 to 2004 is depicted in Fig. 18.6, a log–log plot.

Figure 18.7 shows the temperature scaling behavior at different heights in the global troposphere as a function of altitude. For time lags ranging from four months to six years, it is clear that the DFA-exponent continuously displays persistent long-range power-law correlations, with an overall positive trend in altitude that approximates $1/f$. Notably, at the vicinity of the tropopause level, the DFA exponent can be more than unity. The well-known examples of fractional Brownian motion (fBm) and fractional Gaussian noise (fGn) that are covered in the publications of Varotsos

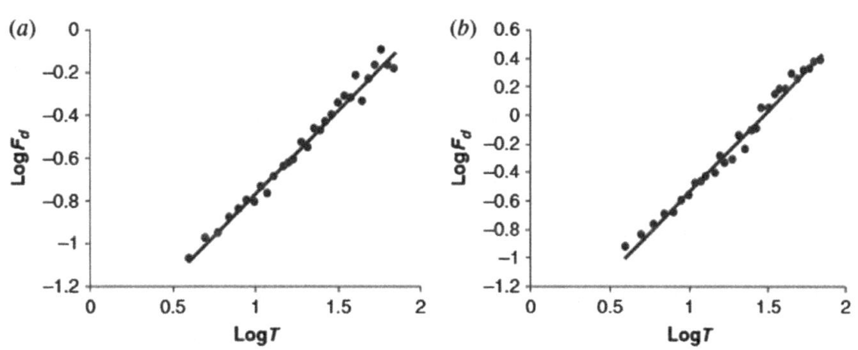

Fig. 18.6 For the detrended and deseasonalized mean monthly and latitudinal averaged temperature values at: **a** 850 hPa ($y = 0.78x - 1.55$ and $R^2 = 0.98$) and **b** 100 hPa ($y = 1.14x - 1.68$ and $R^2 = 0.99$), during 1980–2004, a log–log plot of the temperature rms fluctuation function Fd versus temporal interval τ (in months) and the corresponding best-fit equation and correlation coefficient are shown (Efstathiou and Varotsos 2010)

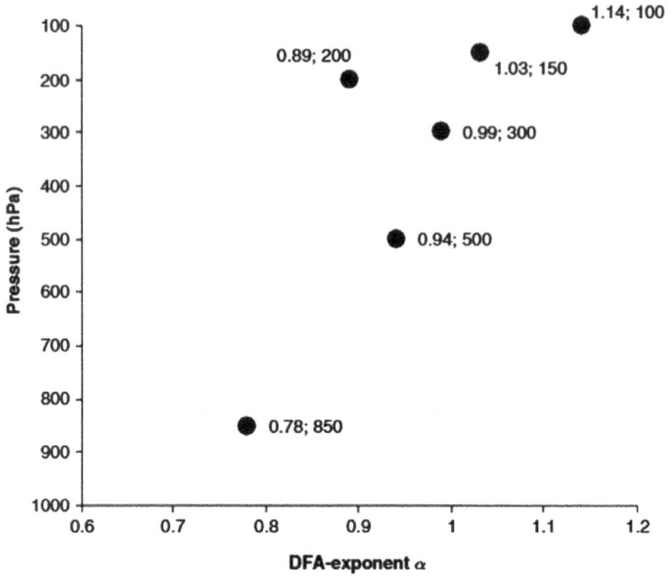

Fig. 18.7 The DFA-exponent's height dependency was observed between 1980 and 2004 for the detrended and deseasonalized mean monthly and latitudinal averaged temperature data at various height layers of the global troposphere (where 100 hPa roughly corresponds to the tropopause level) (Efstathiou and Varotsos 2010)

et al. (2006) and Varotsos et al. (2007) are cited here. In contrast to a fGn, or the time-series of a fBm's increments, which is stationary and has a DFA slope of $\alpha = H$, a fBm with a particular Hurst exponent $H(0 < H < 1)$ is non-stationary and exhibits a DFA slope of $1 + H$.

Two main conclusions can be drawn from the previous analysis:

- First, at time-scales ranging from roughly four months to six years, the mean monthly and latitudinal averaged temperature values at various tropospheric height layers, from 1980 to 2004, exhibit continuous power-law connections. The DFA-exponent values at different pressure points confirm this finding. Second, the long-range power-law persistence of the longitudinal averaged temperature values shows a positive altitude trend, with the regime approaching $1/f$ noise-like behavior close to the tropopause level. This suggests that as we proceed from the surface to the tropopause, the temperature correlation shows increased memory.
- • Secondly, the latitudinal averaged temperature values exhibit a long-range power-law persistence with a positive altitude trend, where the regime approaches $1/f$ noise-like behavior close to the tropopause level. This suggests that as we proceed from the surface to the tropopause, the temperature correlation shows increased memory.

These results imply that fractal behavior and long memory are present in the global tropospheric temperature. It is crucial to examine the output of current climate models to see if the tropospheric temperature shows a similar scaling behavior.

Several factors can be responsible for the vertical distribution of temperature persistence that is depicted in Fig. 18.7. Stronger positive feedbacks or more inertia may be the cause of greater persistence, especially at the tropopause level (Varotsos and Kirk-Davidoff 2006). Therefore, the upward increase in climate sensitivity projected by global greenhouse effect models, which is impacted by height-dependent climate feedbacks, may be related to the decreasing slope of the temperature power distribution in the lower tropospheric altitudes.

References

Efstathiou MN, Varotsos CA (2010) On the altitude dependence of the temperature scaling behaviour at the global troposphere. Int J Remote Sens 31(2):343–349

Efstathiou MN, Tzanis C, Cracknell AP, Varotsos CA (2011) New features of land and sea surface temperature anomalies. Int J Remote Sens 32(11):3231–3238

Maraun D, Rust HW, Timmer J (2004) Tempting long-memory—on the interpretation of DFA results. Nonlinear Process Geophys 11:495–503

Varotsos CA, Efstathiou MN (2015) Symmetric scaling properties in global surface air temperature anomalies. Theor Appl Climatol 121:767–773. https://doi.org/10.1007/s00704-014-1274-0

Varotsos C, Kirk-Davidoff D (2006) Long-memory processes in ozone and temperature variations at the region 60° S–60° N. Atmos Chem Phys 6:4093–4100. https://doi.org/10.5194/acp-6-4093-2006

Varotsos C, Efstathiou M, Tzanis C (2009) Scaling behaviour of the global tropopause. Atmos Chem Phys 9:677–683. https://doi.org/10.5194/acp-9-677-2009

Varotsos CA, Efstathiou MN, Cracknell AP (2013) On the scaling effect in global surface air temperature anomalies. Atmos Chem Phys 13:5243–5253. https://doi.org/10.5194/acp-13-5243-2013

Chapter 19
The Spectral Solar Radiation Variability

19.1 The Long-Range Correlations in the UV Solar Spectral Irradiance

Varotsos et al. (2013) conducted a study on the configuration of solar irradiance (SI) using real observations of high resolution. In more detail, they analyzed the SI discrepancies in the UV spectral region to determine if there is a scaling behavior. Finding out whether and what kind of association exists between the fluctuations in shorter wavelength (WL) intervals and those in longer wavelength intervals was the aim of the investigation.

In their research, Varotsos et al. (2013) presented Fig. 19.1, which displays solar spectra obtained on a horizontal plane at Briançon (French Alps, 1,310 m asl).

Varotsos et al. (2013) aimed to investigate the correlation between solar flux variations in short and long wavelength (λ) bands. They sought to quantitatively analyze these linked variations in the SI spectrum using new tools like the DFA.

It is important to note that Varotsos et al. (2013) focused on the SI structural pattern rather than its temporal evolution. They analyzed real observations of high resolution to examine the scaling behavior of SI fluctuations in the UV spectrum. In essence, their study aimed to determine if there is a correlation between fluctuations in short and long bands and to investigate the nature of this link.

The recorded observations include both atmospheric extinction and diffuse SI band. The measurements were taken on a sunny (17 Sept. 2000) and a cloudy (20 Sept. 2000) day with the same solar elevation (SZA055). The key finding from Fig. 19.1 is that on a sunny day, the diffuse surpasses the direct UVB irradiance.

The analysis of the solar incident flux (SIF) fluctuations across various values of λ in the Earth's atmosphere on May 10, 2001, is illustrated in Fig. 19.2.

Analysis of the SIF-WL dataset using the DFA method, covering wavelengths in the spectral band (278 – 400) nm, showed a consistent long-range power-law self-similarity behaviour. The scaling observed in Fig. 19.3a indicates a power-law relationship between changes in solar incident flux across different WL ranges. This

G. Golitsyn and C. Varotsos, *The Stochastic Nature of Environmental Phenomena and Processes*, https://doi.org/10.1007/978-3-031-77015-9_19

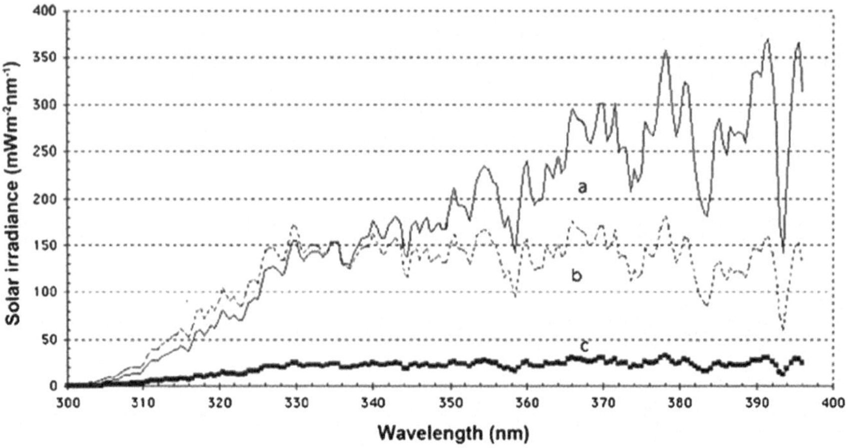

Fig. 19.1 On September 17, 2000, a clear sky day, the direct **a** and diffuse **b** SI spectra were recorded on a horizontal plane at the ground; on September 20, 2000, an overcast sky day, the diffuse **c** SI spectrum was recorded at the same location (Varotsos et al., 2013)

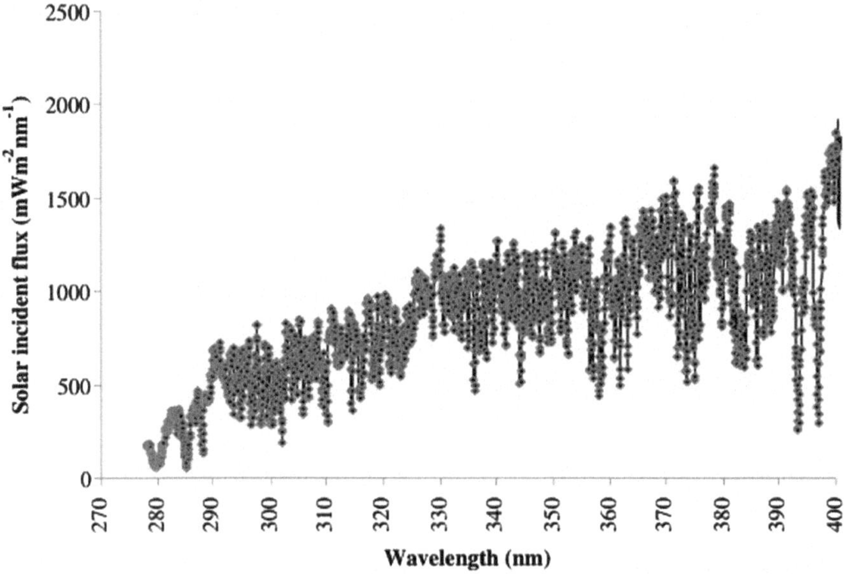

Fig. 19.2 The SIF measurements at the top of the atmosphere against λ (10 May 2001, 7:00 a.m.) in the spectral region (278 – 400) nm (Varotsos et al. 2013)

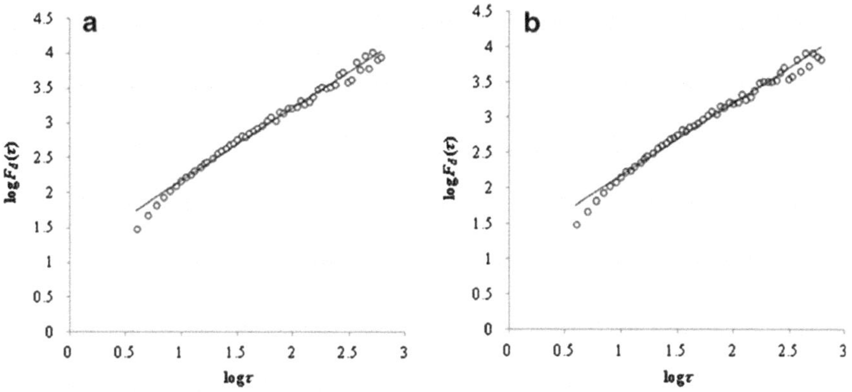

Fig. 19.3 **a** For the **a** SIF-WL data set with best-fit equation $y = 1.05x + 1.11$, $R^2 = 0.99$. For the **b** detrended SIF-WL data set with best-fit equation $y = 1.02x + 1.15$, $R^2 = 0.98$. Double log graph of the root-mean-square fluctuation function $F_d\ (\tau)$ vs λ interval τ (in nm). The slope values near unity signify the existence of scaling, or self-similarity; that is, there is a power-law relationship between the fluctuations in smaller and larger spectral regions, with an exponent around to unity (Varotsos et al. 2013)

suggests that variations in shorter WL ranges are correlated with those in longer WL ranges. In essence, the solar flux in the Earth's atmosphere, as determined by λ in the UV region, exhibits a fractal structure resembling an irregular geometric shape where each part mirrors the whole on a smaller scale. Additionally, the proximity of the a-exponent derived from DFA to unity points towards the presence of $1/f$ noise.

Varotsos et al. (2013) claim that the significant upward trend of the SIF spectral distribution may be the cause of the exponent α's unity value. Varotsos (2013) carried out a specific examination of the DFA approach utilizing the detrended SIF-WL dataset at the top of the atmosphere, within the λ range of 278–400 nm, in order to clarify this point. It's interesting to note that all WL delays between 0.2 and 30 nm (Fig. 19.3b) ($1/f$-type) displayed long-range persistence in the fluctuations of the detrended dataset, with a scaling exponent α of 1.02 ± 0.02. This demonstrates that the significant increase trend of SIF vs UV λ is not the only cause of the unity exponent.

The maximum λ value of 30 nm can be obtained in the manner described below: The greatest value of τ in Fig. 19.3ba, b has a logarithm of 2.8 (=log600). The highest value of τ, given the measurement step of 0.05, is roughly equal to $\tau_0 = 0.05 \times 600$, or $\tau = 0.05 \times 600 = 30$ nm. The minor fluctuations in the solar spectrum inside the Fraunhofer lines, which originate from the outer regions of the Sun and have significant biological implications, could be linked to this maximum λ value (Kondratyev and Varotsos 1995, for example). The λ interval of the $1/f$ behavior seen in this study is of the same order as the λ separation of the major Fraunhofer lines in the solar UV spectrum (299.4, 302.1, 336.1, 358.1, 382.0, 393.4, 396.8, and 410.2 nm). This is quite remarkable.

Varotsos et al. (2013) further proposed that the DFA analysis result is invariant with respect to the sun zenith angle of the UV solar spectral measurement. In particular, from dawn to sunset, there is 1/f behavior in the variations of UV SI on the ground. This result is consistent with the deduction made in Sect. 3.1 that the SIF at the upper atmosphere exhibits $1/f$-type behavior.

It is crucial to stress that the original data series' "roughness" can be gauged by looking at the similarity exponent. Hence, $1/f$ noise can be thought of as a state in between the much smoother landscape of Brownian noise and the total unpredictability of white noise (a very rough "landscape").

The result that the solar spectrum irradiance follows a $1/f$ power-law around UV λ should be interpreted with the knowledge that Planck's law converges to the Wien approximation at short wavelengths. It is possible to express the Wien approximation as:

$$I(\lambda, \mathrm{T}) = \frac{2hc^2}{\lambda^5} e^{-\frac{hc}{\lambda kT}}$$

where $I(\lambda, T)$ represents the energy emitted at a particular λ per unit area, time, solid angle, and λ, T is the black body temperature, h is the Planck's constant, c is the speed of light, and k is Boltzmann's constant.

Varotsos et al. (2013) used the DFA tool to examine the different values of $I(\lambda, T)$ obtained from the measured UV wavelengths λ after holding the temperature constant in the equation above. The analysis's conclusion showed that the computed $I(\lambda, T)$ values deviate from the $1/f$-type scaling. Therefore, in order to match the experimental results reported by Varotsos et al. (2013), the right side of the preceding equation must be multiplied by a $1/f$ noise function of wavelength λ. This function may represent a solar flux-wavelength relationship akin to fBm, which is thought to be the best mathematical model for $1/f$ noise according to Mandelbrot and Wallis (1968).

In conclusion, the UV spectral area shows a strong and permanent structural character in SIF, which has important ramifications for the climate system, as mentioned by Varotsos et al. (2013). It has been noted that a power-law pattern frequently occurs when there is an increase in SIF at one UV λ interval, followed by another increase at a different λ region.

This nonlinearity, as indicated by statistical tools in modern physics, can aid in constructing more reliable weighting functions for satellite observations. These functions are crucial for converting satellite data into accurate measurements. It is important to note that the estimated model values of solar flux versus λ should exhibit the same scaling properties as the real data of SIF across various wavelengths. Interestingly, the pattern of SIF in the UV band obeys a $1/f$ noise type, commonly meeting in nature across different fields such as physics, technology, biology, astrophysics, geophysics, and economics. However, a universally accepted physical explanation for this phenomenon has not yet been proposed.

19.2 The Scaling of the Solar Incident Flux

The work of Varotsos et al. (2013a) to establish the conclusion that the solar spectral irradiance obeys 1/f power-law as a function of UV λ using the well-known Planck's law was briefly mentioned in Sect. 19.1:

$$I(\lambda, T) = \frac{2hc^2}{\lambda^5\left(e^{\frac{hc}{\lambda kT}} - 1\right)}$$

which approaches to the Wien approximation at small wavelengths:

$$I(\lambda, T) = \frac{2hc^2}{\lambda^5} e^{-\frac{hc}{\lambda kT}}$$

where I (λ,T) is the λ-per-unit surface area, duration, solid angle, and energy emitted at a λ, the blackbody temperature is T, the speed of light is c, Boltzmann's constant is k, and Planck's constant is h.

Varotsos et al. (2013) showed that the calculated I (λ,T) values do not follow the $1/f$-type scaling vs. $\Delta\lambda$ (i.e., for fluctuations in (λ,T) throughout a range $\Delta\lambda$ in wavelength) by applying the DFA technique to various I (λ,T) values. Therefore, Varotsos et al. (2013) concluded that the latter may indicate a scaling in fluctuations of the solar incident flux (SIF) that could be associated with the intricate physical processes occurring in the solar atmosphere.

Later, Varotsos et al. (2015) attempted to examine the SIF residues concerning the Planck law in a wider wavelengths window (than in Sect. 19.1), from 115.5 to 629.5 nm.

Figure 19.4a illustrates the SIF observations within the window (115.5–629.5) nm. The main point observed is the presence of obvious non-stationarities vs. λ in the SI distribution and the significant rising trend up to around 450nm. The detrending of this dataset was achieved following the Planck formula:

$$B_1\left(\frac{b_1}{\lambda}\right)^5 / \left[\exp\left(\frac{b_1}{\lambda}\right) - 1\right.$$

In this, $b_1 = 2486.4$ nm based on the Sun's effective temperature (Tsun = 5778K). The derived parameter was found to be $B_1 = 85.8 \pm 0.7$ (0.82%) mWm^{-2}nm^{-1}. Moving forward, Varotsos et al., (2015) focused on the residuals comparing with the Planck function represented by the blue line in Fig. 19.4a. Figure 19.4b shows the results obtained from DFA-n analysis, where the exponent is close to unity.

It should be noted that the DFA tool produces outcomes comparable to those of the Haar tool, yet conceals the discontinuity that is evident in the Haar analysis (see Fig. 19.3 in Varotsos et al. 2015). It is important to emphasize that the $1/f$ scaling dynamics found in SIF pertain.

The main findings are as follows:

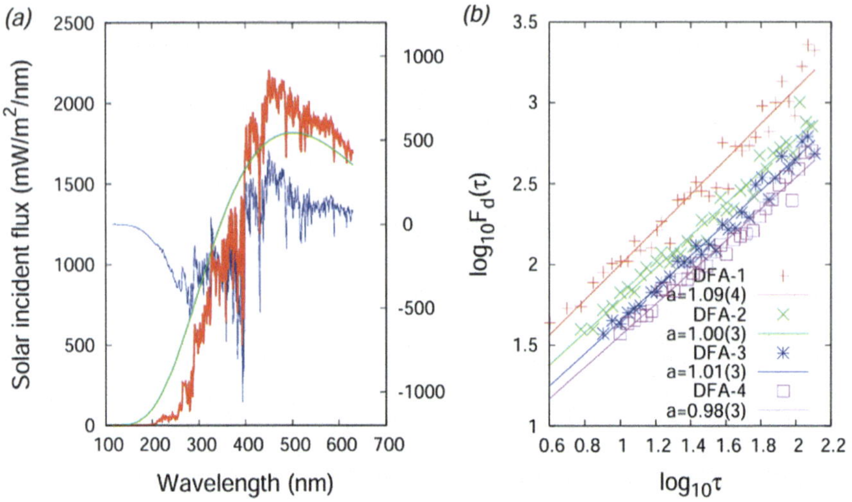

Fig. 19.4 **a** The SIF values (red, left scale) at the top of the atmosphere are shown with the fitting that was employed (green, left scale) and the λ from 115.5 to 629.5 nm. Additionally shown is the detrended SIF data (blue, right scale). **b** a log–log plot, for wavelengths between 115.5 and 629.5 nm, showing the root mean square fluctuation function $F_d(\tau)$ of the detrended SIF against the λ segment size. DFA-1, DFA-2, DFA-3, and DFA-4 have respective values of 1.09 (0.04), 1.00 (0.03), 1.01 (0.03), and 0.98 (0.03)

(1) DFA exponents were found to be around to unity. In simpler terms, the SIF discrepancies around Planck's law exhibit 1/f scaling dynamics.
(2) Analysis of the power spectral density for the detrended SIF dataset revealed a power-law shape with an exponent of 0.99 ($\pm$0.08). The DFA-1 exponent was 1.09 ($\pm$0.04), with DFA-n exponents ranging in (0.98 − 1.01).

References

Kondratyev KY, Varotsos CA (1995) Atmospheric ozone variability in the context of global change. Int J Remote Sens 16:1851–1881

Mandelbrot B, Wallis J (1968) Noah, Joseph and operational hydrology. Water Resour Res 4:909–918

Maraun D, Rust HW, Timmer J (2004) Tempting long-memory—on the interpretation of DFA results. Nonlinear Process Geophys 11:495–503

Varotsos CA, Melnikova I, Efstathiou MN et al (2013) 1/f noise in the UV solar spectral irradiance. Theor Appl Climatol 111:641–648. https://doi.org/10.1007/s00704-012-0697-8

Varotsos CA, Lovejoy S, Sarlis NV, Tzanis CG, Efstathiou MN (2015) On the scaling of the solar incident flux. Atmos Chem Phys 15:7301–7306. https://doi.org/10.5194/acp-15-7301

Chapter 20
Scaling of Near-Ground Spectral Albedo Variability

20.1 Principal Experimental Results on Albedo Features

Varotsos et al. (2013, 2014) revisited the airborne spectral observations of upward and downward irradiances and concentrated on the dependence of near-ground albedo on wavelength across the solar spectrum for different surfaces (snow, water, and sand) and different sky conditions (cloudy or clear). A research aircraft conducting multiple flight paths close to the ground was equipped with a diffraction spectrometer, which was used to determine the radiative upward and downward fluxes. Contrary to popular belief, the results showed that the near-ground albedo does not always increase with longer wavelengths across all surface types. Notably, the albedo of water surfaces in the ultraviolet spectrum holds steady over time. On the other hand, the water albedo exhibits an almost constant power-law relationship with wavelength in the visible and near-infrared spectra. For sand surfaces, the albedo is found to be a quadratic function of wavelength, with greater accuracy when excluding ultraviolet wavelengths. Moreover, despite having different magnitudes—water albedo being lower—the spectral behavior of snow and water are similar, declining by 20–50% from ultraviolet to near-infrared wavelengths. Interestingly, the albedo of snow remains nearly constant in the ultraviolet range, while a second-order polynomial provides the best fit for the visible-near infrared spectrum, similar to sand but with opposite slopes.

20.2 Dependence of Water Albedo on Wavelength in the Entire Solar Spectrum

The near-ground albedo values' variability is displayed in Fig. 20.1, which is based on radiative upward and downward fluxes recorded between May 13 and May 18, 1984, under clear sky conditions, by a research aircraft operating in close proximity

G. Golitsyn and C. Varotsos, *The Stochastic Nature of Environmental Phenomena and Processes*, https://doi.org/10.1007/978-3-031-77015-9_20

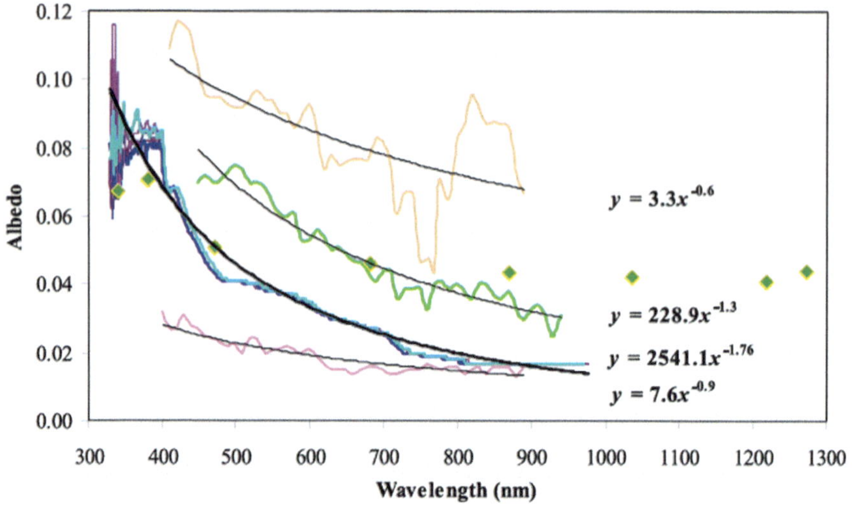

Fig. 20.1 Water albedo's spectrum dependence from the ultraviolet to the near-infrared. Dark green squares: NASA values (Atlantic ocean, 13 September 2000); rose line: Sea of Azov (15 October 1972); green line: Black Sea, Lake Ladoga (lines: pink, cyan, blue, deep pink); burly wood line: Atlantic ocean (12 July 1974) (Varotsos et al. 2014)

to Lake Ladoga's surface. The main information about the carried out experimental campaigns is shown in Table 20.1.

The primary conclusion drawn from Fig. 20.1 is that, throughout the solar spectrum, water albedo decreases with wavelength. The water albedo decay curves recorded over Ladoga between May 13 and May 18, 1984, exhibit a high degree of similarity, as evidenced by the power-law equation $A = 2541.1\lambda^{-1.76}$ with $R^2 = 0.97$, which represents the average fit line. The water albedo has a notable peak in the UV region at 335 nm.

As previously indicated, the power-law fit (illustrated in Fig. 20.1) accurately captures the wavelength variation of water albedo from the UV spectrum to the NIR region. Though the power-law fit is generally accurate for the whole spectral profile of water albedo, a closer look at Fig. 20.1 shows that the spectral water albedo in the UV region does not fit well with the power-law approximation. The albedo profiles obtained from several spectrograms taken over the course of five days are nearly constant over the long run, according to the data. According to this Varotsos et al. (2014) finding, separate research should be done on simulating the spectral dependence of water albedo in the UV and visible-NIR regions. Firstly, it is important to note that the local maxima that are seen in the 330–340 nm range could be explained by a rise in Rayleigh scattering (and consequently, reflectivity) at shorter wavelengths.

Table 20.1 Dates, times, solar elevations, and flight locations over snow, water, and sand

Dates of flights over sand	Lat.(°)	Long.(°)	Moscow time (astr/mic)	Greenwich time	Elevation *h*	Zenith angle,*θ*	cos*θ*
9 Oct 1983	39	60	09:47	07:47	28°34′40”	61.5°	0.477
10 Oct 1983	39	60	09:47	07:47	28°19′24”	61.7°	0.474
11 Oct 1983	39	60	10:06	08:06	30°54′36”	59°	0.515
12 Oct 1983	39	60	09:05	07:05	21°00′19”	69°	0.358
13 Oct 1983	39	60	10:47	08:47	39°35’	52°25’	0.610
14 Oct 1983	39	60	10:12	08:12	41°52’	50°18’	0.639
16 Oct 1983	39	60	10:03	08:03	33°54’	51°	0.629
19 Oct 1983	38:35	63:20	09:47	07:47	25°48′30”	64.2°	0.435
23 Oct 1983	38:35	63:20	10:05	08:05	38°18’	53°42’	0.592
24 Oct 1983	38:35	63:20	08:09	06:09	18°14′40”	71.75°	0.313
26 Oct 1983	38:35	63:20	09:14	07:14	39°17′33”	50.8°	0.632
Dates of flights over sand	Lat.(°)	Long.(°)	Moscow time (astr/mic)	Greenwich time	Elevation *h*	Zenith angle,*θ*	cos*θ*
13 May 1984	60:35	31:31	12:55	10:55	48°26’	51°34’	0.622
14 May 1984	60:38	31:36	12:35	10:35	39°53’	50.2°	0.640
15 May 1984	60:32	31:33	12:44	10:44	47°37’	52°23’	0.611
16 May 1984	60:33	31:35	12:46	10:46	47°22′25’	42.7°	0.735
17 May 1984	60:39	31:32	12:56	10:56	48°31’	51°29’	0.623
18 May 1984	60:35	31:33	12:19	10:19	48°41’	41.4°	0.750
Dates of flights over sand	Lat.(°)	Long.(°)	Moscow time (astr/mic)	Greenwich time	Elevation *h*	Zenith angle,*θ*	cos*θ*
14 Apr 1985	61:30	31:30	10:44	08:44	32°11’	57°49’	0.542
28 Apr 1985	60:49	31:52	13:51	11:51	47°01’	52°59’	0.602
26 Mar 1985	61:53	31:52	12:57	10:57	31°33’	59°27’	0.508

20.3 Water Albedo Versus Wavelength in the Visible-IR Solar Spectrum

The study by Varotsos et al. (2014), on water albedo across different wavelengths in the visible-IR region revealed interesting findings. It was observed that the water albedo remains consistent across the NIR and visible spectra, following a power-law mode. By excluding the UV region, the analysis in Fig. 20.2 (refer to Table 20.2 for errors) highlights this relationship.

A comparison between Figs. 20.1 and 20.2 shows that while the power-law relationship varies across the whole wavelength region from NIR to UV, it remains stable when focusing solely on the visible to NIR regions (excluding UV). Moreover, the information presented in Fig. 20.1 might be misleading concerning the variability in

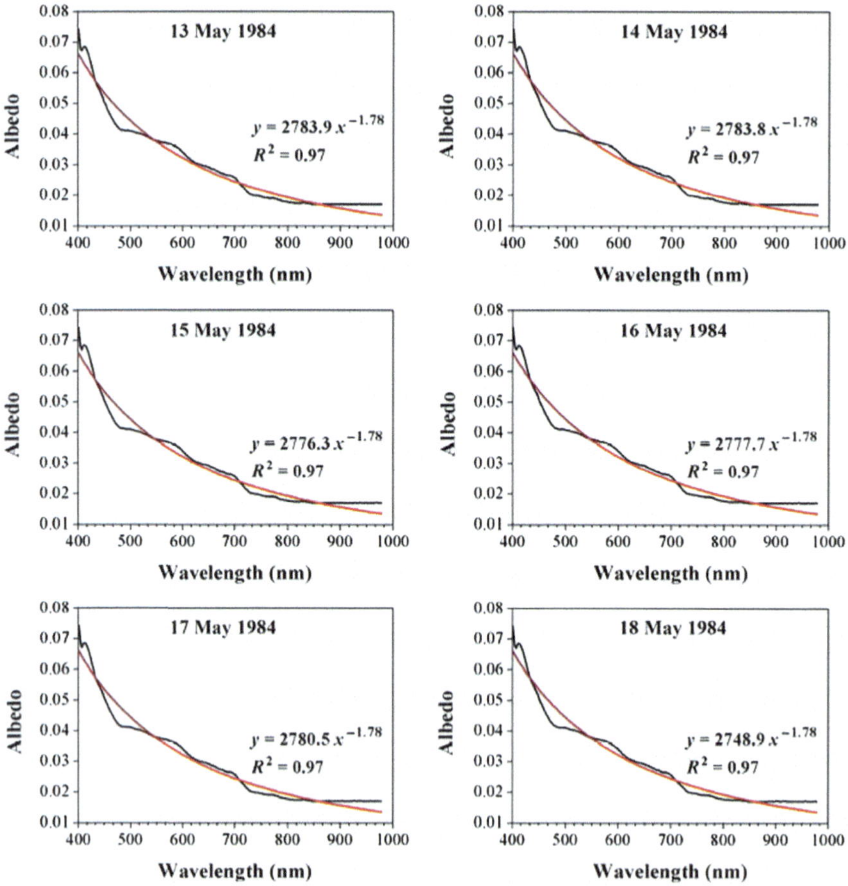

Fig. 20.2 Water albedo (Lake Ladoga) has a spectral dependence in the visible and NIR (UV excluded) bands, as inferred from six days of airborne observations (Varotsos et al. 2014, 2019)

Table 20.2 Fig. 20.2 displays dates, fit equations, and the coefficient errors for water albedo. The given a, and b values have a 95% confidence interval and are statistically significant (Varotsos et al. 2014)

Dates	Power-law fit equation $(y - ax^b)$	Error for a	Error for b
13 May 1984	$Y = 2783.9x^{-1.78}$	327.2	0.02
14 May 1984	$Y = 278.38x^{-1.78}$	327.2	0.02
15 May 1984	$Y = 2776.3x^{-1.78}$	326.7	0.02
16 May 1984	$Y = 2777.7x^{-1.78}$	327.1	0.02
17 May 1984	$Y = 2780.5x^{-1.78}$	327.6	0.02
18 May 1984	$Y = 2748.9x^{-1.77}$	322.1	0.02

the UV of the spectrum of water albedo, where water albedo is independent over a broad UV spectrum.

In summary, the albedo of water surfaces shows a weak dependence on wavelength in the UV region. However, in the NIR and visible wavelengths, the albedo of water follows a consistent power-law mode with wavelength. This finding is important for climate and radiation studies utilizing ground-based, airborne, or satellite observations (Katsambas et al. 1997; Tzanis and Varotsos 2008; Varotsos 1995, 2005; Varotsos et al. 2001).

References

Katsambas A, Varotsos CA, Veziryianni G, Antoniou C (1997) Surface solar ultraviolet radiation: a theoretical approach of the SUVR reaching the ground in Athens, Greece. Environ Sci Pollut Res 4:69–73

Tzanis C, Varotsos CA (2008) Tropospheric aerosol forcing of climate: a case study for the greater area of Greece. Int J Remote Sens 29(9):2507–2517

Varotsos C (1995) Simulation of broad-band and spectral solar ultraviolet radiation. Int J Solar Energy 16(3):203–216

Varotsos C, Alexandris D, Chronopoulos G, Tzanis C (2001) Aircraft observations of the solar ultraviolet irradiance throughout the troposphere. J Geophys Res: Atmosph 106(D14):14843–14854

Varotsos CA, Melnikova IN, Cracknell AP, Tzanis C, Vasilyev AV (2013) New spectral functions of the near-ground albedo derived from aircraft diffraction spectrometer observations. Atmos Chem Phys 13:16211–16245

Varotsos CA, Melnikova IN, Cracknell AP, Tzanis C, Vasilyev AV (2014) New spectral functions of the near-ground albedo derived from aircraft diffraction spectrometer observations. Atmos Chem Phys 14(13):6953–6965

Varotsos CA, Krapivin VF, Chukhlantsev AA (2019) Microwave polarization characteristics of snow at 6.9 and 18.7 GHz: estimating the water content of the snow layers. J Quant Spectrosc Radiat Transfer 225:219–226

Varotsos C (2005) Airborne measurements of aerosol, ozone, and solar ultraviolet irradiance in the troposphere. J Geophys Res: Atmospheres 110(D9)

Chapter 21
Scaling Properties of Air Pollution

21.1 Long-Memory in Air Pollution in Megacities

Varotsos et al. (2005) performed a detrended fluctuation analysis (DFA) on the hourly readings of O_3, NOx, and particulate matter from five monitoring stations from 1987 to 2003 in the Athens air pollution dataset (Fig. 21.1).

With lag durations varying from one week to five years, persistent power-law relationships were seen in the swings of NOx concentrations and daytime and nighttime O_3 concentrations. During the day, stronger relationships were seen (Fig. 21.2).

Additionally, from roughly 4 h to 9 months, Athens' PM10 anomalies showed consistent power-law connections. Long-range correlations were likewise discovered for PM2.5 changes in a 6-month data set obtained from the "Supersite" of University of Maryland in East Baltimore, over lag durations varying from around 4 h to 2 weeks (Fig. 21.3).

To comprehend the fundamental causes of air pollution's long-range behavior, more research is required. These results most likely represent short- and long-term sources and meteorological elements related to the atmosphere's self-organized critical pattern. This study's power-law connections may prove useful in creating more precise simulation models for changes in particulate matter, O_3, and NOx.

The following is a summary of the primary conclusions made by Varotsos et al. (2005):

(1) At temporal periods ranging from one week to five years, strong power-law correlations were discovered between daytime and nighttime O_3 changes, with stronger correlations occurring during the day. Similar patterns in the fluctuation of NOx were noted.
(2) Over time periods spanning from 4 h to 9 months, persistent power-law correlations in PM10 changes were identified, representing both short- and long-term source/meteorological aspects. In Baltimore, similar persistence was observed in PM2.5 during periods of 4 h to 15 days.

G. Golitsyn and C. Varotsos, *The Stochastic Nature of Environmental Phenomena and Processes*, https://doi.org/10.1007/978-3-031-77015-9_21

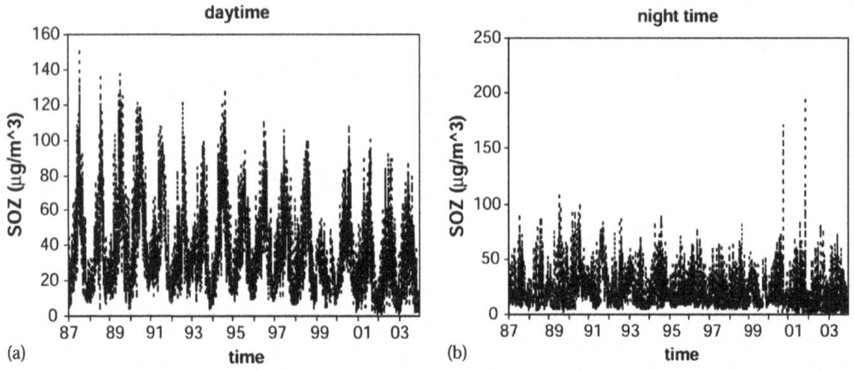

Fig. 21.1 Daily average values of surface O_3 (SOZ) dataset (1987–2003) for **a** daytime, **b** nighttime at the Patision air-pollution site (center of Athens) (Varotsos et al. 2005)

The short- and long-term source/meteorological aspects associated with the atmospheric self-organized critical pattern are probably the source of these long-range correlations in the temporal evolution of air pollution. To comprehend the cause of this long-term persistence in air pollution, more research is required. The results may be utilized to evaluate current models for their scaling behavior in the temporal evolution of air pollutants and to create new models for projecting future concentrations of air pollutants under various scenarios. Moreover, Varotsos et al. (2012a, b) found that the persistence of surface ozone concentration (SOC) in the early 1900s and early 2000s shared similar features.

This suggests that the SOC fractal behavior was unaffected by the industrialization and development of in situ photochemistry in the Athens basin.

Furthermore, there were only slight differences in the SOC variations between day and night. The accuracy of simulation models for variations in air pollution near surface and the predictability of their time scale variations could be improved by these findings.

21.2 Long-Memory in Aerosols Content

Varotsos et al. (2006) examined self-similarity traits by applying DFA to zonal average daily Aerosol Index (AI) values obtained from satellite data covering the period from 1979 to 2003. The results showed that on time scales longer than 4 days but less than 2 years, detrended and deseasonalized AI discrepancies show persistent long-range correlations of power-law mode in northern/southern hemispheres and globally. This suggests that oscillations in AI within shorter time frames are power-law correlated with fluctuations in AI within longer time frames (about 4 days to 2 years). The quasi-biennial oscillation (QBO) likely originates on a time scale of

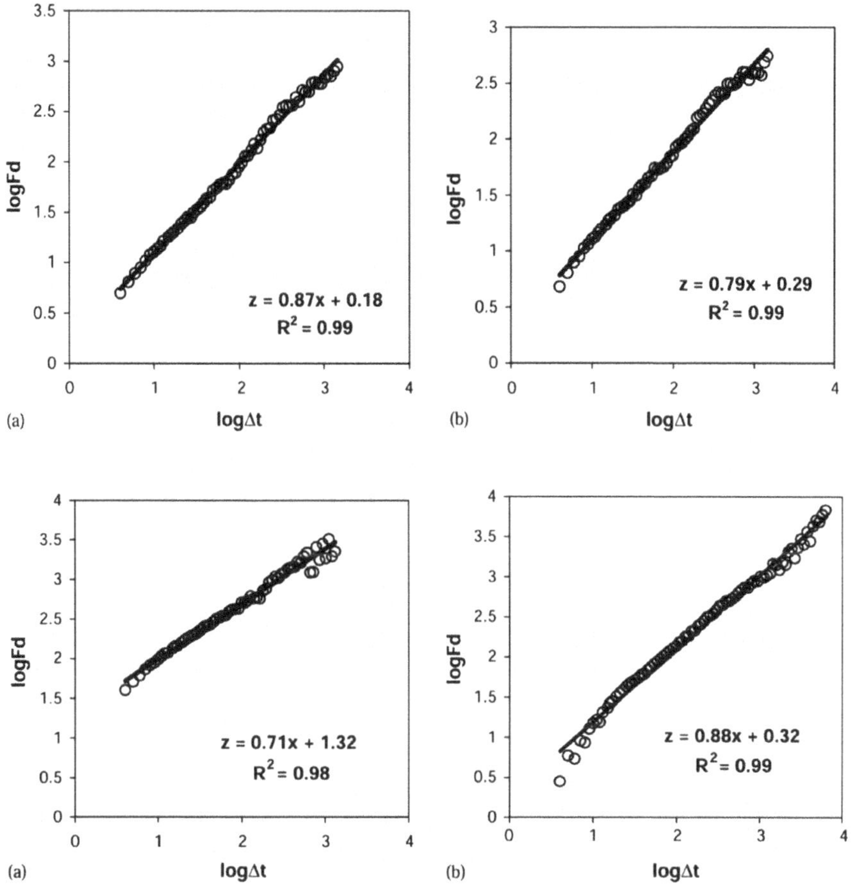

Fig. 21.2 (**Up**) The deseasonalized daily average values of **a** daytime surface O_3 (SOZ) and (**b**) nighttime SOZ at the Patision air-pollution site between 1987 and 2003 (Δt in days) are shown using the DFA-function in a log–log plot. The deseasonalized daily average values of NOx throughout the day at Patision air-pollution site from 1988 to 2003 (Δt in days) and the hourly average PM10 values at Zografou air-pollution site from 2000 to 2003 (Δt in hours) are plotted in a log–log plot with the DFA-function (**Down**) (Varotsos et al. 2005)

around two years, while the weather systems of the synoptic-scale likely give rise to a time scale of approximately four days.

In essence, AI anomalies show a power-law evolution as they go from one-time domain to the next. Discussion is held regarding the effects of 12-month and 6-month cycles on the scaling behavior of AI time series in both hemispheres. The 2-year temporal scale in AI time series could possibly be explained by the QBO modulating the Brewer-Dobson cell in the zonal wind of the equatorial stratosphere. The 4-day temporal scale is also probably influenced by synoptic-scale weather

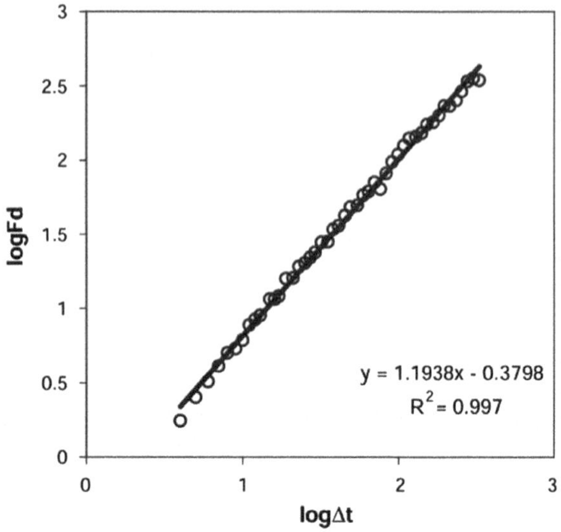

Fig. 21.3 The deseasonalized hourly mean PM2.5 data at the "Supersite" Station at Baltimore from July 4, 2002, to December 26, 2002, are plotted in a log–log format using the DFA-function (Varotsos et al. 2005)

processes. These findings may be useful for verifying current models to see if they display the above-described scaling tendency.

The time series data does not offer any clear proof during times of volcanic eruptions. In particular, there was no O_3 shortage in the southern hemisphere as a result of El Chichon's 1982 eruption or Pinatubo's 1991 eruption. The dominant winds in the lower stratosphere during the post-El Chichon era can account for this. The volcanic effect is responsible for the reported ozone shortage, which varies from 2 to 4% at equatorial latitudes to up to about 5% at middle and high latitudes. These inadequacies, which include the noise term, last for several months following the eruption and are more significant than the predicted inaccuracy brought on by radiance tainted by aerosols.

The aforementioned results are of interest for various applications in air pollution research (Christodoulakis et al., 2017, 2022; Tzanis et al. 2011; Varotsos 2004; Varotsos et al. 1992, 2001, 2012a, b, 2014a, b, 2015, 2021).

References

Christodoulakis J, Tzanis CG, Varotsos CA, Ferm M, Tidblad J (2017) Impacts of air pollution and climate on materials in Athens, Greece. Atmos Chem Phys 17(1):439–448

Christodoulakis J, Karinou F, Kelemen M, Kouremadas G, Fotaki EF, Varotsos CA (2022) Assessment of air pollution from Athens International Airport and suggestions for adaptation to new aviation emissions restrictions. Atmos Pollut Res 13(6):101441

Tzanis C, Varotsos C, Christodoulakis J, Tidblad J, Ferm M, Ionescu A, Lefevre RA, Theodorakopoulou K, Kreislova K (2011) On the corrosion and soiling effects on materials by air pollution in Athens, Greece. Atmos Chem Phys 11(23):12039–12048

Varotsos C (2004) Atmospheric pollution and remote sensing: implications for the southern hemisphere ozone hole split in 2002 and the northern mid-latitude ozone trend. Adv Space Res 33(3):249–253

Varotsos C, Cartalis C, Feidas C, Gerasi E, Asimakopoulos DN (1992) Relationship of ozone and its precursors in the West Coast Air Basin of Athens: a statistical model for the assessment of air quality in an urban area. Atmos Res 28(1):41–47

Varotsos C, Kondratyev KY, Efstathiou M (2001) On the seasonal variation of the surface ozone in Athens, Greece. Atmos Environ 35(2):315–320

Varotsos C, Ondov J, Efstathiou M (2005) Scaling properties of air pollution in Athens, Greece and Baltimore, Maryland. Atmos Environ 39(22):4041–4047

Varotsos CA, Ondov JM, Cracknell AP, Efstathiou MN, Assimakopoulos MN (2006) Long-range persistence in global Aerosol Index dynamics. Int J Remote Sens 27(16):3593–3603

Varotsos C, Efstathiou M, Tzanis C et al (2012a) On the limits of the air pollution predictability: the case of the surface ozone at Athens, Greece. Environ Sci Pollut Res 19:295–300. https://doi.org/10.1007/s11356-011-0555-8

Varotsos C, Ondov J, Tzanis C, Öztürk F, Nelson M, Ke H, Christodoulakis J (2012b) An observational study of the atmospheric ultra-fine particle dynamics. Atmos Environ 59:312–319

Varotsos CA, Ondov JM, Efstathiou MN, Cracknell AP (2014a) The local and regional atmospheric oxidants at Athens (Greece). Environ Sci Pollut Res 21:4430–4440

Varotsos C, Christodoulakis J, Tzanis C, Cracknell AP (2014b) Signature of tropospheric ozone and nitrogen dioxide from space: a case study for Athens, Greece. Atmos Environ 89:721–730

Varotsos C, Tzanis C, Efstathiou M, Deligiorgi D (2015) Tempting long-memory in the historic surface ozone concentrations at Athens, Greece. Atmos Pollut Res 6(6):1055–1057

Varotsos CA, Mazei Y, Saldaev D, Efstathiou M, Voronova T, Xue Y (2021) Nowcasting of air pollution episodes in megacities: a case study for Athens, Greece. Atmos Pollut Res 12(7):101099

Chapter 22
Scaling Effect in Greenhouse Gasses

22.1 Long-Memory in the Atmospheric Carbon Dioxide Content

In order to identify potential scaling behavior in the temporal history of the monthly mean values of the atmospheric carbon dioxide (CO_2) concentration at Mauna Loa Observatory in Hawaii, USA, Varotsos et al. (2007) employed the detrended fluctuation analysis (DFA). This dataset is the longest continuous record that is currently available worldwide, running from 1958 to 2004.

According to their research, there is a long memory because the variations in CO_2 concentrations show long-range relationships in a power-law fashion. The lag durations of these associations range from 4 months to 11 years. The study also showed that noise is produced by random variations in CO_2 concentrations. This noise has an exponent that is close to one and a power-law frequency spectrum. This implies that there is a notable increase in the correlation times. In addition, a properly rescaled subset of the original CO_2 concentration time series closely mimics the original dataset, the researchers found. In the end, the power-law connection generated from actual CO_2 concentration data may strengthen the trust in global climate models and the transport of atmospheric chemical species.

More specifically, Varotsos et al. (2007) assumed that Wiener filtering is the most effective way to deseasonalize the CO_2 dataset and that the 10th order polynomial trend is used to detrend it. As a result, the α-values of DFA-1 have an average of 1.08 and a standard deviation of 0.03 (Fig. 22.1). The DFA-1 α-values have an average of 1.20 and a standard deviation of 0.03 after the detrending is completed as previously and the deseasonalization is treated using the discrepancies from the regular data (Fig. 22.2). It is important to emphasize that there is close agreement between the α-values acquired in the two situations stated above.

G. Golitsyn and C. Varotsos, *The Stochastic Nature of Environmental Phenomena and Processes*, https://doi.org/10.1007/978-3-031-77015-9_22

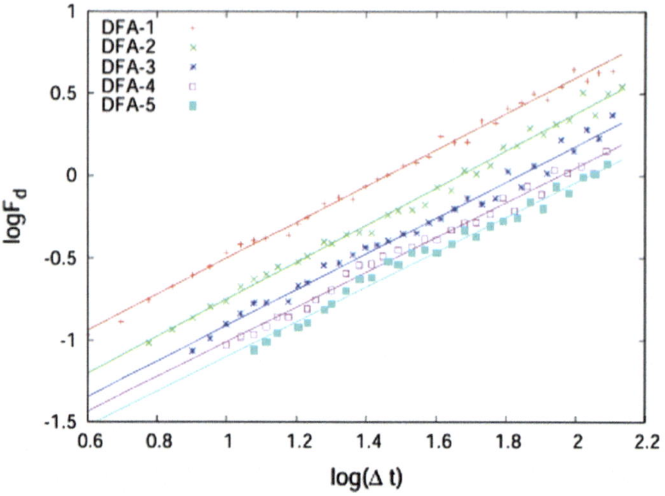

Fig. 22.1 A log–log graph of the DFA-l against the period Δt (measured in months) for CO_2 content values that were detrended (using the 10th order polynomial tool) and deseasonalized (using the Wiener filter tool) between 1959 and 2004. DFA-1, DFA-2, DFA 3, DFA-4, and DFA-5 have α-values of 1.10 (±0.02), 1.13 (±0.02), 1.09 (±0.02), 1.06 (±0.02), and 1.06 (±0.03) (Varotsos et al. 2007)

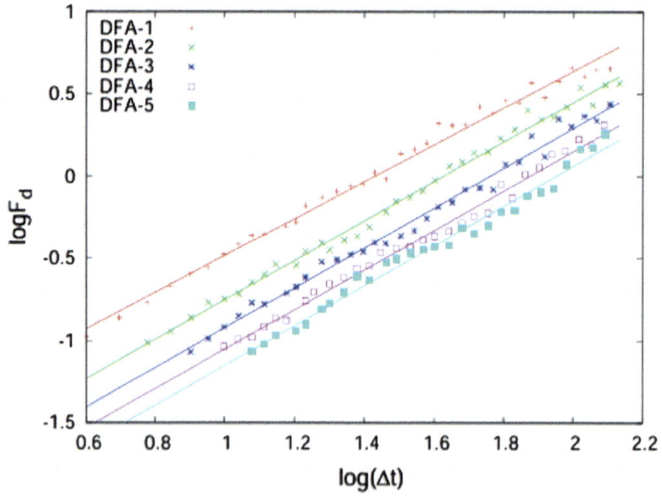

Fig. 22.2 The connection between DFA-l and the time region Δt (measured in months) for the CO_2 data that has been detrended and deseasonalized is shown in a log–log graph. The monthly average values' departures from their regular values were used to analyze the CO_2 data. The analysis is done for the years 1959 through 2004. For DFA-1, DFA-2, DFA-3, DFA-4, and DFA-5, the calculated α values are 1.12 (±0.02), 1.20 (±0.02), 1.21 (±0.02), and 1.21 (±0.03), (Varotsos et al. 2007)

It is also worth mentioning that filtering out the seasonal variability by the regular "deviations from the normal" technique results in α-values varying in the region 0.91–1.03. This confirms the α-values that were obtained in cases when the detrending and deseasonalization techniques were marginally distinct.

The primary conclusion drawn from Varotsos et al. (2007)'s numerous analyses is that, irrespective of the deseasonalization and detrending techniques employed, the DFA α-value is between 0.91 and 1.21. This suggests that, like a 1/f-type pattern, the variations in CO_2 content show persistent long-range behaviour. The significant persistence seen indicates that there is a power-law positive correlation between the variations in CO_2 content over both short and extended time periods (up to eleven years). Put more simply, a power-law pattern tends to indicate that a rise in CO_2 content is accompanied by another rise at another time. This conclusion suggests that the correlations between the CO_2 content variations exhibit slower declining correlations rather than the typical stochastic pattern of Markov-type, which is an exponential drop with time.

One interesting point to note is that the persistence observed can potentially serve as a forecast for CO_2 concentration. It assumes that the CO_2 content in the "next time segment" (up to eleven years) will be comparable with the corresponding "present time segment". However, it is important to recognize that this differs from the classical climatological prediction, according to which the CO_2 content in the "next" eleven years will be comparable with the climatological mean content.

Additionally, an effort has been made to compare the results of DFA analysis on the Mauna Loa CO_2 data with the CO_2 observations from the Antarctic region (89∘59′ S, 24∘48′ W) from 1973 to 2004. The South Pole data shows less seasonal dependence compared to Mauna Loa. By applying DFA-1 to the detrended and deseasonalized CO_2 dataset at the southern polar regions, it was found that the exponent is 1.22, indicating the persistent long-range correlations, such as observed over Mauna Loa.

Additionally, we looked into whether the CO_2 content dataset's intense persistent behaviour was caused by the content's actual levels or by how they changed over time. We accomplished this by shuffling the detrended and deseasonalized CO_2 content at random. If the shuffled CO_2 content exhibits random (white) noise, the previously observed persistence is due to the time evolution of the data rather than the data itself. The exponent $\alpha = 0.49 \pm 0.02$ was obtained by applying DFA-1 to the shuffled CO_2 data, showing that the shuffled data is essentially uncorrelated.

Therefore, the power-law behaviour obtained from the actual observations of CO_2 content is ultimately a result of their temporal course and correlations. The scalability performance of climate prediction models under various CO_2 scenarios can also be evaluated using these correlations.

In conclusion, the CO_2 fluctuations show strong long-range persistence, indicating positive correlations in a power-law manner across different time intervals. These correlations are attributed to the time evolution rather than the actual CO_2 values. The analysis of these correlations can aid in identifying human-induced changes due to increased CO_2 emissions against the backdrop of natural variations in the atmosphere. The scaling behaviour observed in real CO_2 data could be utilized to assess the effectiveness of climate modeling and enhance atmospheric chemistry-transport models on a global basis.

22.2 The Necessity to Establish the Power-Law: The Two Criteria

The autocorrelation function and the constancy of the individual gradients of fluctuation functions must be examined to determine whether long-range correlations exist in a dataset. It is required since a power-law behavior is not implied by the DFA single straight line for a dataset under study. The following subsections go over three examples.

22.2.1 Do the Thermospheric CO_2 and nO Power Exhibit Power-Law Behavior?

An investigation into the energy released globally by nitrogen monoxide and CO_2 from the thermosphere was carried out by Varotsos and Efstathiou (2018). They examined the development of this energy using data obtained through observation and empirical analysis. They started by looking at the daily power observations of nitrogen monoxide and CO_2 that were gathered from the NASA TIMED satellite's SABER sensor between 2002 and 2016. They then examined the recently obtained empirical daily power released by nitrogen monoxide and CO_2 from the thermosphere's IR energy budget, which covered the years 1947 to 2016.

The power emitted by nitrogen monoxide and CO_2 from the thermosphere was examined for power-law behavior using the DFA. The results of the empirical and observational measurements did not corroborate the existence of power-law pattern. This implies that the intrinsic qualities of the empirically acquired data and the observational data are comparable, which strengthens their dependability.

Figure 22.3a illustrates the evolution of nitrogen monoxide and CO_2 daily power over time from 2002 to 2016. Meanwhile, Fig. 22.3b shows the matching root-mean-square fluctuation functions $F_d(\tau)$ of the DFA technique versus time scale τ (in days).

The classical Wiener method (Maraun et al. 2004) and polynomial regression analysis are used to remove the strong polynomial trend and seasonality in the dataset of nitrogen monoxide and CO_2 daily power. Additionally, the autocorrelation function and the technique of the individual gradients of fluctuation functions are used to confirm the existence of long-range correlations. To ensure accuracy, the individual gradients of $\log F_d\ (\tau)$ against $\log\tau$ are examined for steadiness within an adequate extent. Similarly, the profile of the power spectral density is analyzed to determine if it fits better algebraically (power-law) or exponentially. The derived

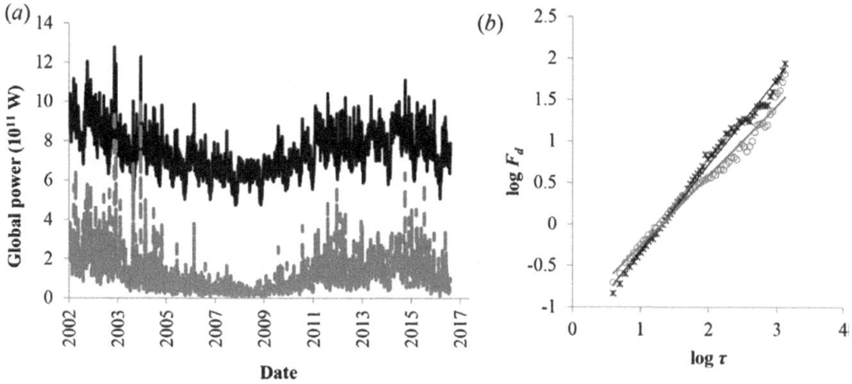

Fig. 22.3 **a** The daily power of nitrogen monoxide and carbon dioxide from 2002 to 2016. **b** The DFA root-mean-square fluctuation functions $F_d(\tau)$ are shown in a log–log plot with best-fit equations for both CO_2 and NO. Varotsos and Efstathiou (2018)

DFA scaling exponent suggests long-range persistence in the initial datasets of CO_2 (NO nitrogen monoxide) daily power $\alpha = 1.04 \pm 0.01$ ($\alpha = 0.84 \pm 0.02$). However, further investigation is needed to validate long-range correlations and scaling of power-law type.

The power spectral density profiles for the detrended and deseasonalized datasets of daily power of CO_2 and NO, from 2002 to 2016, are specifically shown in Fig. 22.4a and b. This indicates that the exponential decay could only be ruled out in the case of CO_2, where a fit of the power-law type appears to provide statistical reliability. However, Fig. 22.4a and b and 22.5(a and b) clearly show that the requirements of Maraun et al. (2004) are not met, indicating that the detrended and deseasonalized datasets of the daily power of CO_2 and NO cannot be validated using power-law scaling. Remarkably, there appears to be a stability of the individual gradients against logτ in a narrow range only for the case of nitrogen monoxide; this, however, is insufficient to validate the long-range correlations (Fig. 22.5).

Outlining the previously mentioned work, Varotsos and Efstathiou (2018) investigated the changes in the daily power of CO_2 and NO during 2002–2016. Power-law scaling and long-range correlations were not found for CO_2 and NO daily power, even if the DFA scaling exponent for the original datasets of both quantities displayed persistent behavior.

Analyzing the scaling pattern of detrended and deseasonalized datasets of carbon dioxide and nitrogen monoxide daily power over the same time revealed similar results. Using data from the thermosphere's infrared energy budget, Varotsos and Efstathiou (2018) also examined the historical evolution of radiated CO_2 and NO daily power during 1947- 2016. For CO_2 and NO daily power, power-law mode and long-range correlations were not apparent, despite the scaling DFA exponent exhibiting stable behavior for the initial datasets of both gases.

Analyzing the scaling dynamics of detrended and deseasonalized datasets of carbon dioxide and nitrogen monoxide daily power during 1947–2016 produced

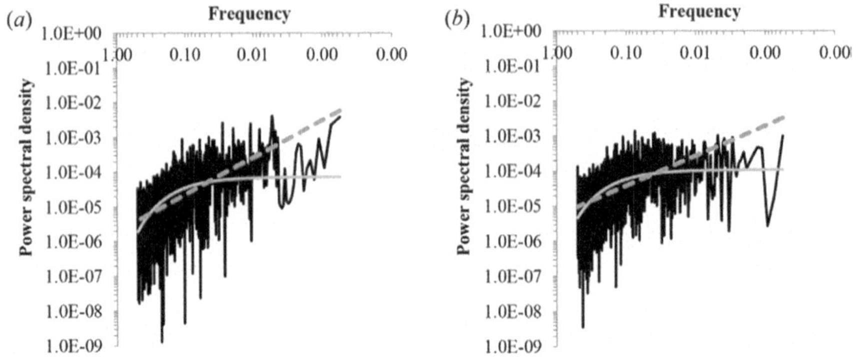

Fig. 22.4 Power spectral density for the detrended and deseasonalized of daily power datasets (during 2002–2016) for **a** carbon dioxide and **b** nitrogen monoxide with the respective power-law (dashed line) and the exponential (solid line) fit (carbon dioxide: $y = 1.95 \cdot 10^{-6}\ x^{-1.06}$ with $R^2 = 0.36$ and $y = 7.31 \cdot 10^{-5}\ e^{-7.31x}$ with $R^2 = 0.37$, nitrogen monoxide: $y = 5.5 \cdot 10^{-6}\ x^{-0.85}$ with $R^2 = 0.28$, $y = 1.1 \cdot 10^{-4}\ e^{-6.31x}$ with $R^2 = 0.33$) (Efstathiou and Varotsos 2018)

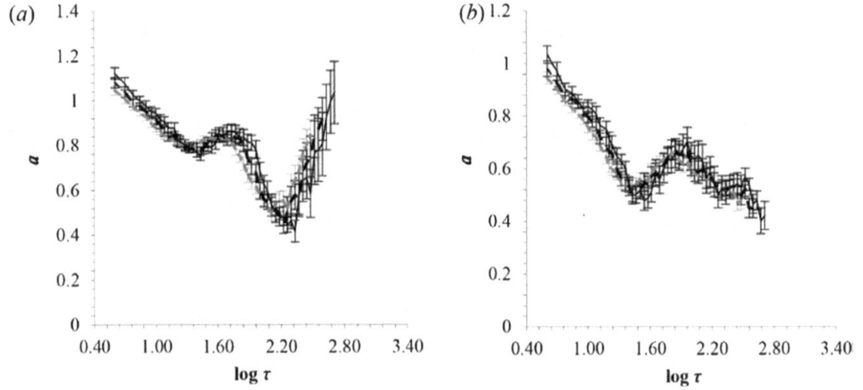

Fig. 22.5 Local slopes of the detrended and deseasonalized datasets of daily power of **a** carbon dioxide and **b** nitrogen monoxide were determined within a segment of 18 points (dashed grey line), 12 points (solid thin black line), and 15 points (dashed black line). The $\log F_d(\tau)$ against $\log\tau$ were plotted. The relevant 1.96·sa–segments of the gradients across all scales under consideration are indicated by the error bars

results that were comparable. The most important finding from the DFA results discussed above is that the IR power datasets observed by the SABER instrument during 15 years and its extension back to 1947 have consistent intrinsic features. This implies that the observations and the extended datasets are in agreement. To fully comprehend the impact of sustained variations in rising carbon dioxide and nitrogen monoxide contents on IR power, more research is necessary.

22.2.2 Does the Tropical Ozone Exhibit Power-Law Behavior?

Based on model simulations at heights of 30–50 km, Varotsos et al. (2017) examined the tropical monthly average O_3 volume mixing ratios (VMR) during the period 1980–2014. Whether the vertical O_3 VMR shows long-range power-law persistence was the main focus of the inquiry. The acquired results demonstrated that there is no long-range power-law connection in the modeled O_3 data, indicating that the data are not influenced by a persistent signal. The information is displayed below.

Checking for the satisfaction of the following two criteria is crucial in order to establish the previously mentioned persistence at long time scales: (1) whether the scaling exponent's individual gradients are steady over a particular segment, and (2) whether the autocorrelation function's exponential decay is rejected (Maraun et al. 2004). In reference to the initial criterion, Fig. 22.6b displays the individual gradients as a function of $\log\tau$ for two distinct segment widths, namely 12 and 13 points. The results indicate that the individual gradients diminish without attaining a stable period. With respect to the second requirement, the spectral density function of the O_3 VMRs' detrended and deseasonalised datasets, fitted by power-law and exponential curves, is shown in Fig. 22.2c. These fits' statistical metrics (such as the R^2 value) demonstrate that the exponential approach outperforms the algebraic (power-law) fit. As a result, the previously indicated requirements for the long-range correlations' validation are not met. Using models B_SAT and C_SOR, we also applied the DFA approach to the detrended and deseasonalized datasets of O_3 VMRs, at 35 km, over the years 1980–2014. The findings were comparable to those previously obtained. Therefore, the calculations employing all three models indicate that the O_3 VMRs at 35 km do not appear to be characterized by long-range correlations of the power-law fashion, that is, with $a > 0.5$.

22.2.3 Does the Solar and Volcanic Forcing Exhibit Power Law? DFA and Haar Tools

There is a serious flaw in the conclusions drawn by a number of climatic studies on long-range reliance on the memory of climate conditions. These investigations contradict the straightforward exponential decay of the autocorrelation function and are unable to demonstrate power-law scaling. Below is an example that illustrates this problem: from insufficient data analysis, a strong long-range reliance was mistakenly inferred.

Varotsos and Efstathiou (2017) used the DFA approach on the solar and volcanic forcing (SF, VF) datasets for the past 10^3 years in the tropical Pacific to show this. A straight-line approach with a gradient greater than 0.5 was demonstrated by the results in a log–log representation. This, however, should not be interpreted as proof of enduring long-range relationships.

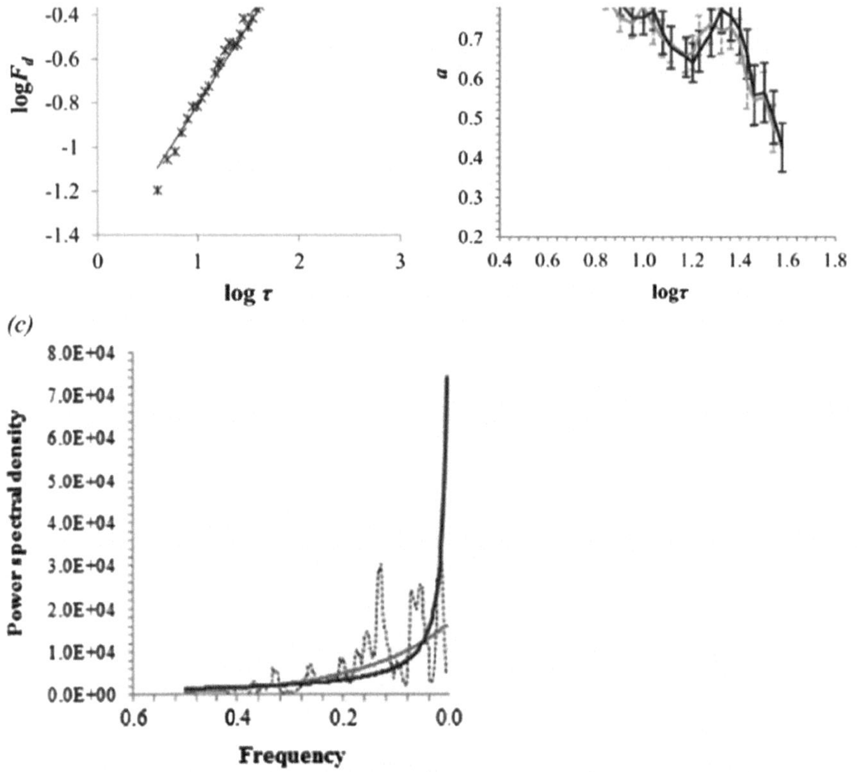

Fig. 22.6 **a** The root-mean-square fluctuation function $F_d(\tau)$ is plotted in a log–log fashion against the time scale τ (in years) for the dataset of the monthly average O_3 VMRs at 35 km that were detrended and deseasonalized between 1980 and 2014. The best-fit equations for each series are y = 0.72x − 1.53, $R^2 = 0.98$. **b** For the same dataset, the individual gradients of the $F_d(\tau)$ against $\log\tau$ estimated inside a segment of 13 points (dashed grey line) and 12 points (solid thin black line). The matching 1.96·sa—segments of the gradients over all scales under consideration are indicated by the grey and black error bars. The power spectral density of the monthly average O_3 VMRs that were detrended and deseasonalized was obtained using the power-law (black line) and exponential (grey line) fits, with $y = 8.0{\cdot}10^2{\cdot}x^{-0.92}$ with $R^2 = 0.53$ and $y = 1.7{\cdot}10^4{\cdot}e^{-6.1\times}$ with $R^2 = 0.64$, respectively (Varotsos et al. 2017)

According to Varotsos and Efstathiou (2017), this straight-line fit alone cannot be used to determine the presence of long-range dependence. It necessitates validating power-law scaling and rejecting exponential decay in the autocorrelation function. Their analysis showed that the existence of long-term correlations in the SF and VF over the past millennium cannot be explained by a DFA exponent of more than 0.5.

Stated differently, empirical investigations predicated on these two requirements ought to be seen as corroborating evidence for the plausibility of the scaling hypothesis rather than as an absolute demonstration of scaling. Varotsos and Efstathiou (2017) also explore the scaling behavior of SF and VF data using the Haar tool, which has proven its reliability in detecting the scaling property in climate datasets.

Table 22.1 The DFA-l a values for the VF from 1000 to 1999 (Varotsos and Efstathiou (2017)

	No detrending	6th order polynomial detrending
DFA-1	$a = 0.73 \pm 0.02$	$a = 0.73 \pm 0.03$
	with $R^2 = 0.96$	with $R^2 = 0.96$
DFA-2	$a = 0.64 \pm 0.03$	$a = 0.62 \pm 0.03$
	with $R^2 = 0.93$	with $R^2 = 0.93$
DFA-3	$a = 0.55 \pm 0.02$	$a = 0.56 \pm 0.02$
	with $R^2 = 0.94$	with $R^2 = 0.94$

Table 22.2 The DFA-l a values of for the SF from 1000 to 1999 (Varotsos and Efstathiou (2017)

	No detrending	6th order polynomial detrending
DFA-1	$a = 0.9 \pm 0.04$	$a = 0.76 \pm 0.03$
	with $R^2 = 0.92$	with $R^2 = 0.93$
DFA-2	$a = 0.76 \pm 0.05$	$a = 0.64 \pm 0.04$
	with $R^2 = 0.86$	with $R^2 = 0.88$
DFA-3	$a = 0.58 \pm 0.04$	$a = 0.55 \pm 0.03$
	with $R^2 = 0.86$	with $R^2 = 0.89$

Without deleting their periodicities, Tables 22.1 and 22.2 present the DFA-l values that were obtained for the original and detrended VF (SF) datasets from 1000 to 1999.

The results show that a straight line fits well in a log–log representation of the fluctuation function, with a gradient larger than 0.5.

Nevertheless, it is clear from looking at the standards put forth by Maraun et al. (2004) that for every examined dataset, neither the stability of individual gradients nor the rejection of ACF exponential decay is satisfied (absence of long-term memory).

Importantly, the Zebiak-Cane model dataset, which is based on the Haar method, was also used to validate all the inherent aspects of the VF and SF datasets that were previously discussed (Lovejoy and Schertzer 2013). To elaborate, it is found that, when the RMS cubic DFA is increased by the proper factor, the RMS Haar and the RMS cubic DFA are almost equal to the Haar fluctuations.

Long-range scaling typically denotes the presence of significant long-range statistical relationships. There is, however, an exception in the case of Gaussian white noise $T(t)$, where $a = 0.5$, $K(q) = 0$, and $H =$ -0.5. However, the discovery of $a \approx 0.5$ in an empirical study simply suggests that $1 + 2H \approx K(2)$, demonstrating the existence of long-range interdependence. In order to confirm their absence, one must show that $a \approx 0.5$, but it is also rare to show that $K(q) = 0$, especially in the geophysical domain.

Different interpretations of the DFA exponent a rely on the quasi-Gaussian approximation. For example, the terms "persistence" and "antipersistence" can be interpreted as variations on "wandering" or "cancelling" behaviors, although they are more concerned with Gaussian white noises than with the average discrepancies that the sign of H indicates. When the variance of a process is growing more quickly than that of Gaussian white noise ($a > 0.5$), it is said to be "persistent" (Gaussian), whereas the variance of a "antipersistent" process increases more slowly ($a < 0.5$). For strongly non-Gaussian processes, this classification might not be particularly helpful, but the sign of H is still quite important.

The capacity of the DFA approach to remove nonstationarities is a topic of significant discussion. The original series' $l - 1$ order polynomials are eliminated by alth order DFA analysis, which is defined using l th order polynomial regressions. This has resemblance to polynomial extensions of fluctuations based on the Haar wavelet as presented in Lovejoy and Schertzer (2012), or to wavelets such as the "Mexican hat" and their higher-order derivatives. It is incorrect to state that it alone eliminates nonstationarities because it eliminates trends of all sizes, not just the largest ones. Multifractal processes, for instance, are generally specified on finite regions with finite outer scales, where they can maintain statistical homogeneity/stationarity regardless of their spectral slope or H value. Strong gradients may be seen throughout the region in each realization, and linear trends may be seen in each subregion, but these are all random features of a statistically stationary/homogeneous process. In polynomial regression, the DFA depends on a strong stationarity requirement for the residuals. It is recommended to treat trends as a "pretreatment" by deleting them over the entire series instead than at intervals, in order to eliminate nonstationarities. Although measures are employed to reduce geophysical nonstationarities such as diurnal and yearly cycles, oscillations are still present.

Anticipate notable oscillations and features outside of quasi-Gaussian processes if the process is multifractal. Treating them as nonstationarities or inhomogeneities can result from misinterpretations. Without making any presumptions, it is impossible to derive statistical stationarity or homogeneity from actual data. These are theoretical characteristics of stochastic processes. Common pretreatments for eliminating nonstationarities include DFA, which estimates long-range correlations impacted by gaps, trends, spikes, and harmonics (Chen et al. 2002). The following succinctly describes the primary conclusions of Varotsos and Efstathiou's (2017) analysis:

1. No obvious long-range correlations were found, despite the fluctuation function of solar data seeming linear in a log–log form. The main reason for this is that, for bigger scales, the local slopes do not converge to a constant value within a wide enough range. As a result, it's critical to use power-law scaling and the analysis of autocorrelation function decay in addition to the DFA method for examining extended memory. Empirical analyses, like the one by Maraun et al. (2004), should be interpreted cautiously nevertheless, as they might not offer concrete evidence of scaling but rather imply the plausibility of the scaling hypothesis.
2. The scaling properties of the Zebiak-Cane model dataset were found to be almost same when the RMS cubic DFA method and the RMS Haar-tool were applied. It is important to remember that the latter method's fluctuations must be corrected by a suitable factor.

The earlier works in this field (Kondratyev and Varotsos 1995, 1997; Cracknell et al. 2009; Krapivin and Varotsos 2008; Varotsos 2013; Varotsos et al. 2014, 2019, 2020, 2023; Varotsos and Ghosh 2017; Varotsos and Efstathiou 2019) contain additional information about the atmospheric greenhouse effect and its components. The phenomenon's previously noted nonlinearity is a development that can be applied to a fresh understanding of the underlying mechanics.

References

Chen Z, Ivanov PC, Hu K, Stanley HE (2002) Effect of nonstationarities on detrended fluctuation analysis. Phys Rev E 65(4):041107

Cracknell AP, Varotsos CA, Krapivin VF, Zakharov VI (2009) Regarding greenhouse explosion. In: Global climatology and ecodynamics: anthropogenic changes to planet earth, pp 107–132

Kondratyev KY, Varotsos C (1997) A review on greenhouse effect and ozone dynamics over greece. In: Atmospheric ozone dynamics: observations in the mediterranean region, pp 175–228

Kondratyev KY, Varotsos C (1995) Atmospheric greenhouse effect in the context of global climate change. Il Nuovo Cimento C 18:123–151

Krapivin VF, Varotsos CA (2008) Modeling the interactive cycles of greenhouse gases and other chemicals. In: Biogeochemical cycles in globalization and sustainable development, pp 213–289

Lovejoy S, Schertzer D (2012) Haar wavelets, fluctuations, and structure functions: convenient choices for geophysics. Nonlinear Proc Geoph 19:513–527

Maraun D, Rust HW, Timmer J (2004) Tempting long-memory– on the interpretation of DFA results. Nonlinear Process Geophys 11:495–503

Varotsos CA (2013) The global signature of the ENSO and SST-like fields. Theoret Appl Climatol 113:197–204

Varotsos CA, Efstathiou MN (2017) On the wrong inference of long-range correlations in climate data; the case of the solar and volcanic forcing over the tropical Pacific. Theor Appl Climatol 128:761–767. https://doi.org/10.1007/s00704-016-1738-5

Varotsos CA, Efstathiou MN (2018) The observational and empirical thermospheric CO_2 and NO power do not exhibit power-law behavior; an indication of their reliability. J Atmos Solar Terr Phys 168:1–7

Varotsos CA, Efstathiou MN (2019) Has global warming already arrived? J Atmos Solar Terr Phys 182:31–38

Varotsos CA, Ghosh S (2017) Impacts of climate warming on atmospheric phase transition mechanisms. Theoret Appl Climatol 130:1111–1122

Varotsos C, Assimakopoulos MN, Efstathiou M (2007) Long-term memory effect in the atmospheric CO_2 concentration at Mauna Loa. Atmos Chem Phys 7(3):629–634. https://doi.org/10.5194/acp-7-629-2007

Varotsos CA, Krapivin VF, Soldatov VY (2014) Modeling the carbon and nitrogen cycles. Front Environ Sci 2:8

Varotsos CA, Efstathiou MN, Cracknell AP (2017) On the temporal evolution of the tropical stratospheric ozone. J Atmos Solar Terr Phys 157:1–5. https://doi.org/10.1016/j.jastp.2018.01.006

Varotsos CA, Efstathiou MN, Christodoulakis J (2019) Abrupt changes in global tropospheric temperature. Atmos Res 217:114–119

Varotsos C, Mazei Y, Efstathiou M (2020) Paleoecological and recent data show a steady temporal evolution of carbon dioxide and temperature. Atmos Pollut Res 11(4):714–722

Varotsos CA, Krapivin VF, Mkrtchyan FA, Xue Y (2023) Global climate change and hydrogeochemistry. Constructive processing of microwave and optical data for hydrogeochemical applications. Springer International Publishing, Cham, pp 399–474

Wiener N (1950) Extrapolation, interpolation and smoothing of stationary time series. MIT Technology Press and John Wiley and Sons, New York

Chapter 23
A New Tool to Study Complex Systems: The Natural Time

23.1 Introduction

The largest problem facing science is time, not space. The representation of traditional time in real numbers is a 1D continuum, but this continuity is not predicated on any core idea. On the other hand, a new time field known as "natural time" (NT) was introduced more than 20 years ago. It is not continuous and its values fall into countable sets. Datasets can reveal hidden dynamic features by analyzing complex systems in NT (e.g. Abe et al. 2005).

This method aids in lowering uncertainty and extracting useful signal data. Reversing NT fluctuations allows one to measure long-range dependence in datasets. We can also study the dynamic evolution of complex systems and pinpoint critical states with the help of "natural time analysis" (NTA).

When the normalized power spectrum $\Pi(\omega)$ is brought in NT, the expression $\Pi(\omega) \approx 1 - \kappa_1 \omega^2$ results from its Taylor expansion at low frequencies, ω ($\omega \rightarrow 0$). When determining the approach to a critical point, the coefficient κ_1, which represents the variance of NT $\kappa_1 = <\chi^2> - <\chi>^2$ is helpful. Moreover, NTA aids in differentiating between the two origins of self-similarity: infinite variance in the process's increments or long-range temporal correlations. Nonetheless, self-similarity can generally originate from both of these sources, and NTA can also detect such instances.

The following sections will discuss several ways that NTA has been applied to the study of a few geophysical complex systems.

G. Golitsyn and C. Varotsos, *The Stochastic Nature of Environmental Phenomena and Processes*, https://doi.org/10.1007/978-3-031-77015-9_23

23.2 Ozone Hole as a Complex System Using the Natural Time Analysis

Varotsos and Tzanis (2012) conducted an analysis of the dataset for the maximum daily O_3 hole area (OHA) over Antarctica during 1979–2009. In order to capture the dynamics of the O_3 hole complex system, this analysis concentrated on the novel time field known as the NT and using the entropy S. The results show that the S under time reversal ($S_{\_}$) for all scales (3–15 years) and in NT for those ranging from 3 to 7 years have stabilized recently. Varotsos (2002, 2003, 2004) first noted the notable event of the major, sudden stratospheric warming and the subsequent split of the Antarctic O_3 hole into two mini-holes in September 2002. However, distinct characteristics of this entropy were observed prior to this event. In particular, the antecedent modifications listed below have been found: First, after about 1999, the S in NT gradually increased for scales larger than 8 years. Second, during 2000–2001, all scales (3–15 years) exhibit an increase in S under time reversal (S_), with the exception of the 13-year scale. Finally, for shorter scales of 3–7 years, the values of the entropy change (ΔS) in NT nearly coincide at 2000 and then decline. The eddy heat flux, which is connected to the vertically propagating wave activity that impacts the O_3 hole over Antarctica, is also analyzed in the NT field. The outcomes validated the inferences drawn from the diagnostics of the OHA.

23.2.1 *The Natural Time Analysis*

Varotsos and Tzanis (2012) used the NTA to analyze the dataset of the maximum daily ozone hole area (MD-OHA) over Antarctica for each year during 1979–2009. The OHA, which is defined as the area south of 40°S with O_3 values below 220 Dobson Units (DU), is derived from measurements made by all O_3 measuring satellites.

Additionally, Varotsos and Tzanis (2012) used the same analysis to examine the eddy heat flux (EHF) annual average values in the belt 45°–75°S on average during 1979–2010 at 10 and 100 hPa.

As a quick review of the NTA process, let's look at a random sample of the MD-OHA events that is shown in Fig. 23.1a. It is noteworthy that this time series may also be representative of other geophysical events, like volcanic eruptions or earthquakes. The steps involved in the NTA process are as follows (Varotsos 2005):

(1) *First Step*

Converting the initial dataset into a new one without taking the event occurrences' chronological order into account.

When N events make up a time series known as MD-OHA, the k-th event's occurrence is indicated by the NT, χ_k. It has the following definition:

Fig. 23.1 A stochastic segment of the MD-OHA event time series **a** in conventional time and **b** in natural time χ_k (evolution of the pair (χ_k, Q_k)). The k-th event's intensity is quantified by Q_k (Varotsos and Tzanis 2012)

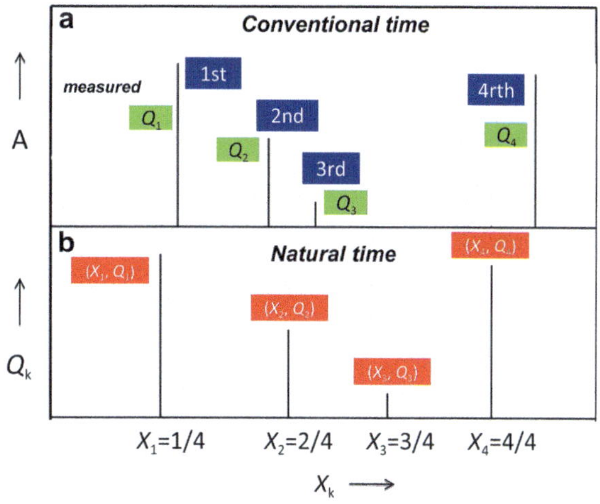

$$\chi_k \equiv \frac{\text{Order of occurrence of an event}(k)}{\text{Total number of the events}(N)}$$

By definition, χ_k progresses with each subsequent k-th event with intensity Q_k, and it ranges from zero to unity.

Figure 23.1b displays a new dataset that represents the development of the pair (χ_k, Q_k), or in simpler terms, the original dataset transformed into the NT field.

(1) *Step 2*

Calculating the S:

Alternatively, alongside Q_k, the quantity p_k can be considered and assessed as:

$$p_k = Q_k / \sum_{n=1}^{N} Q_n$$

Here, p_k represents a normalized intensity of the k-th event, describing the "probability" of observing it. Therefore, the progression of the pair (χ_k, Q_k) in the above-said NT can be equally substituted by the progression of (χ_k, p_k).

Then the mean value of χ, which is denoted by $<\chi>$, can be calculated as:

$$\langle \chi \rangle = \sum_{k=1}^{N} p_k \chi_k;$$

likewise, the mean value of the quantity $f(\chi)$, is denoted by $<f(\chi)>$, and can be determined by:

$$\langle f(\chi)\rangle = \sum_{k=1}^{N} p_k f(\chi_k),$$

and therefore

$$\langle \chi \ln \chi \rangle = \sum_{k=1}^{N} p_k \chi_k \ln \chi_k .$$

As a result, the entropy which by definition is: S ≡ <χlnχ> − <χ> ln <χ> , transforms in the NT field as shown below:

$$S \equiv \sum_{k=1}^{N} p_k \chi_k \ln \chi_k - \left(\sum_{k=1}^{N} p_k \chi_k\right) \ln\left(\sum_{m=1}^{N} p_m \chi_m\right) \quad (23.2)$$

Every time a subsequent event happens, the NT value χ_k advances as previously stated, and p_k likewise changes according to Eq. (23.1); as a result, all the values on the right side of Eq. (23.2), and thus *S*, change.

Going on, we will take the following tack: A length-*i* segment moves across the whole MD-OHA dataset over Antarctica during 1979–2009, one event (year) at a time.

Every time, the entropy *S* is calculated (Fig. 23.2).

Now let's dissect how we calculate *S* for a sequence of successive MD-OHA events. To be clear, if we take a segment of three events (years), then the following 3-year sequences (1998, 1999, 2000), (1999, 2000, 2001), (2000, 2001, 2002) corresponding to S_3 (2000), S_3 (2001), and S_3 (2002), respectively, will have their MD-OHA value of 2000 taken into account when calculating *S*. S_3 evolution year by year will be examined in detail as the *S* value for each of these sequences will essentially be found by analyzing them chronologically.

(2) *Step 3*

Calculating the entropy S_ while considering the time reversal:

In the next step, we will repeat the computation from step 2 using the *S* that we obtained by taking the time reversal into account, which is represented by the symbol S_ (Fig. 23.3). The last event is now interpreted as the first one, the second-to-last event as the second, and so on, according to the definition of the operator *T*′, which is $\mathrm{Tp_k}' = \mathrm{p_{N-k+1}}$.

We wish to emphasize that the entropy *S* employed in the aforementioned analysis is a dynamic entropy that takes the order of events into account, rather than merely a statistical one. Additionally, S_ generally deviates from *S*, pointing to the disruption of time-reversal symmetry and emphasizing the significance of taking the true time arrow into account when dividing similar signals with distinct dynamics. *S* and S_ are both less than the value *Su* of a "uniform" distribution for critical dynamics. We will examine the findings of the NTA of the MD-OHA and EHF datasets in the following sections (Varotsos and Tzanis 2012).

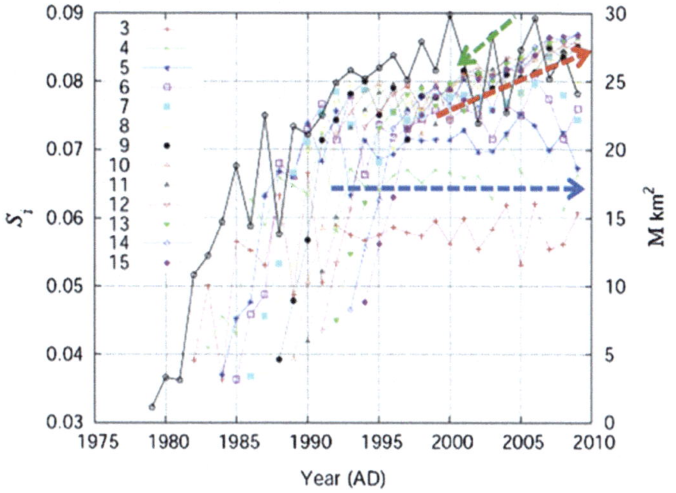

Fig. 23.2 The entropy S_i in NT is shown for different segment lengths i ranging in 3–15 years (on the left scale). The data, measured in Million km^2, is represented by black pentagons. For segment lengths in 3–7 years, the entropy S stabilizes in recent years (indicated by the blue arrow). However, for segment lengths greater than eight years, there has been a slow rise in entropy since 1999 (as shown by the red arrow). At longer scales, during 2000–2001, the entropy is approximately 0.080 (as indicated by the green arrow), which is similar to this obtained by a fBm with a DFA exponent 1. For critical dynamics, the entropy is less than S_u (equal to ln2/2–1/4, approximately 0.0966), indicating a "uniform" distribution with long-range correlations. (Varotsos and Tzanis 2012)

23.2.2 *The Natural Time Analysis in the Study of the Ozone Hole Dynamics*

First, we will look at the results for entropy in natural time S_i over various segment lengths ranging in 3–15 years, moving annually through the complete MD-OHA dataset during 1979–2009 (Fig. 23.2). As we can see, the value of S stabilizes in the last few years for i = 3–7 years, but after 1999, there is a gradual increase for i > 8. Notably, all S_i values are less than the standard deviation (Su) of a "uniform" distribution. This is consistent with the long-range correlations that were covered in the previous section. Furthermore, the S_i values during 2000–2001 are near or vary near the value $S \approx 0.080$ on longer scales.

The S value obtained from a fractional Brownian motion (fBm) with a DFA near unity is comparable to this (see Fig. 23.4 of Varotsos et al. 2006).

Next, we will examine the entropy results in NT under time reversal $(S\_)_i$ for different segment lengths ranging in 3–15 years, with annual shifts over the whole period (1979–2009). These results are shown in Fig. 23.3, where we can observe that the value of $(S\_)_i$ almost completely stabilizes in recent years for all scales i = 3–15 years (with the exception of a transient period observed in 2002–2007). On the other hand, prior to 2002, a significant precursor shift is visible: from 2000 to 2001, $S\_$

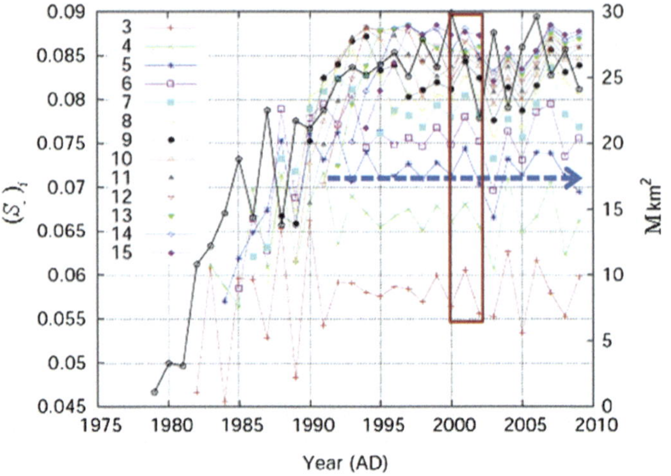

Fig. 23.3 Over the entire MD-OHA time series from 1979 to 2009, the entropy in NT under time reversal $(S\_)i$ for different segment lengths $i = 3$–15 years (left scale) is shown sliding by one year each time. The black pentagons represent the data (in million km^2; right scale); With the exception of a brief period 2002–2007, the $(S\_)i$ nearly stabilizes for $3 \le i \le 15$ years during the last few years (see the blue arrow). The values of $(S_)I$continue to be less than those of (Su) in a "uniform" distribution. Remember that both S and $S\_$ should be less than Su for critical dynamics (The reader is directed to the online version of this article for an interpretation of the color references in this figure legend.) (Tzanis and Varotos 2012)

values rise on all scales, with the exception of $i = 13$. From 2001 to 2002, all $(S\_)_i$ values fall, with the exception of $i = 13$, and these $(S\_)_i$ values continuously stay below that (Su) of a "uniform" distribution. Recall that S and $S_$values for critical dynamics should both be less than Su, as was previously discussed.

Now, we are concentrating on the outcomes of the entropy change $\Delta S_i = S_i - (S\_)_I$ under time reversal for various segment lengths i (3–15 years), sliding through the whole time series (1979–2009) by one year each time. These findings are shown in Fig. 23.4. A close inspection of the figure reveals that, prior to 2002, a scale invariance existed, albeit on very small scales. In particular, we observe that the ΔS_i values nearly coincide at 2000 and then decline for scales i 3–7 years. On the other hand, the ΔSi values rise for longer scales (i.e., i – 10) between 1990 and 2005.

Given that precursory changes can be detected on both shorter and longer scales in this complex system, the 11-year solar cycle seems to be important. For example, we discovered that for scales I 3–7 years, the ΔS_i values coincide in 2000. In contrast, the S_i values exhibit a slow increase after 1999 for scales longer than eight years. Consequently, given the complexity of the system under study, it is crucial to examine S_i, $(S\_)_i$, and ΔS_i at scales close to the 11-year cycle in order to pinpoint precursory effects. It is important to note that the 2000 signal may have been related to the OHA phase transition that occurred in 2001 (the largest OHA) and 2002 (the smaller OHA).

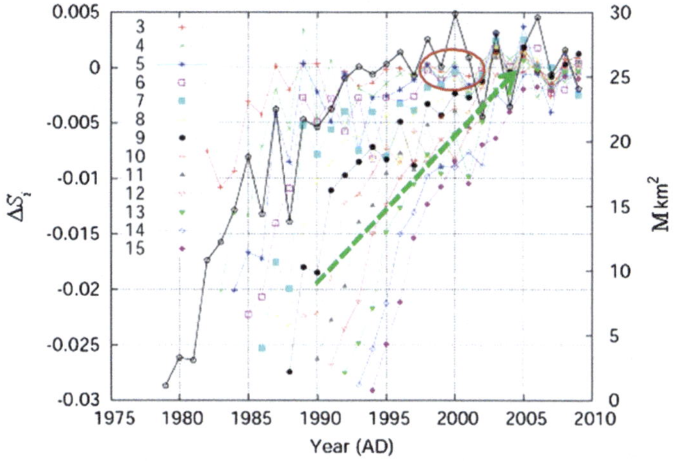

Fig. 23.4 The evolution of S under time reversal in NT. The $\Delta Si = Si - (S\_)_I$ is the representation of the S change. The segment lengths (i) span a period of $3 - 15$ years. The data in million km^2 is represented by the right scale, while the values of ΔSi are represented by the left scale. Black pentagons represent the data points, and linear interpolation was applied for the year 1995. It is important to remember that scale invariance existed prior to 2002, but only on small scales. In particular, the values of ΔSi are nearly the same in 2000 and then decrease (shown by the brown oval) for the scales $i = 3$–7 years. On the other hand, ΔSi rises for longer scales in 1990–2005 (shown by the green arrow) (Varotos and Tzanis 2012)

Lastly, we will talk about the year 1988, when it appeared as though the Antarctic O_3 hole would experience a similar abrupt shift to what was seen in 2002. Nonetheless, a somewhat powerful event occurred in 1988, though it was not a significant one (without a reversal of the zonal average zonal wind at 600). This explains why the 1988 small Antarctic O_3 hole did not rupture. Furthermore, there were differences between the 5–7 year period characteristic features of El Niño in 1988 and 2002. Sadly, no precursory changes, if any, could be found because of the scant observations of the OHA prior to 1988 (only 9 years of data are available). We will now proceed to the next subsection, where we will examine the temporal evolution of a proxy parameter for the dynamics of the O_3 hole, in an effort to further corroborate these findings.

The eddy heat flux, which is directly related to the vertically propagating wave activity that affects the O_3 hole over Antarctica, was also examined in the NT field. The results of the eddy heat flux NTA at 10 and 100 hPa, averaged in the belt 45-75^{0}S, support the conclusions drawn from the analysis of the OHA (Varotsos and Tzanis, 2012).

In conclusion, three initial changes have been found through the examination of the maximum daily OHA dataset. The following are the modifications:

(1) After about 1999, there is a slow rise in S in NT for scales larger than 8 years.
(2) Between 2000 and 2001, theS in NT under time reversal rose for all scales (3–15 years), with the exception of the nearly 13-year scale.

(3) For short time scales of 3–7 years, the ΔS values in NT nearly coincide at 2000 and then decline.

These three facts together point to the system's approach to a dynamic phase transition, or critical point, almost two years before the O_3 hole split over Antarctica (Varotsos 2002, 2005).

We used the same analysis to look into the dynamical evolution of the eddy heat flux in the NT field because we were confident in the relationship between the eddy heat flux and the O_3 hole's longevity. The obtained results demonstrated a noteworthy degree of coherence between the two variables.

References

Abe S, Sarlis NV, Skordas ES, Tanaka HK, Varotsos PA (2005) Origin of the usefulness of the natural-time representation of complex time series. Phys Rev Lett 94:170601

Varotsos C (2002) The southern hemisphere ozone hole split in (2002). Environ Sci Pollut Res 9:375–376

Varotsos C (2003) What is the lesson from the unprecedented event over Antarctica in 2002? Environ Sci Pollut Res 10:80–81

Varotsos C (2004) The extraordinary events of the major, sudden stratospheric warming, the diminutive Antarctic ozone hole, and its split in 2002. Environ Sci Pollut Res 11:405–411

Varotsos C (2005) Power-law correlations in column ozone over Antarctica. Int J Remote Sens 26:3333–3342

Varotsos CA, Tzanis C (2012) A new tool for the study of the ozone hole dynamics over Antarctica. Atmos Environ 47:428–434

Chapter 24
El Niño Southern Oscillation; A New Prediction Tool

24.1 Present Understanding and Questions

The central and eastern tropical Pacific experience vast and erratic warming or cooling when El Niño/La Niña, an oceanic event, occurs. There are important ramifications to this phenomenon, including a discernible change in weather regime throughout the Pacific. Along with variations in the strength of trade winds over the Pacific, these changes include increased convection and cloudiness over the region of the central tropical Pacific. Numerous research works (Cordero et al. 2024) have emphasized the extensive effects of El Niño/La Niña.

The Southern Oscillation, named for the seesaw pattern of the pressure of the air over the surface among the eastern and western parts of the South Pacific, is a crucial link between the atmosphere and the oceanic phenomenon of El Niño/La Niña. The Southern Oscillation Index (SOI), which is a measurement of the intensity of this atmospheric phenomenon, is the name given to this oscillation. The difference in the surface air pressure on a monthly basis between Tahiti (17° 40′ S, 149° 25′ W) and Darwin (12° 27′ S, 130° 50′ E) is used to calculate the SOI.

It is commonly known that El Niño episodes are linked to negative SOI values, whereas La Niña episodes are linked to positive SOI values. This results in the composite oceanic-atmospheric phenomenon known as the El Niño/La Niña Southern Oscillation (ENSO) (Fig. 24.1). Wunsch (1999) proposes that the ocean's reaction to random atmospheric forcing accounts for a part of the El Niño phenomenon. Because ENSOs have climatological effects in areas known as teleconnections that extend beyond the tropical Pacific, forecasting ENSOs is a matter of global concern. Severe weather events like droughts and floods, shifts in the prevalence of epidemiological illnesses like malaria, extreme coral bleaching, civil unrest, and other events can all have an impact.

As a result, climatologists have found ENSO to be a useful climate indicator (Stenseth et al. 2003). For instance, Eastern Australia suffered greatly as a result of the most powerful La Niña ever recorded from 2010 to 2011, which led to the

G. Golitsyn and C. Varotsos, *The Stochastic Nature of Environmental Phenomena and Processes*, https://doi.org/10.1007/978-3-031-77015-9_24

Fig. 24.1 The warm phase of the Southern Oscillation (ENSO) in the tropical Pacific region is known as El Niño. Usually, it starts with the water's temperature along South America's western coast rising significantly

nation's second-wettest year ever (National Climate Centre, Bureau of Meteorology 2012).

The objective in climate prediction is to create methods that produce the most precise forecasts for specific cases, like one location at a given time, while simultaneously attaining the highest level of proficiency over a set of instances. It has been highlighted that because of possible disruptions brought on by climate noise, the best estimate might not match exact observed results in individual cases or show consistent skill across a number of cases. Recent advances have improved the measurement of the climatic effects associated with the ENSO phenomenon by demonstrating that seasonal predictions of the three-month mean temperature at the surface or precipitation can be accurate in specific seasons, locations, and situations. The Seasonal Diagnostics Consortium's efforts (Barnston et al. 2005) have helped dynamical approaches to better understand climate variability and forecast seasonal to decadal timescales become more widely understood.

There is no doubt that there is a limit to the accuracy of ENSO extreme prediction using statistical and hydrodynamic combined atmosphere–ocean models. As a result, geophysics continues to face the challenge of improving ENSO forecasting. Several studies have brought attention to this limitation (e.g., Cheng et al. 2024). The complexity stems from the lack of current knowledge about ENSO, a quasi-periodic interannual variability in global oceanic-atmospheric circulation that is nonstationary and nonlinear.

Although there has been progress in the understanding of ENSO, much remains to be learned.

Furthermore, ENSO happens erratically every 3–7 years or roughly every 4–5 years. This begs the fundamental question of whether there is a lead time for this climate anomaly if it is easy to predict. Numerous coupled models have been used to analyze and forecast ENSO phenomena in order to answer this question. The special issue (vol 103, 1998) of the J Geophys. Res., which concentrated on the Tropical Ocean Global Atmosphere (TOGA) program, reviewed these models in great detail. Interestingly, these models take the equilibrium between atmosphere–ocean for granted, but they ignore many extratropical teleconnections as well as intraseasonal oscillations. In evaluating the predictions of various models, Anderson and Davey (1998) take into account the variability in the features of ENSO severe events, which can have a substantial impact on their effects.

A multivariate linear stochastic differential equation with a nonnormal linear operator appears to be able to describe seasonal variability of the sea surface temperatures (SSTs) in the tropics, including El Niño events, based on a substantial body of evidence. Research like that done by Penland and Sardeshmukh (1995) has proven this. However, resolved nonlinearities might also be significant at finer resolutions, like monthly scales. In spite of this, general circulation tools are now able to forecast the pattern, amplitude, and temporal progress of tropical SST more accurately than linear statistical models thanks to recent developments. Notably, Saha et al. (2006) have demonstrated these models' efficacy. It is crucial to remember that these linear stochastic tools do not disregard nonlinearities; rather, as Ruelle (1991) explains, the stochastic behavior results from rapidly fluctuating, unresolved chaos. It is important to note that there is disagreement over this description of El Niño and that there is still discussion about whether or not it is a nonlinear system with resolved nonlinearities.

El Niño is an irregular occurrence that is occasionally called quasi-periodic, though there isn't much data to back up this assertion. The SOI webpage of the Climate Prediction Center of the National Oceanic and Atmospheric Administration correctly reports that the ENSO cycle normally lasts four years, with historical records indicating variations between 2 and 7 years. Numerous observational studies have provided evidence that the ENSO is a wide-ranging phenomenon with a broad peak period. On the other hand, the ENSO component and its intensity vary significantly over decadal and interdecadal temporal frames. Using a basic delayed oscillator tool, Eccles and Tziperman (2004) demonstrated that in a weakly nonlinear pattern, the period increases with amplitude of the ENSO cycle, whereas in a highly nonlinear regime, a larger amplitude of the ENSO cycle results in a limited period.

An and Wang (2000) discovered that the ENSO period changed from two to four years between 1962 and 1975 to four to six years between 1980 and 1993. This change was highly dependent on both the meridional scale of wind stress and the zonal phase lag between sea surface temperature and wind stress. The North Pacific experienced an interdecadal climate shift at the same time as this ENSO period change. After examining observed data, Lin (2007) proposed that there were four specific ENSO periods: three years before 1910, four to seven years during 1910–1965, three to four years during 1965–1980, and four to five years after 1980. These results are consistent with many earlier studies (e.g., Mann et al. 2005).

Kirtman (1997) ran simulations with limited, wide, and control scenarios and found that, although the control simulation had a 5-year period, the period decreased to roughly 3 years with a narrower anomaly and increased to roughly 9 years when a wider wind stress anomaly was used. The presence of increased meridional mode Rossby waves, which have reduced phase speeds and thereby result in longer delays, could be one explanation for the lengthy periods linked to the wider meridional structure in the wind stress anomaly. The large variations in the forcing of the gravest Rossby mode brought about by the widening and contracting of the wind stress anomaly structure offer yet another possible explanation.

24.2 A New El Niño Prediction Model; The 1982–1983 and 1997–1998 Events

Varotsos et al. (2016a) conducted a study to examine the temporal changes in the ENSO using a new time field known as natural time (NT). The investigation spanned from January 1876 to November 2011 and was detailed in Chap. 23. The results of this exploration revealed that the major ENSO extremes exhibit precursor signals that reach their maximum intensity within a temporal interval of approximately 2 years. This significant result has the potential to increase the precision of forecasting models for extreme ENSO events, thereby facilitating measures to mitigate the detrimental impacts associated with this phenomenon.

In their study, Varotsos et al. (2016a) analyzed the SOI monthly averaged data from January 1876 to November 2011. The SOI calculation in this study followed Troup's formula (Troup 1965), which is expressed as the deviation among the Tahiti (PA Tahiti) pressure anomaly and the Darwin (PA Darwin) pressure anomaly divided by the standard deviation of the deviation in air-pressure between Tahiti and Darwin (SDD):

$$10 \times [PA(Tahiti) - PA(Darwin)]/SDD \tag{24.1}$$

According to the definition of a pressure anomaly, it is the monthly average sea level (air) pressure difference between Tahiti (P Tahiti) and Darwin (P Darwin) from the long-term mean of the pressure difference for each month (relatively the period from 1887 to 1989). It is significant to remember that the normalized average sea level pressure difference between Tahiti and Darwin is the source of Troup's monthly SOI data starting in 1876.

The oscillation in surface air pressure among the eastern and western parts of the South Pacific is reflected in the SOI, as was previously mentioned. The largest region in the world with warm water temperatures (above 28 °C) is the western South Pacific, which is also thought to be a major cause of air warming that drives large-scale convective circulation processes. As a result, in the exchange of energy between the combined ocean–atmosphere systems, the SOI is essential.

24.2.1 Application of the Natural Time Analysis to the Southern Oscillation Index

The natural time analysis (NTA) is used in the subsequent steps to analyze the SOI dataset that was discussed in the preceding section. First, the initial SOI dataset of N events, designated Q_k (k = 1, 2,... N) $\mathrm{Q_k}$, is converted into a new dataset by keeping the temporal order of extremes intact but ignoring their incidence times. In order to do this, the ratio (χ_k) among the number of SOI values (N) and the order in which they occur (k) is calculated for each SOI value (Q_k) as follows:

$$\chi_k \equiv \frac{\text{Order of occurrence of a maximum}}{\text{Total number of the maxima}} = \frac{k}{N} \tag{24.2}$$

This process introduces a new series consisting of pairs (Q_k, χ_k). The new parameter (χ_k), referred to as "natural time," replaces the conventional time (t) and ranges from zero to unity. Unlike the continuous conventional time, natural time (NT) is not continuous. It serves as a marker of the occurrence of the kth maximum, increasing in intensity (Q_k) with each subsequent kth maximum. Furthermore, since Q_k in NT denotes an intensity (i.e., a positive value), we take $Q_k = \mathrm{SOI}_k + \min(\mathrm{SOI})$, where min(SOI) is the lowest value that the SOI achieved during the period of the study.

To provide a concise explanation of the NTA methodology, let us consider a segment of the SOI maxima dataset (illustrated in Fig. 23.1a of Chap. 23). The transformed dataset, which represents the progression of the (Q_k, χ_k) pair or, in simpler terms, the initial dataset converted into the NT field, is displayed in Fig. 23.1b (Chap. 23).

In the scenario of a "uniform" pattern (such as when the system is in a steady condition and releases energy in the form of uncorrelated pulses), the Q_k takes positive unbiased and stochastic values. As the number of observations (N) approaches infinity, the probability value p(χ) of the NT variable χ tends towards $p(\chi) = 1$, producing the following average NT value:

$$\langle \chi \rangle = \int_0^1 \chi p(\chi) d\chi = \frac{1}{2} \tag{24.3}$$

Furthermore, as stated by Varotsos et al. (2005), the following formula is used to determine the entropy (S) of SOI values in the NT field is $S \equiv \langle \chi \ln \chi \rangle - \langle \chi \rangle \ln \langle \chi \rangle$, where:

$$\langle \chi \rangle = \sum_{k=1}^{N} p_k \chi_k, \langle \chi \ln \chi \rangle = \sum_{k=1}^{N} p_k \chi_k \ln \chi_k$$

$$p_k = Q_k / \sum_{n=1}^{N} Q_n$$

and p_k represents the normalized intensity of the kth maximum. It is possible to think of p_k as a probability because it is less than unity and the sum of all p_k values equals unity, p_k. Hence, the entropy (S) is given by the equation:

$$S = \sum_{k=1}^{N} p_k \chi_k \ln \chi_k - \left(\sum_{k=1}^{N} p_k \chi_k \right) \ln \left(\sum_{m=1}^{N} p_m \chi_m \right) \tag{24.4}$$

This entropy calculation is performed for a moving segment of SOI maxima (monthly averages) with a size (i) that increases by 1 month each time, covering the entire dataset of the SOI. The values are computed using a moving segment with length three months ($i = 3$) at first, and then it is repeated using segments that are longer up to 7 years (84 months) ($i = 3, 4, \ldots, 84$). The maximum ENSO period of seven years is closely aligned with the upper limit of 84 months.

Furthermore, identical entropy computations are made, as described previously, but this time taking into account time reversal. For instance, in reference to Fig. 23.1, the last maximum is now considered to be the first, the second-to-last maximum as the second, and so on. It was found that this entropy, represented as (S –), generally differs from S. As a result, the entropy in NT highlights the breaking of time reversal symmetry and emphasizes how important it is to take the (true) time direction into account when differentiating between signals that appear to be similar but have different dynamics. It is important to emphasize that both stochastic and deterministic processes can be understood in terms of entropy in NT. Last but not least, a moving segment i (which represents the number of consecutive maxima, or the number of months in the sliding segment) is used to calculate the discrepancy $\Delta S_i = S_i - (S -)_i$ for each (S) and (S –).

For example, Fig. 24.2 illustrates the change in entropy over NT with time reversal ΔS_i for a segment length of $i = 36$ months (red, left scale) moving by 1 month each time throughout the entire dataset (spanning from January 1876 to November 2011) of SOI (blue, scale on the right). A shorter subset of the dataset that shows the relationship between $\Delta\ S_i$ (e.g. ΔS_{20}) and SOI is demonstrated in the following.

Physically speaking, it was shown that a negative ΔS and likewise are correlated with an increasing trend in the dataset. The physical significance of ΔS is further explained by looking at tiny increasing and decreasing trends to clarify its definition. In this case, the expression for $p(\chi;\varepsilon)$ is as follows, assuming that ε is the trend parameter (that is, $\varepsilon > 0$ denotes a rising trend and $\varepsilon < 0$ denotes a falling trend):

$$p(\chi; \varepsilon) = 1 + \varepsilon(\chi - 1/2) \tag{24.5}$$

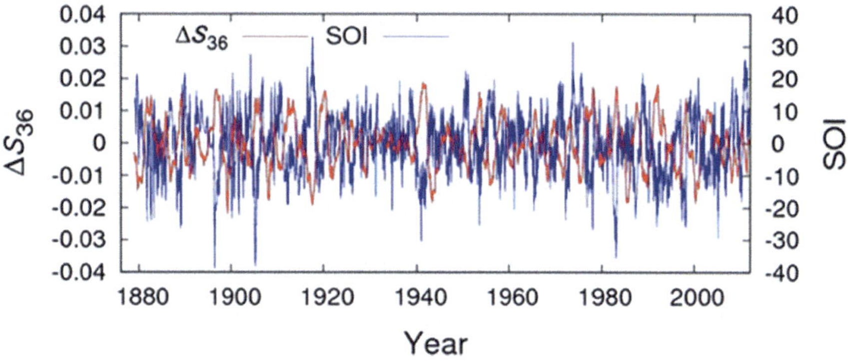

Fig. 24.2 The temporal evolution of the entropy change ΔS_{36} over 36 months (in red) on the left scale, alongside the SOI values (in blue) on the right scale. To facilitate comparison, the ΔS_{36} values are shifted by one month, allowing for a direct comparison with the corresponding SOI values on the same abscissa. This indicates that a window one month ahead of the SOI value is used to calculate the ΔS_{36} value (Varotsos et al. 2016a)

where $p(\chi;\varepsilon)$ denotes the point probabilities discussed earlier. Subsequently, the analytical calculation of entropy is determined as follows:

$$S(\varepsilon) = -\frac{1}{4} + \frac{\varepsilon}{72} - \left(\frac{1}{2} + \frac{\varepsilon}{12}\right)\ln\left(\frac{1}{2} + \frac{\varepsilon}{12}\right) \tag{24.6}$$

It is clear from Eq. (24.5) that the entropy in NT under time reversal $S-$ is obtained by changing the sign of ε in Eq. (24.5). Hence, $S(\varepsilon) - S - (\varepsilon) = S(\varepsilon) - S(-\varepsilon)$ is equal to $\Delta S(\varepsilon)$. $\Delta S(\varepsilon)$ as a function of the slight trend parameter ε (i.e., $\Delta S(\varepsilon)$ where $-1/2 \leq \varepsilon \leq 1/2$) is depicted in Fig. 24.3, demonstrating that a negative (positive) $\Delta S(\varepsilon)$ trend corresponds to a rising (declining) trend.

Fig. 24.3 The change in entropy, $\Delta S(\varepsilon)$, expressed against the slight trend parameter ε (Varotsos et al. 2016a)

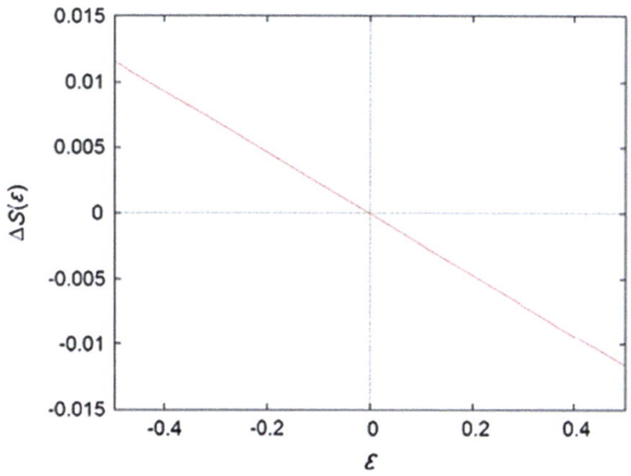

It is important to highlight that NT has been utilized in analyzing intermittent dataset such as earthquakes, where specific maxima can be simply spotted. Moreover, NT has also been employed in datasets, where the values of the dataset itself (after adjusting for a suitable constant to ensure positivity) constitute the quantity Q_k; in this scenario, the average values per month form the dataset.

The experimental robustness or stability of S and $S-$ has been proven (Varotsos et al. 2005). The Lesche stability criterion, which was first proposed by Lesche, states that an entropic measure, like the one used in this demonstration, is stable if its variation due to a small perturbation in the distribution p_k (which represents fluctuations in experimental data) remains minimal. This was the basis for the demonstration.

24.2.2 Results of the Natural Time Analysis of the Southern Oscillation Index

In the NT field, the change in the entropy is computed at different segment lengths, ranging from three to eighty-four months. Utilizing the dynamics of ΔS_i for the purpose of forecasting future SOI maxima, the receiver operating characteristics (ROC) tool has been employed to assess the efficacy of various prediction algorithms, as demonstrated by Sarlis et al. (2011). Figure 24.4 shows the ROC graph that was produced by using ΔS_i as a binary predictor for SOI values in the following month with various values of i (e.g., i = 6, 12, 20, 24, 48).

In essence, the windows utilized are one month ahead of the SOI, and this value is transformed into a binary value according to whether it exceeds a particular threshold

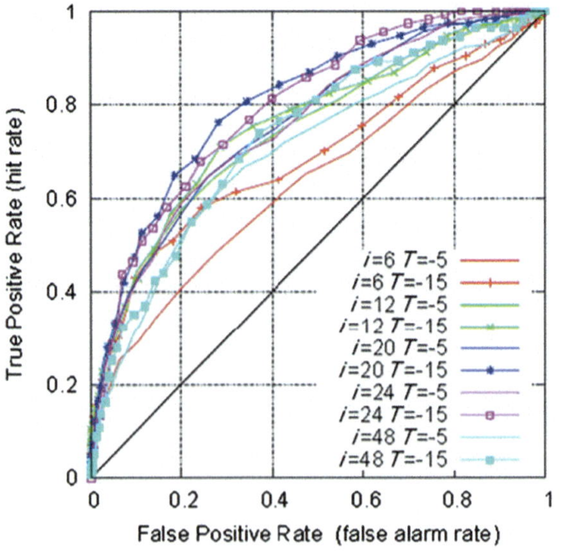

Fig. 24.4 The hit rate compared to the false alarm rate considering varying ΔS_i and maintaining a steady target value T (Varotsos et al. 2016a)

T. A time increased probability (TIP) is triggered when ΔS_i is greater than or equal to a threshold ΔS_T. The prediction is deemed accurate if the monthly mean SOI value in the subsequent month is less than or equal to a target value T. The false alarm rate is the ratio of false alarms to the total number of cases where SOI exceeded T. The ROC curve plots the true positive rate (hit rate) against the false positive rate (false alarm rate). The hit rate is the percentage of correctly predicted large negative monthly mean SOI using data from prior months to calculate ΔS_i.

We can observe different curves for $T = -5$ (weak El Niño maximum) or $T = -15$ (strong El Niño maximum) as shown in Fig. 24.4 by varying the change in ΔS while maintaining the target value T constant. For example, Table 24.1 shows the outcomes of predicting maxima with $T = -15$ by using ΔS_{20} as a predictor and a threshold of $\Delta S_T = 0.00326$. It is important to remember that the black line on the ROC graph represents the diagonal, which would be produced by a random predictor (Table 24.1).

Based on the findings obtained using the ROC analysis, it is obvious that the highest skill or best-hit rate is achieved when considering ΔS_{20} to ΔS_{24} (approximately 24 months before) as predictors based on SOI data. In this context, Fig. 24.5 illustrates the change in entropy in NT under time reversal ΔS_i for a segment size of $i = 20$ months. The red curve represents the scale on the left, while the blue curve represents the scale on the right, corresponding to the entire time series of SOI from January 1876 to November 2011.

Next, we attempt to forecast the maximum value of SOI using the proposed NTA technique, taking into account that the reliability of SOI data before 1935 was questioned by Trenberth and Hoar in 1997. We focus on predicting the most intense El Niño episode of the previous century, which occurred in 1982–1983 and exhibited a distinct pattern in terms of warming and onset time.

The SOI value (blue, scale on the right) and ΔS_{20} (red, scale on the left) are shown in Fig. 24.6. The time increased probability (TIP) activation is shown by the black line. The TIP is activated for the following month when ΔS_{20} exceeds the value ΔS_T, which corresponds to a false positive rate (false alarm rate) of 50% for $T = -5$.

The outcomes derived from NTA analysis of the intense ENSO maximum in 1997–1998 are illustrated in Fig. 24.7. The progression of ΔS_{20} depicted in this figure indicates that the occurrence of a more intense ENSO maximum could be anticipated in advance through the utilization of the NTA tool. It is important to highlight that comparable successful outcomes, as shown in Figs. 24.6 and 24.7, were achieved when testing the NTA tool for predicting the more intense El Niño maximum in 2009–2010. There were comparable outcomes for segment sozes $20 < i \leq 24$ as well.

The quasi-biennial oscillation (QBO) in the tropical stratosphere's zonal wind is a plausible mechanism that could impact the roughly two-year time window. The average meridional circulation is driven by this phenomenon, which causes warm or cold anomalies to arise during the declining zonal average westerly or easterly shear. In essence, the upper tropical troposphere's deep convection is modulated by the ENSO and QBO.

Table 24.1 Results obtained when using ΔS_{20} as a predictor with a threshold $\Delta S_T = 0.00326$ to predict maxima with $T = -15$ provide an example of the binary prediction method that has been proposed

	Positive predictor(Number of cases $\Delta S_{20} \geq \Delta S_T$)	Negative predictor(Number of cases $\Delta S_{20} \geq \Delta S_T$)	Totals
Strong El Niño maximum (number of cases SOI ≤ -15)	87 (correct positive prediction)	27 (false prediction)	114
Not Strong El Niño maximum (number of cases SOI > -15)	422 (false prediction)	1074 (correct negative prediction)	1496
Totals	509	1101	1610

Roughly speaking, the ratio of accurate predictions to incorrect ones is 2.6, or (87 + 1074)/(27 + 422) (Varotsos et al. 2016a)

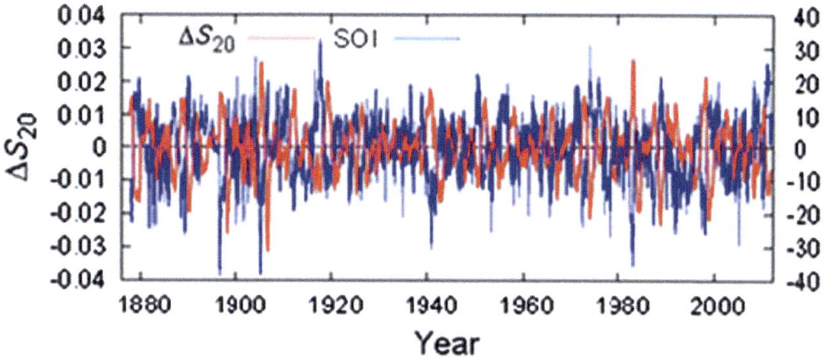

Fig. 24.5 The time march of the entropy change ΔS_{20} in NT for the segment size $i = 20$ months (red, scale on left) along with SOI (blue, scale on right) (Varotsos et al. 2016a)

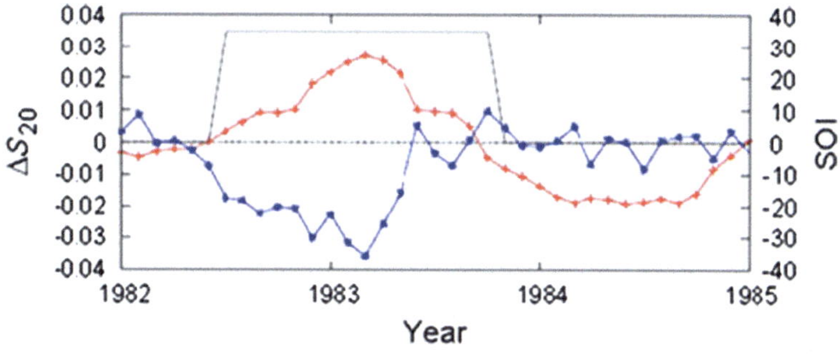

Fig. 24.6 The time rising probability is placed on 1982–1983, the major ENSO maximum of the twentieth century, when ΔS20 (red line) surpasses the threshold ΔST value (Varotsos et al. 2016a)

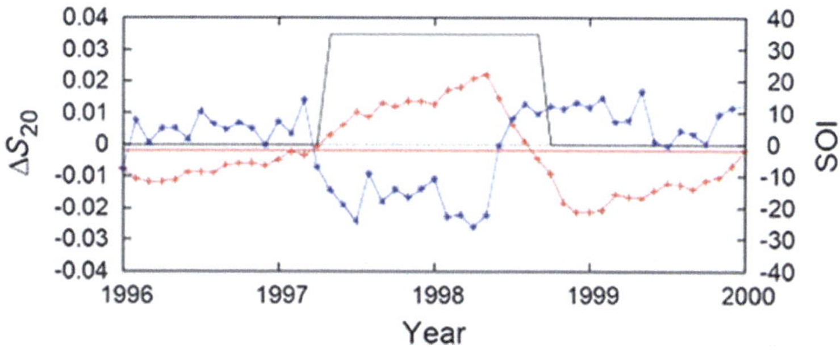

Fig. 24.7 As the NTA results for correctly predicting the 1997–1998 ENSO peak, one of the strongest events of the twentieth century, are shown in Fig. 24.6. The horizontal red line denotes the threshold ΔS_T for a 50% false positive rate. Table 24.1 shows that 422 out of 1496 cases, or about 28% of cases, are false positives. Thus, in order to attain a 50% false positive rate, it is necessary to reduce the value of ΔS_T until 748 false positive predictions are made (Varotsos et al. 2016a)

Further exploration of this distinct relationship is the primary focus of a forthcoming publication. It is worth noting that the utilization of different segment sizes is reminiscent of similar approaches employed in existing literature, such as the "optimal climate normal" method proposed by Huang et al. (1996). However, the method presented here differs significantly in terms of the timescale utilized and the overall approach mentioned and implemented above.

As previously clarified, the interpretation of a rising trend in a dataset is represented by a negative change in entropy (ΔS), while a decreasing trend corresponds to a positive change in entropy. Therefore, the predictive power of ΔS_{20} for the next month's value can be seen as an indication of the persistence of the SOI dataset. However, ΔS also captures intrinsic dynamics.

Figure 24.8 illustrates this by comparing three predictors for the value of SOI after a 24-month lag (SOI($k + 24$)). The negative correlation between SOI(k) and SOI($k + 24$) is made clear in this example. The value of $-$ SOI(k) (represented by magenta and cyan lines with points, where $T = -5$ and $T = -15$) and $-\Delta S_{40}(k)$, which are represented by red and blue lines with points where $T = -5$ and $T = -15$, and $-\Delta S_{48}(k)$, which are represented by red and blue lines without points, are the predictors taken into consideration.

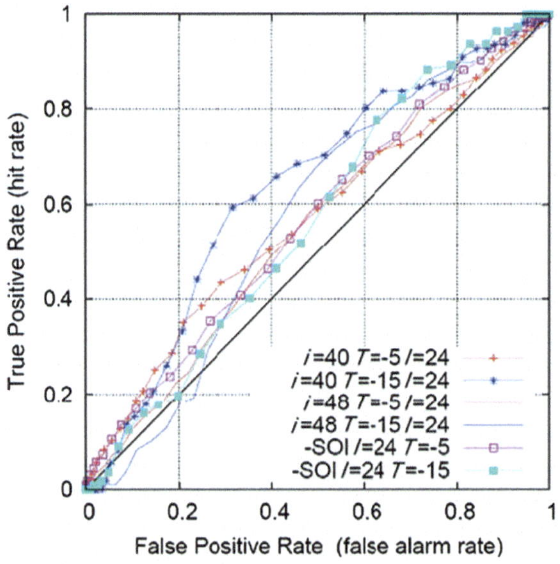

Fig. 24.8 Comparison of 3 predictors for the SOI future value, or SOI($k + 24$), following a 24 month lag, denoted as is presented in this investigation. The three predictors under consideration are $-\Delta S_{40}(k)$ (represented by red lines with points for $T = -5$ and blue lines with points for $T = -15$), $-\Delta S_{48}(k)$ (represented by red lines without points for $T = -5$ and blue lines without points for $T = -15$), and the value of $-$ SOI(k) (represented by magenta lines with points for $T = -5$ and cyan lines with points for $T = -15$). The study also includes the depiction of Receiver Operating Characteristics (ROCs) associated with these predictors (Varotsos et al. 2016a)

$-\Delta S_{40}(k)$ is found to perform better than $-$ SOI(k) in terms of false positive rates, which range from 0.1 to 0.9. This outcome validates that the suggested approach employing ΔS permits sophisticated forecasting, beyond its prior accomplishments.

The idea of a critical point provides another intuitive way to interpret the previously mentioned results. Based on the number of consecutive maxima within the NT segment size used in the computation, we can think of S as an indicator of "disorder." When we get close to a phase change or critical point, the difference ΔS between the disorder in the near future (S) and the disorder in the earlier past (S −) becomes much more significant than the difference in ordinary circumstances.

This phenomenon is a direct consequence of the maximum amplitude of ΔS. It is crucial to note the successive series of these maxima is important to observe when a dynamic system goes through a phase transition. To illustrate this point, if we reverse the order, the system will transition from the new phase back to the old one. To further support this argument, we conducted a ROC analysis using S or $S-$ as predictors. The findings indicated that ΔS performed better than both of these predictors.

In conclusion, the characteristics of the change under time reversal of the NT entropy explain the skill discovered during this analysis. The differentiated pathways that add to the SOI's value are quantified by this entropy.

24.2.3 The Precursory El Niño Signals

In the realm of weather and climate literature, new forecast methodologies undergo rigorous testing before being introduced to the public. This testing entails evaluating the forecast expertise with that of current methods in real-time over the course of months or even years.

However, when it comes to long-term variability, such as phenomena like El Niño which exhibit extreme values over extended periods, testing a forecasting method in live time urges for decades of observational data. Rather than holding off until the required long-term dataset is collected, it is more practical to assess the forecasting tool using existing time series of observations.

Our primary objective is to determine whether major ENSO peaks serve as precursory signals. The results of our analysis show that these peaks do produce early SOI warnings, which peak after about two years. Based on recorded SOI values during ENSO maxima, these signals are used to forecast if an upcoming SOI peak surpasses a specific target value. This prediction is binary, signaling either an impending event or not, without specifying a precise value for the upcoming SOI peak.

This approach sets our method apart from traditional statistical methods, which typically offer expected monthly values for the SOI. While a direct comparison with existing tools is not feasible, an indirect comparison can be made by analyzing ROC curves, concentrating especially on the region under the curve (A). A value of A greater than 0.5 indicates a more effective predictor, as highlighted in previous research by Mason and Graham (2002).

Focusing on the most robust maxima, specifically T = − 15, the subsequent outcomes are as follows: For the 24-month delay (refer to Fig. 24.8), It may be estimated that the predictor $-\Delta S_{40}(k)$ has an area A beneath the ROC curve of 0.63, while the predictor − SOI(k) has an area A of 0.57. As a result, $-\Delta S_{40}(k)$ predictor is greater than − SOI(k). Considering the SOI value for the next month, the value of A for $\Delta S_{20}(k)$ is 0.80, indicating a strong forecast quality based on the data presented in Fig. 24.4.

Therefore, a forecasting tool can be envisioned as follows: The most efficient $\Delta S_i(k)$ predictor can be initially found based on the forecasting time lag (which, as previously shown, can be selected to exceed − SOI(k)). and subsequently, by amalgamating the respective outcomes, future predictors can be estimated for various timeframes reaching forward several seasons. Nonetheless, such a prediction scheme exceeds the scope of the current effort and will be explored in a future publication alongside an extensive comparison of our approach with the existing ENSO predicting models and techniques.

24.2.4 Conclusions

The examination of the dataset of SOI using entropy in the NT field enables the identification of certain features of the complicated atmosphere–ocean system's dynamics that could be utilized for detecting precursor signals of significant ENSO peaks. Varotsos et al. (2016a) demonstrated that the computation of entropy change when the SOI is reversed over the period from January 1876 to November 2011, with varying window lengths ranging from three to eighty-four months, provides the most robust precursor signal when the segment length is approximately two years.

This implies that calculating the SOI entropy change during the preceding 2 years allows for a warning of ENSO maximum occurrence one month in advance. By investigating the features of entropy change in the NT field of the SOI, we have identified strong precursor signals for the major ENSO peaks in 1982–1983 and 1997–1998. Based on these findings, particularly the results of the ROC analysis, we propose that combining our suggested method with existing forecasting tools may enhance the precision of nowcasting ENSO extreme events.

24.3 The Prediction of the 2015–2016 El Niño Event

Varotsos et al. (2016b) conducted a prediction of the strong El Niño event that occurred during 2015–2016. This was done by employing the Natural Time Analysis (NTA) methodology, presented in Sect. 24.2. Thus Figs. 24.9 and 24.10, illustrate the monthly values of the SOI and the entropy change in NT under time reversal (S_{20}) for a segment size of 20 months.

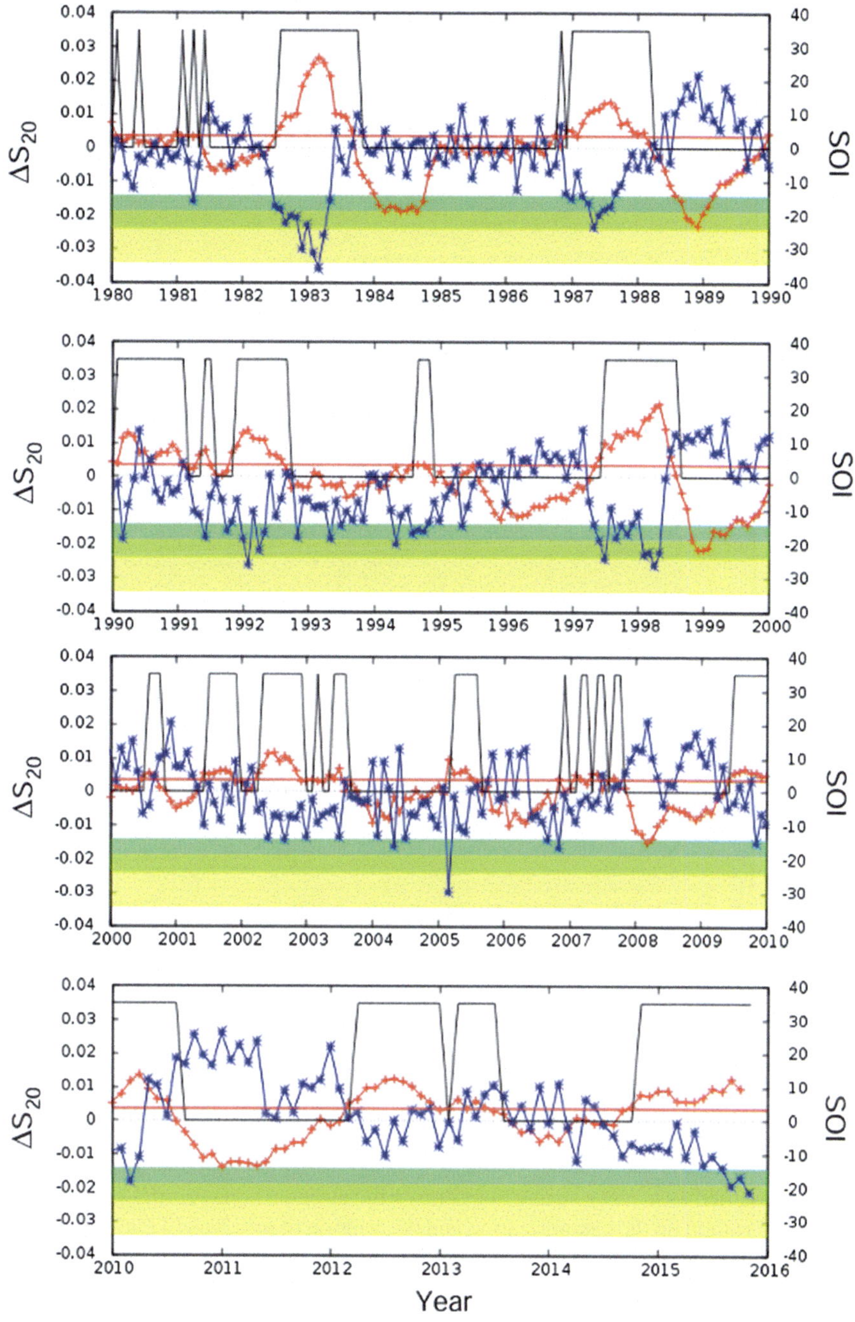

Fig. 24.9 The ΔS_{20} in NT for the segment size $i = 20$ months (red line, left scale) with SOI monthly values (blue line, right scale) from January 1980 to October 2015. The alarm is activated on (black line) when S_{20} is greater than the limit (threshold value) $S_{thres} = 00.035$ (Varotsos et al. 2016b)

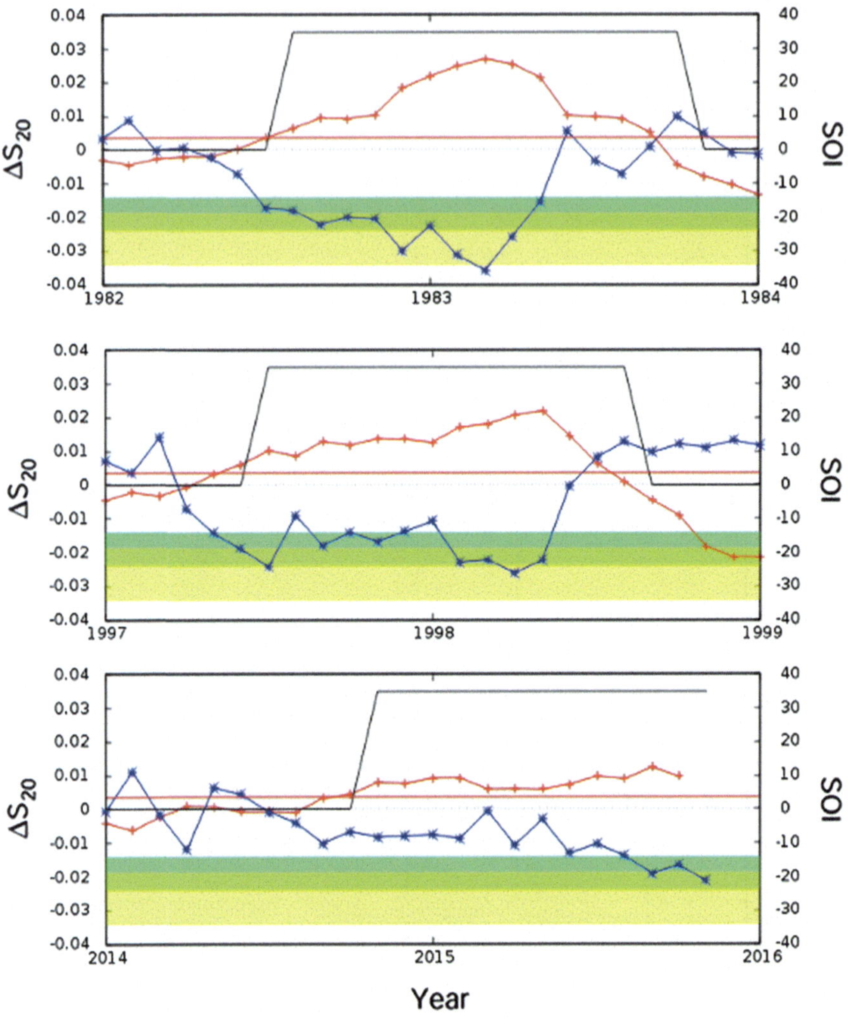

Fig. 24.10 Similar to Fig. 24.9, but only for the El Niño occurrences of 1982–1983, 1997–1998 (the two strongest in the past century), and 2015–2016

To further assess the intensity of the 2015–2016 El Niño event and determine if it can be classified as a "very strong" or "one of the strongest on record," a number of scientists brought up the categorization and description of previous El Niño episodes. The colored areas in Figs. 24.9 and 24.10 correspond to the mean minimum negative values of SOI for the categories of "weak, weak to moderate, moderate, moderate to strong" (green band) and "strong, very strong" (yellow band) El Niño occurrences. These values also include the 1 standard deviation band.

The SOI values for the previous three months fall inside the green band and at the boundaries of the yellow band, as Fig. 24.10 makes evident. It would be more accurate to describe the 2015 El Niño event as "moderate to strong" or perhaps "strong" than as "one of the strongest on record." This conclusion is further supported by comparing it with the El Niño events of 1982–1983 and 1997–1998. Additionally, the variation of S_{20} during the 2015 El Niño event, compared to the events of 1982–1983 and 1997–1998, is not as pronounced. Hence Varotsos et al. (2016b) concluded that this confirms that the 2015–2016 El Niño event was not "one of the strongest on record." As other models predicted. To quantify this variation, Varotsos et al. (2016b) calculated the probability density function (PDF) of S_{20}, obtained from the estimator:

$$f_N(\Delta S_{20}) = \frac{1}{N}\sum_{i=1}^{N}\frac{1}{b_N}K\left(\frac{\Delta S_{20} - O_i}{b_N}\right),$$

Figure 24.11 displays this estimator with a black curve. The values of S_{20} observed since the commencement of the study are denoted by Oi, where i represents the individual observations. The total number of observations is represented by N. The kernel K (x) is only non-zero for$|x| < 1$, with a value of $K(x) = 3/4(1 - x^2)$. The relationship between b_N and the standard deviation of the observed S_{20} values is given by $b_N = 10.25\ \sigma/\mathrm{N}^{0.34}$, as proposed by Mercik et al. (1999). Analysis of Fig. 24.11 reveals that S_{20} is rarely greater than the value of 0.02, as evidenced by the red histogram generated using the TISEAN package and displayed in Fig. 24.11.

The histogram displays the bar's lowest non-zero height, which corresponds to S_{20} = 0.002, and it spans the range up to roughly 0.0205. Blue crosses on the right axis of

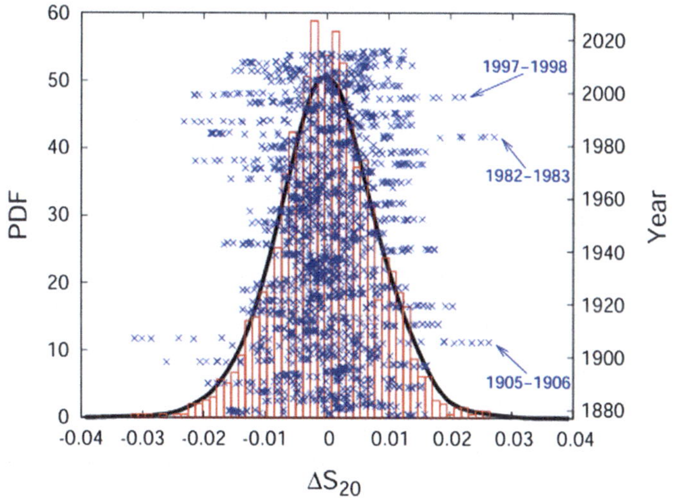

Fig. 24.11 The PDF of S_{20} (black curve, left scale) and the matching histogram (red bars, left scale) derived from the S_{20} dataset, which is also illustrated along the vertical axis against time (blue crosses, right scale). The arrows, which are termed with the corresponding El Niño events from 2014 to 2016, show when S_{20} is greater than 0.0205

Fig. 24.11 represent a plot of the dataset of S_{20} against time, which is used to identify the instances where S_{20} exceeds this threshold. The plot (blue arrows in Fig. 24.11) makes it clear that $S_{20} > 0.0205$ is only seen during the three major El Niño events that occurred in 1905–1906, 1982–1983, and 1997–1998. But in the current scenario (2015–2016 El Niño), this imbalance is not met because the observed values are significantly lower than 0.0205, close to 0.01.

In summary, rather than being referred to as "one of the strongest on record," the El Niño event of 2015–2016 should be categorized as "moderate to strong" or even "strong."

24.4 Forecasting the 2023–2024 El Niño Event

Varotsos et al. (2024) have released a feature article predicting the El Niño event for 2023–2024. The colored areas in Fig. 24.12 represent the mean minimum negative SOI value.

The 1σ standard deviation bands for two categories of El Niño events—"strong, very strong" (yellow band) and "weak, weak to moderate, moderate, moderate to strong" (green band)—accompany these areas. Figure 24.12 shows that the monthly SOI events that occurred during 2015–2016 are located close to the yellow limit and inside the green zone.

Varotsos et al. (2024) set out to further investigate claims of a very strong El Niño occurring in 2023–2024 based on their prior experience in forecasting major El Niño events. The monthly SOI events from January 2021 to May 2023, as shown in Fig. 24.12, do, however, continuously stay above the green zone following an upward trend. There was a sharp drop in June 2023 without any sign of a strong El Niño. Moreover, the changes in ΔS during the El Niño event of 2023–2024 are not as noticeable as the previous variations during the El Niño events of 1982–1983 and 1997–1998. Only in May and June 2023 is an alarm (i.e., $S \geq \Delta S_{\text{thresh}}$) detected.

Varotsos et al. (2024) presented the ΔS histogram and the probability density function (PDF) (Fig. 24.13) obtained using the kernel estimation tool outlined in Sect. 24.3 (Fig. 24.11) in order to evaluate the degree of ΔS variation associated with SOI variation.

Varotsos et al. (2024) summarized the NTA model and found that the El Niño event of 2015–2016 had been classified as a "moderate to strong" event, contradicting the claims made by several models that it would be one of the strongest on record. It should be noted that this method has already been applied in other climatic components (e.g., Varotsos 2020a, b).

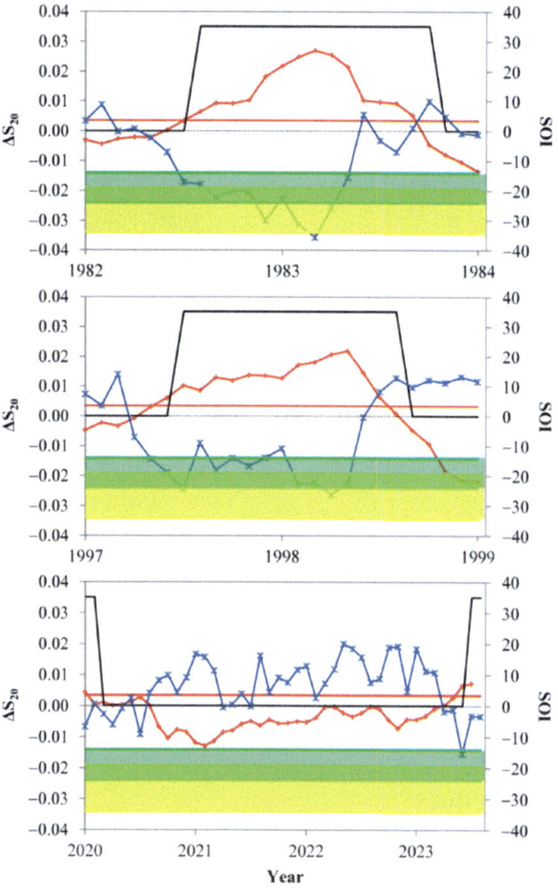

Fig. 24.12 Monthly SOI events showcasing a blue line on the right scale, plotted against time. The change in entropy (ΔS_{20}) is illustrated in a time-reversed scheme with a red line on the left scale, covering during January 2010–July 2023. The graph includes the mean minimum negative SOI values, complemented by 1σ standard deviation zones indicating the intensity of El Niño events, ranging from "weak" to "very strong" as represented by green and yellow bands, respectively. An alarm is triggered when the entropy change surpasses the cutoff point of $\Delta S_{\text{thresh}} = 0.0035$, marked by a red straight line. It is important to note that the SOI values have been shifted by one month to facilitate a more straightforward comparison

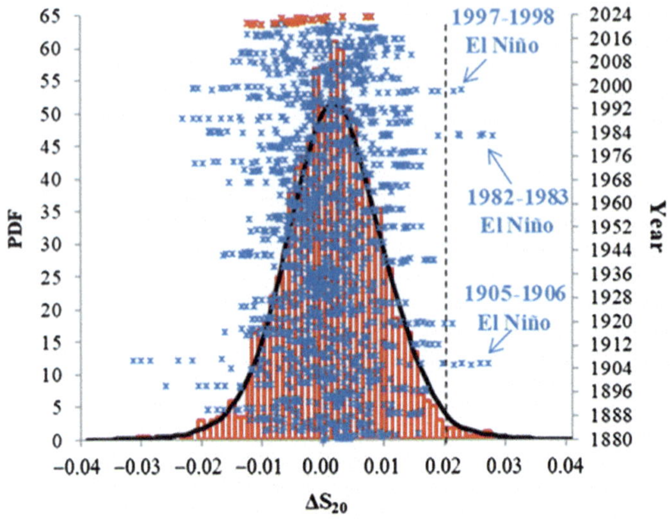

Fig. 24.13 The black curve on the left scale represents the PDF of ΔS, and the red bars on the same scale represent the relevant histogram. The ΔS dataset, spanning during January 1880–July 2023, provided these values. The blue points on the right scale show how ΔS_20 changes over time along the vertical axis. The vertical dashed black line represents the entire dataset; blue arrows indicate the ΔS values that exceed the 99% percentile, $p_{99\%} = 0.02$ of the total dataset. Strong El Niño events are linked to these extreme values. During January 2021–July 2023, the ΔS values are represented by the red points and label (Varotsos et al. 2024)

References

Anderson DL, Davey MK (1998) Predicting the El Niño of 1997/98. Weather 53(9):303–310

An S-I, Wang B (2000) Interdecadal change of the structure of the ENSO mode and its impact on the ENSO frequency. J Clim 13:2044–2055

Barnston AG, Kumar A, Goddard L, Hoerling MP (2005) Improving seasonal prediction practices through attribution of climate variability. Bull Am Meteorol Soc 86:59–72

National Climate Centre, Bureau of Meteorology (2012) Australia's wettest two-year period on record; 2010–2011. Special Climate Statement 38, Melbourne, Australia

Cheng L, Abraham J, Trenberth KE, Boyer T, Mann ME, Zhu J, Wang F, Yu F, Locarnini R, Fasullo J et al (2024) New record ocean temperatures and related climate indicators in 2023. Adv Atmos Sci 1–15

Cordero RR, Feron S, Damiani A, Carrasco J, Karas C, Wang C, Kramwinkel CT, Beaulieu A (2024) Extreme fire weather in Chile driven by climate change and El Niño-Southern Oscillation (ENSO). Sci Rep 14:1974

Eccles F, Tziperman E (2004) Nonlinear effects on ENSO'speriod. J Atmos Sci 61:474–482

Huang J, van den Dool HM, Barnston AG (1996) Long-lead seasonal temperature prediction using optimal climate normals. J Clim 9:809–817

Kirtman BP (1997) Oceanic Rossby wave dynamics and the ENSO period in a coupled model. J Clim 10:1690–1704

Lin J-L (2007) Interdecadal variabilityof ENSOin21 IPCCAR4coupled GCMs. Geophys Res Lett. https://doi.org/10.1029/2006GL028937

Mann ME, Cane MA, Zebiak SE, Clement A (2005) Volcanic and solar forcing of the tropical Pacific over the past 1000 years. J Clim 18:447–456

Mason SJ, Graham NE (2002) Areas beneath the relative operating characteristics (ROC) and relative operating levels (ROL) curves: statistical significance and interpretation. Q J R Meteorol Soc 128:2145–2166

Mercik S, Weron K, Siwy Z (1999) Statistical analysis of ionic current fluctuations in membrane channels. Phys Rev E 60:7343–7348

Penland C, Sardeshmukh PD (1995) The optimal growth of tropical sea surface temperature anomalies. J Clim 8(8):1999–2024

Ruelle D (1991) Chance and chaos. Princeton University Press, Princeton

Saha S, Nadiga S, Thiaw C, Wang J, Wang W, Zhang Q, Van den Dool HM, Pan HL, Moorthi S, Behringer D, Stokes D, Pena M, Lord S, White G, Ebisuzaki W, Peng P, Xie P (2006) The NCEP climate forecast system. J Clim 19:3483–3517

Sarlis NV, Skordas ES, Varotsos PA (2011) The change of the entropy in natural time under time-reversal in the Olami-Feder-Christensen earthquake model. Tectonophysics 513:49–53

Stenseth NC, Ottersen G, Hurrell JW, Mysterud A, Lima M, Chan KS, Yoccoz NG, Adlandsvik B (2003) Studying climate effects on ecology through the use of climate indices: the North Atlantic Oscillation, El Niño Southern Oscillation and beyond. Proc R Soc Lond B 270:2087–2096

Troup AJ (1965) The southern oscillation. Q J R Meteorol Soc 91:490–506

Varotsos PA, Sarlis NV, Tanaka HK, Skordas ES (2005) Some properties of the entropy in the natural time. Phys Rev E. https://doi.org/10.1103/PhysRevE.71.032102

Varotsos CA, Tzanis C, Cracknell AP (2016a) Precursory signals of the major El Niño southern oscillation events. Theor Appl Climatol 124:903–912

Varotsos CA, Tzanis CG, Sarlis NV (2016b) On the progress of the 2015–2016 El Niño event. Atmos Chem Phys 16:2007–2011

Varotsos CA, Efstathiou MN, Christodoulakis J (2020a) The lesson learned from the unprecedented ozone hole in the Arctic in 2020 A novel nowcasting tool for such extreme event. J Atmos Sol Terr Phys 207:105330

Varotsos CA, Mazei Y, Novenko E, Tsyganov AN, Olchev A, Pampura T, Mazei N, Fatynina Y, Saldaev D, Efstathiou MA (2020b) New climate nowcasting tool based on paleoclimatic data. Sustainability 12:5546

Varotsos C, Sarlis NV, Mazei Y, Saldaev D, Efstathiou M (2024) A composite tool for forecasting el niño: the case of the 2023–2024 event. Forecasting 6(1):187–203

Wunsch C (1999) The interpretation of short climate records, with comments on the North Atlantic and Southern oscillations. Bull Am Meteorol Soc 80:245–255

Chapter 25
Other Applications of Natural Time to Extreme Phenomena

25.1 The Variance Properties of Natural Time

In the Natural Time Analysis described in Sect. 23.2.1 an interesting scenario arises when the increments of the timeseries of Q_k are positive independent and identically distributed (p.i.i.d.) random variables with finite variance. Initially, we focus on the case where the increments of the Q_k timeseries are p.i.i.d. random variables r_n with finite variance. In this particular case Q_k can be expressed as the sum of r_n from $n = 1$ to k:

$$Q_k = \sum_{n=1}^{k} r_n \quad (25.1)$$

Consequently, Q_k exhibits a linear relationship with k on average. As a result, it is anticipated that the continuous distribution $p(\chi)$, which corresponds to p_k can be represented as $p(\chi) = 2\chi$. By utilizing the expression for variance κ_l of natural time given by Equation:

$$\kappa_1 = \int_0^1 p(\chi)\chi^2 \mathrm{d}\chi - \left(\int_0^1 p(\chi)\chi \mathrm{d}\chi\right)^2 \quad (25.2)$$

a direct computation yields $\kappa_l = 1/18 \approx 0.056$. This value significantly deviates from $\kappa_u \approx 0.083$, which corresponds to the "uniform" distribution.

Considering that the increments possess finite variance, the distribution of Q_k for a given N also possesses finite variance. Therefore, based on the findings of the previous Sect. 23.2.1, it is expected that when Q_k is randomly shuffled, the resulting κ_l values will exhibit a scattering effect around κ_u. A numerical example illustrating this phenomenon for exponentially distributed increments is depicted in Fig. 25.1.

G. Golitsyn and C. Varotsos, *The Stochastic Nature of Environmental Phenomena and Processes*, https://doi.org/10.1007/978-3-031-77015-9_25

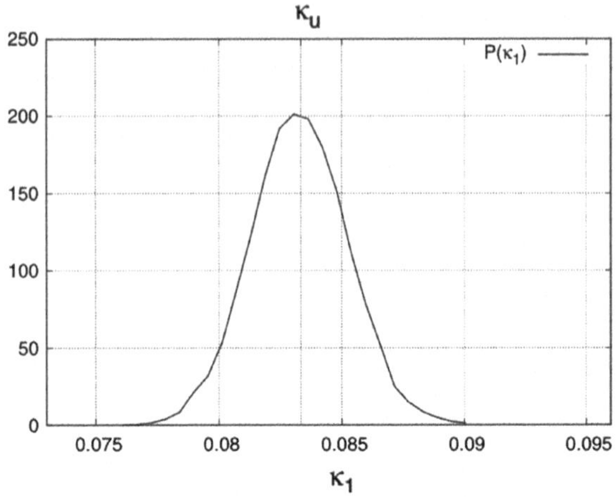

Fig. 25.1 The pdf of κ_1 that has been obtained by shuffling the Q_k randomly in the case of exponential increments, i.e., r_n are randomly drawn from an exponential distribution. Here, $N = 500$, and the original time series results in $\kappa_1 = 0.055$

In the following, we focus on the value of κ_1 when a (natural) time window is sliding through power-law distributed energy bursts.

25.2 Prediction of Earthquakes, Icequakes, Solar Flares, and Microfractures

We will now examine a scenario of self-similarity that arises from the infinite variance of the increments. Our focus is on systems that are driven slowly and produce energy bursts following a power law distribution with a constant γ:

$$P(E) \sim E^{-\gamma} \tag{25.3}$$

Through an analysis of experimental data from various fields, it is observed that the γ exponent typically falls within a narrow range of $1.5 \leq \gamma \leq 2.1$, with an even narrower range of 1.5–1.8. To illustrate the range of phenomena displaying this characteristic, we present some representative examples in Table 25.1.

Crystalline materials under external stress exhibit bursts of activity due to the nucleation and movement of dislocations. These localized changes generate acoustic emission waves, indicating cooperative movement of numerous dislocations intermittently. For instance, acoustic emission experiments on stressed ice single crystals undergoing viscoelastic deformation reveal a power law distribution of energy burst intensities with $\gamma = 1.6$.

Similarly, intermittent plastic flow observations in nickel microcrystals also show an exponent of $\gamma = 1.60 \pm 0.02$.

Solar flares, representing sudden energy releases in the solar corona, exhibit similar power law distributions in energy sizes, inter-occurrence times, and temporal

Table 25.1 Compilation of the experimental values of the power law exponent γ determined in different physical processes (Varotsos et al. 2006)

Process/type of measurement	γ
Dislocation glide in hexagonal ice single crystals (acoustic emission)	1.6
Intermittent plastic flow in nickel microcrystals	1.6
Solar flares	1.5–2.1
Microfractures before the breakup of wood (acoustic emission)	1.5
Microfractures before the breakup of fiberglass (acoustic emission)	2.0
Earthquakes	1.5–1.8
Icequakes	≈ 1.8

clustering as earthquakes. The statistical analysis of these events demonstrates power law distributions with exponents γ ranging from 1.5 to around 2.1.

Other examples include acoustic emissions from microfractures before the breakup in heterogeneous materials (wood, fiberglass), icequakes, and earthquakes.

The subsequent method is currently implemented. A substantial quantity (500,000) of synthetic data following Eq. (25.3) for a specific γ value with energy $E \geq 1$ is generated and then randomly rearranged. This process was iterated for different γ values while maintaining a constant total number of events (indicating that altering γ also changes the maximum energy considered in the computation). The randomized ("shuffled" data are subsequently examined in the natural time domain: the computation of the variance κ_1 is conducted for an event encompassing time windows ranging from $l = 6$–40 consecutive events.

The determination of the specific upper limit value for l is not considered crucial, as similar outcomes are achieved regardless of whether the range of consecutive events is adjusted from 6–40 to 6–100. Additionally, this procedure was executed for all events (across all l values, such as between $l = 6$ to $l = 40$) by thoroughly examining the entire dataset. Figure 25.2 illustrates the plot of the probability density function $P(\kappa_1)$ against κ_1 for various γ values. Furthermore, Fig. 25.3 displays the plotted most probable value $\kappa_{1,p}$ (for a constant γ) in relation to the corresponding γ value.

The relationship between κ_1 and γ for the shuffled data leads to the destruction of a process's memory, resulting in the plotted $\kappa_{1,p}$ values reflecting self-similarity due to the heavy-tailed distribution, distinct from κ_u.

In brief, the source of self-similarity can be differentiated in the following manner: If self-similarity arises solely from the memory of the process, the κ_1 is expected to change to $\kappa_u = 1/12$ for the (randomly) shuffled data. This scenario applies to Seismic Electromagnetic Signals (SES). On the contrary, if self-similarity is a result of the "infinite" variance in the process's increments, the most likely value $\kappa_{1,p}$ should be the same for both the original and the (randomly) shuffled data, although different from κ_u. When both sources of self-similarity coexist, the relative influence of one source compared to the other can be quantified by the expectation value of κ_1.

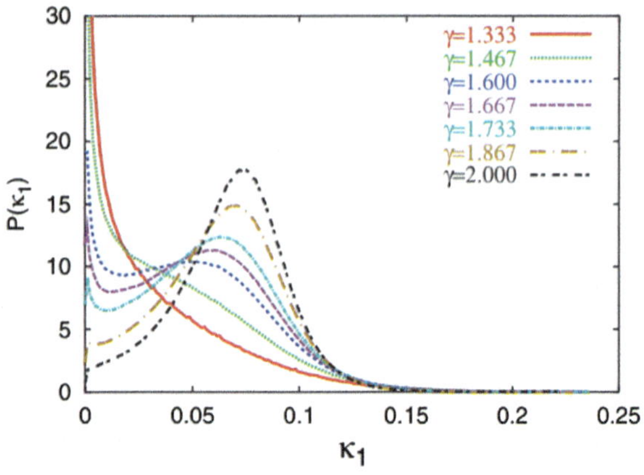

Fig. 25.2 The probability density function $P(\kappa_1)$ versus κ_1 for several γ (Varotsos et al. 2006)

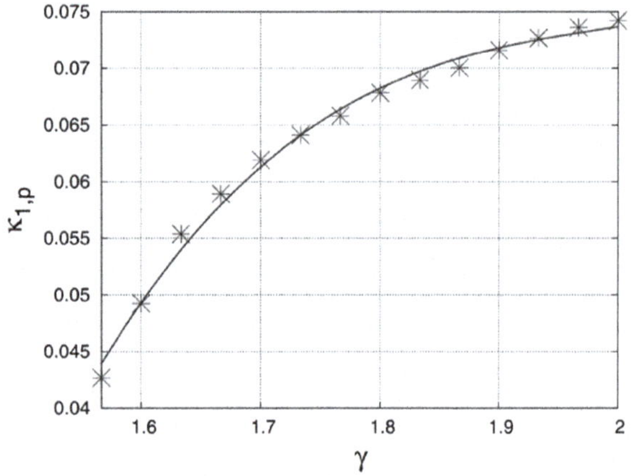

Fig. 25.3 The values of $\kappa_{1,p}$ as a function of γ for power-law distributed data. The continuous line has been drawn as a guide to the eye. Note that $\kappa_{1,p} \approx 0.070$ for $\gamma \approx 1.87$, see also Fig. 25.2 (Varotsos et al. 2006)

The values of the coefficient κ_1, denoting the variance of natural time, can be utilized to determine the approach towards a critical point, of a dynamical system. The κ_1 value is observed to be 0.070 for several dynamical systems approaching criticality. Thus, it provides a prediction parameter for the arrival of an extreme event that might be disastrous.

Moreover, natural time analysis allows for the differentiation between the two sources of self-similarity, namely whether it arises solely from long-range temporal correlations (the process's memory alone) or solely from the infinite variance of the process's increments (heavy tails in their distribution). However, in general, self-similarity may arise from both of these sources, a scenario that can also be identified through natural time analysis.

25.3 The Cosmic Rays Spectrum Properties Revealed in the Natural Time Domain

Varotsos et al. (2024) explored the temporal evolution of the CR power spectral density derived from long-term terrestrial observational data. This investigation aimed to shed light on the validity of the '5/3 law' in the CR spectrum.

Both the conventional and natural time domains have been considered using the daily CR data of the real-time neutron monitoring stations at the Athens Neutron Monitor Station (A.NE.MO.S) (http://cosray.phys.uoa.gr), Jungfraujoch (Switzerland), and Oulu (Finland) sites provided by the High-Resolution Neutron Monitor DataBase-NMDB (http://www.nmdb.eu/) (Christodoulakis et al. 2019; Mavromichalaki 2010; Mavromichalaki et al. 2011, 2018; Varotsos et al. 2023a, b; Xaplanteris et al. 2021). These stations were selected based on the length of their data time-series. The characteristics of these stations are listed in Table 25.2.

The effective vertical cutoff rigidity denotes the minimum rigidity a charged particle requires to reach the middle atmosphere in vertical directions (20 km altitude). Research has shown that it almost remains stable for European CR stations.

These averages were computed over the period from 2000 to 2017 and then subtracted from the corresponding CR time series for that specific day. Additionally, the long-term trend of the CR time series was eliminated through a 6th-degree polynomial regression analysis.

In the data mentioned above, Varotsos et al. (2024) applied the natural time analysis described briefly in Sect. 23.2.1. Using Eq. (25.4)

$$S_k = \sum_{j=1}^{k} P_j N_j ln(N_j) - \left(\sum_{j=1}^{k} P_j N_j\right) ln\left(\sum_{i=1}^{k} P_i N_i\right) \tag{25.4}$$

Table 25.2 The list of the stations used in this study for the period 11/2000–10/2023

Station (Neutron monitor)	Coordinates	Altitude (above sea level-asl)	Effective vertical cut-off rigidity (GV)	Organization
Athens NM64 (A.Ne.Mo.S)	37.97°N, 23.78°E	260 m	8.53	National and Kapodistrian University of Athens, GR
Jungfraujoch IGY (JIGY) (NM64)	46.55°N, 7.98°E	3570 m	4.5	Physikalisches Institut of the Univ. of Bern, CH Int Foundation & High-Altitude Research Stations Jungfraujoch Gornergrat (HFSJG), Bern CH
Oulu NM64 station (OULU)	65.05°N, 25.47°E	15 m	0.8	Sodankyla Geophysical Obs. of the University of Oulu, FI

the entropy $S = S_k$ is calculated for a sliding window of k-length, each time by 1 day, running the entire CR time series of the N-events. Then, we consider the window size $k = 730$ days (2-years) and we calculate S_{730} for the past 730 days and this window is sliding, each time by 1 day, and ran the entire CR time series of N-events.

The selection of the 2-year window size was not arbitrary. The equatorial lower and middle stratosphere goes through a regular cycle known as the quasi-biennial oscillation (QBO). This cycle lasts for 2–3 years (average period 28 months) and is characterized by alternating patterns of westward and eastward zonal wind. Along with the zonal wind, the QBO also brings changes in temperature, trace constituents, and mean meridional circulation.

The QBO has a significant impact on the equatorial lower and middle stratosphere and also affects other parts of the atmosphere, such as the tropical troposphere, tropical upper stratosphere and mesosphere, the extratropical middle atmosphere during winter, and sudden stratospheric warmings of the high latitudes. Therefore, the QBO influences climate phenomena beyond the tropical stratosphere, including ozone transport, the North Atlantic, and the Madden–Julian Oscillations (Anstey et al. 2022). However, simulating the QBO is challenging due to uncertainties surrounding the waves that drive the oscillation, particularly the momentum fluxes from small-scale gravity waves caused by deep convection.

Despite the complexity and unpredictability of these wave motions, the predictability of the QBO is remarkable, considering their wide range of spatial and temporal scales. Hopefully, by improving our understanding of the processes that control the QBO, we can also gain insight into unexpected events like the two QBO disruptions observed since 2016.

Varotsos et al. (2018) investigated the unusual equatorial QBO event in the zonal wind in 2016 and suggested that it was not related to any previous events. They used NTA to analyze the QBO data and found a precursor behavior before the increase in zonal wind velocity, indicating a possible connection with the strong El Niño event in 2015–2016.

In addition, their previous research on the dynamics of the ozone hole in the Antarctic and the El Niño phenomenon showed that a window of approximately 2–3 years was optimal for detecting critical system states (Varotsos and Tzanis 2012; Varotsos et al. 2010, 2016, 2024).

Since the CR data utilized in this study were collected from instruments on the ground, it is logical to consider that they may be influenced by natural atmospheric fluctuations, like the QBO (Kang et al. 2024). This is why we opted for a 2-year window size as the threshold for our NTA analysis.

Consequently, considering the potential impact of QBO on the CR flux reaching the ground, we calculated the exponent γ of the power-law fit to the power spectral density of Ej values ($j = 1,2,\ldots, k$) as a function of frequency, compared to S_k using the sliding window of length k. This analysis was performed on the entire CR time series of the N-events at OULU station.

Bearing in mind the afore-mentioned discussion on the potential influence of the QBO to the CR flux reaching the ground, we proceed calculating the exponent γ of the power-law fit to the power spectral density of E_j values ($j = 1, 2, \ldots, k$) as a

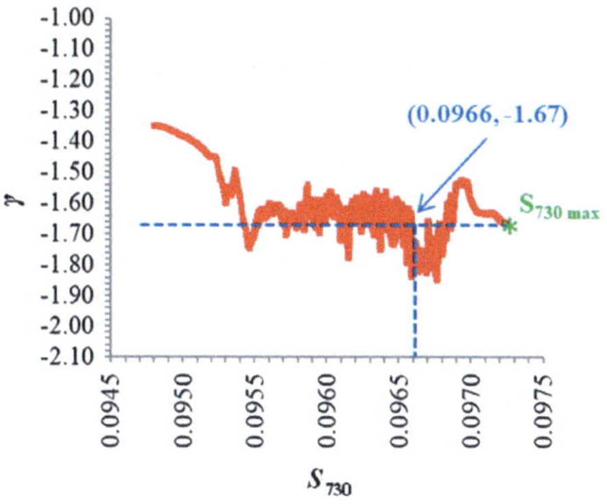

Fig. 25.4 Power law exponent γ of the E_j values ($j = 1, 2, \ldots, 730$) as a function of frequency versus S_{730}, for the above-mentioned sliding window of 730-length, running the entire CR time series of the N-events at OULU station, expressed as 30-day running mean of the data (red line). Interestingly, we find that one of the points with $S_{730} = S_u$ (=0.0966) corresponds to $\gamma = -5/3 \approx -1.67$ (blue coordinates and green star)

function of frequency, versus S_k employing the above-mentioned sliding window of k-length, running the entire CR time series of the N-events at OULU station.

The results obtained are shown in Fig. 25.4 where plots of the γ-exponent versus S_{730} for the CR dataset at OULU station smoothed by applying the 30-day running mean (red line) are presented. It can be seen that the γ-value varies from -1.9 to -1.3, as the sliding window of 730-length runs the entire CR time series. Moreover, it is noteworthy from Fig. 25.4 that the maximum value of S_{730} is observed for $\gamma = -1.67 = -5/3$.

It should be emphasized that for the uniform distribution (u), (e.g. when our system is in a stationary state emitting uncorrelated bursts of energy) Eq. (25.4) gives the entropy S_u of the uniform distribution (Varotsos et al. 2004):

$$S_u = (\ln 2)/2 - 1/4 \approx 0.0966, \text{ where } \gamma = -1.67 \text{ (see Fig. 25.4)}$$

This is reminiscent of the fact that according to Varotsos et al. (2006) the experimental data of systems emitting energy bursts, when analyzed by NTA, show that they obey a power-law distribution with an exponent γ between 1.5 and 2.1. For example, this exponent is between 1.5 and 2.1 in solar flares, and 1.5 and 1.8 in earthquakes and takes the value 1.8 in the case of icequakes (Fig. 25.2). Consequently, this very important result may be used for the prediction modelling of the CR extreme events and their interplay with other geophysical phenomena.

Next, the same analysis was conducted on the data collected at JIGYNM and Athens stations (see Fig. 25.5). The results indicate that the power law exponent corresponding to the maximum S_{730} value is $\gamma = -1.59$ and $\gamma = -1.54$ at JIGYNM and Athens stations, respectively. Both values deviate from the -1.67 ($-5/3$) that was found at OULOU. It is noticeable that the exponent's value displays the most

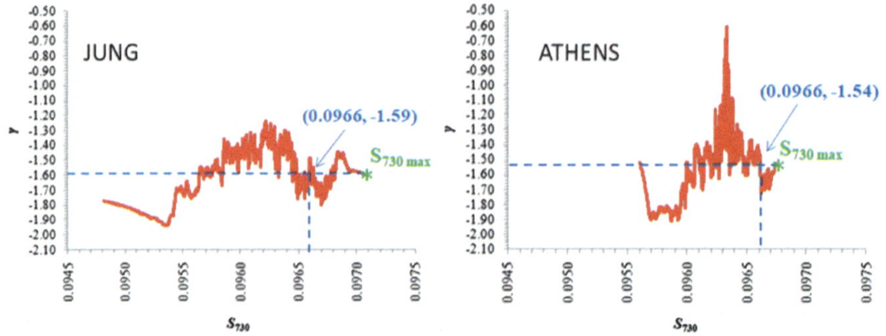

Fig. 25.5 As in Fig. 25.4, but for the Stations JUNG and ATHENS

significant deviation from − 5/3 in the data from Athens' station, the southernmost station.

This variation could be attributed to the notably higher maximum effective vertical cut-off rigidity at the Athens station, which exceeds that of OULU by more than tenfold. It is essential to emphasize that magnetic rigidity is defined as the momentum of a charged particle divided by its electric charge, then multiplied by the speed of light.

This fundamental quantity plays a critical role in studying the motion of charged particles within a magnetic field, as particles with the same rigidity and initial conditions will follow identical paths in a specific magnetic field. This does not hold in the case of the selected three stations, in our study.

It should be stressed that variations in atmospheric parameters impact the generation multiplicity of CR secondary particles, with barometric and temperature effects playing a significant role. While the neutron component is mainly influenced by barometric effects (taken into account in the data used here), there is evidence of a humidity effect that has been historically overlooked but may have an impact on CR neutron intensity near Earth's surface (Dorman 1957, 1972, 1975). Neutrons formed in the atmosphere through interactions with CR undergo elastic collisions, losing energy and eventually being absorbed.

The rate of neutron production in the atmosphere remains constant, but variations in water vapor content cause intensity variations in the detected neutron component (Zrenda et al. 2012).

Furthermore, the neutron flux density near Earth's surface is inversely proportional to air and soil humidity (Yanchukovsky et al. 2024). Although the humidity effect is much less than the barometric one, the neutron energy spectrum displays information about moisture.

As highlighted in the Introduction, CR flux is connected to climate change. Hence, climate modeling must incorporate our findings for enhancement. For instance, CRs ionize the atmosphere, forming cloud condensation nuclei leading to cloud formation acting like an "umbrella". For example, when the CR flux decreased (because of the increased solar wind from sunspots), there was a decrease in cloud cover, which could potentially lead to warming (Svensmark et al. 2007, 2017). Some scientists, however, have raised doubts about the connection between CR and cloud cover and argue that just because there is a correlation doesn't mean there is causation (Erlykin et al. 2011).

By summarizing, the above analysis and discussion, demonstrate that the observed temporal evolution of CR deduced from several stations of the terrestrial network exhibits a power spectral density consistent with the finding that the energy spectrum of CR obeys the Kolmogorov-Obukhov 5/3 law, which has already been confirmed in several geophysical quantities (Golitsyn 2023, 2024; Varotsos et al. 2006).

It is observed even more so when the frequency is less than 10^{-1} day^{-1}. Furthermore, with the natural time analysis we showed that the 5/3 Kolmogorov-Obukhov power law is also related to the maximum value of CR entropy in which the QBO was revealed as a leading index.

This CR behaviour is consistent with the finding that systems emitting energy bursts (such as solar flares, earthquakes and icequakes) obey a power-law distribution with an exponent between 1.5 and 2.1. This may be used for the prediction of the CR extreme events, as in prediction problems maximized entropy gives the maximum room for the data to reveal secrets hidden and ensures that no unconscious arbitrary assumptions are introduced into the method used (Jaynes 1957, 2003).

Consequently, there are several unanswered questions in the CR domain, such as the origin and mass composition of ultra-high energy CR and how they achieve extreme energies (Alves Batista et al. 2019). These questions drive ongoing research and exploration, pushing the boundaries of our understanding of CR and the universe. It is increasingly imperative considering that CR have made a lasting mark on early life and could have affected certain crucial biological aspects with potential consequences.

References

Anstey JA, Osprey SM, Alexander J et al (2022) Impacts, processes and projections of the quasi-biennial oscillation. Nat Rev Earth Environ 3:588–603. https://doi.org/10.1038/s43017-022-00323-7

Alves Batista R, Biteau J, Bustamante M, Dolag K, Engel R, Fang K et al (2019) Open questions in cosmic-ray research at ultrahigh energies. Front Astron Space Sci 6:23

Christodoulakis J, Varotsos CA, Mavromichalaki H, Efstathiou MN, Gerontidou M (2019) On the link between atmospheric cloud parameters and cosmic rays. J Atmos Solar Terr Phys 189:98–106

Dorman LI (1957) Variatsii kosmicheskikh luchey [Variations of Cosmic Rays]. Gostechizdat Publication, Moscow, p 285 (In Russian)

Dorman LI (1972) Meteorologicheskie effekty kosmicheskikh lu-chei [Meteorological Effects of Cosmic Rays]. NaukaPublications, Moscow, p 211 (in Russian)

Dorman LI (1975) Eksperimentalnye i teoreticheskie osnovy as-trofiziki kosmicheskikh luchei [Experimental and Theoretical Foundations of Cosmic Ray Astrophysics]. Nauka Publications, Moscow, 462 p (in Russian)

Erlykin A, Sloan T, Wolfendale A (2011) Cosmic rays and climate. Phys World 24(10):25

Golitsyn GS (2023) Random moves equation Kolmogorov-1934. A unified approach for description of statistical phenomena of nature. arXiv preprint arXiv:2310.06057

Golitsyn GS (2024) AN Kolmogorov's 1934 paper is the basis for explaining the statistics of natural phenomena of the macrocosm. UspekhiFizicheskikhNauk 194(1):86–96

Jaynes ET (1957) Information theory and statistical mechanics. Phys Rev 106(4):620

Jaynes ET (2003) Probability theory: the logic of science. Cambridge University Press, Cambridge

Kang MJ, Kim H, Son SW (2024) QBO modulation of MJO teleconnections in the North Pacific: impact of preceding MJO phases. NPJ Clim Atmos Sci 7:12. https://doi.org/10.1038/s41612-024-00565-w

Mavromichalaki H (2010) World-wide integration of neutron monitors—the NMDB project. EOS AGU 31(8):305–306

Mavromichalaki H et al (2011) Applications and usage of the real-time neutron monitor database. Adv Space Res 47:2210–2222

Mavromichalaki H, Gerontidou M, Paschalis P, Paouris E, Tezari A, Sgouropoulos C et al (2018) Real-time detection of the ground level enhancement on 10 September 2017 by A.Ne.Mo.S. System report. Space Weather 16(11):1797–1805

Svensmark H, Bondo T, Svensmark J (2009) Cosmic ray decreases affect atmospheric aerosols and clouds. Geophys Res Lett 36:L15101. https://doi.org/10.1029/2009GL038429

Svensmark H, Enghoff MB, Shaviv NJ et al (2017) Increased ionization supports growth of aerosols into cloud condensation nuclei. Nat Commun 8:2199. https://doi.org/10.1038/s41467-017-020 82-2

Varotsos CA, Golitsyn GS, Efstathiou M, Sarlis N (2023a) A new method of nowcasting extreme cosmic ray events. Remote Sens Lett 14(6):576–584

Varotsos CA, Golitsyn GS, Xue Y, Efstathiou M, Sarlis N, Voronova T (2023b) On the relation between rain, clouds, and cosmic rays. Remote Sens Lett 14(3):301–312

Varotsos CA, Tzanis C (2012) A new tool for the study of the ozone hole dynamics over Antarctica. Atmos Environ 47:428–434. https://doi.org/10.1016/j.atmosenv.2011.10.038

Varotsos CA, Cracknell AP, Tzanis C (2010) Major atmospheric events monitored by deep underground muon data. Remote Sens Lett 1(3):169–178

Varotsos CA, Tzanis CG, Sarlis NV (2016) On the progress of the 2015–2016 El Niño event. Atmos Chem Phys 16(4):2007–2011

Varotsos C, Sarlis NV, Efstathiou M (2018) On the association between the recent episode of the quasi-biennial oscillation and the strong El Niño event. Theor Appl Climatol 2018(133):569–577. https://doi.org/10.1007/s00704-017-2191-9

Varotsos C, Sarlis NV, Mazei Y, Saldaev D, Efstathiou M (2024) A composite tool for forecasting El-Nino. The case of the 2023–2024 event

Varotsos PA, Sarlis NV, Skordas ES, Lazaridou MS (2004) Entropy in the natural time domain. Phys Rev E 70:011106. https://doi.org/10.1103/PhysRevE.70.011106

Varotsos P, Sarlis N, Skordas E, Lazaridou M (2006) Attempt to distinguish long-range temporal correlations from the statistics of the increments by natural time analysis. Phys Rev E 74:021123. https://doi.org/10.1103/PhysRevE.74.021123

Yanchukovsky V, Kalyuzhnaya MA, Khisamov RZ (2024) Intensity of the neutron component of cosmic rays and air humidity. Sol-Terr Phys 10(1):34–39. https://doi.org/10.12737/stp-101 202405

Zreda M, Shuttleworth WJ, Zeng X, Zweck C, Desilets D, Franz T, Rosolem R (2012) COSMOS: the COsmic-ray Soil Moisture Observing System. Hydrol Earth Syst Sci 16:4079–4099. https://doi.org/10.5194/hess-16-4079-2012

Xaplanteris I et al (2021) Improved approach in the coupling function between primary and ground level cosmic ray particles based on neutron monitor data. Sol Phys 296:91. https://doi.org/10.1007/s11207-021-01836-y

Chapter 26
The Climate Linear and Non-linear Regime

26.1 Current and Missing Knowledge

26.1.1 The Concept of Linearity and Non-linearity

The general circulation model (GCM), a method for modeling climate change, is predicated on the idea that, in contrast to weather, climate is a boundary value problem (Pielke 1998). This indicates that while weather shows a sensitive dependence on initial conditions, making it unpredictable over short timescales, averaging over a large enough sample size causes the weather to converge towards the climate of the model. The climate, also known as control runs, is the state at which specified atmospheric shapes and boundary conditions cause model outputs to converge.

It becomes interesting to see how the system reacts to slight modifications in the boundary conditions. For example, over longer timescales, the magnitude of solar forcing (SF) and volcanic forcing (VF) is almost the same or less, and anthropogenic forcings are generally smaller than 2 W/m^2. Given that these values represent about 1% of the average radiative flux from the sun, the atmospheric response appears to be largely linear. This presumption supports the reduction of potentially intricate forcings to general radiative forcings at annual scales in climate algorithms (Meehl et al. 2004; Hansen et al. 2005, for GHGs). On longer timescales, however, linearity diminishes. Intense nonlinear interactions and feedbacks among global albedos and temperatures are evident in the "Daisyworld" algorithm (Watson and Lovelock 1983) and later research on energy balance models. Over multimillennial timescales, temperature-albedo feedbacks play a particularly significant role during glacial-interglacial transitions. The methods used by different authors to ascertain the critical ice-albedo feedback timescales vary. Some writers take into account short timescales such as 200 years, including Roques et al. (2014). Conversely, over the previous millennium, Østvand et al. (2014) and Rypdal and Rypdal (2014) assume a fairly linear temperature response to SF and VF forcings. Pelletier (1998) and Fraedrich et al. (2009) prolong this linearity assumption to even greater timescales.

G. Golitsyn and C. Varotsos, *The Stochastic Nature of Environmental Phenomena and Processes*, https://doi.org/10.1007/978-3-031-77015-9_26

Determining the timescales over which linear responses are valid is essential. It should be noted, though, that if the forcing is volcanic or volcanic-like, exhibiting spiky or intermittent behavior, the response may become nonlinear, even over scales where normal responses to minor forcings obey linearity.

26.1.2 Scaling Regimes in the Atmosphere

Prior to discussing models, it is important to examine the empirical findings. Higher frequencies exhibit a highly nonlinearity of the atmosphere, as seen in the weather patterns. However, after approximately ten days, there is a significant shift to a lower-frequency regime in the atmosphere, known as the "macroweather" regime, which exhibits potentially quasi-linear responses. The identification of the fundamental scaling regimes in the atmosphere was initially achieved through spectral analysis, as demonstrated by Lovejoy and Schertzer (1986), Pelletier (1998), Shackleton and Imbrie and Huybers and Curry (2006). Nevertheless, a clearer and simpler interpretation was obtained by analyzing real space fluctuations. This approach revealed that the conventional understanding of atmospheric variability, characterized by a series of limited-range processes like nonlinear oscillators, has largely overlooked the scaling "background spectrum," which is the main cause of variability. The analysis showed that for nearly all atmospheric domains, there is a shift from the macroweather regime-corresponding lower-frequency scaling regime with $H < 0$ for scales $t > 10$ days, to the scaling behavior of mean temperature fluctuations, represented as $\Delta T(t) \approx \Delta t^H$ with $H > 0$. The theoretical prediction of this transition scale near ten days is based on the turbulent wind scaling induced by SF, specifically through the required density of energy rate (Lovejoy and Schertzer 2010, 2013; Lovejoy et al. 2014). The lower-frequency $H < 0$ pattern is characterized by fluctuations that tend to cancel out, effectively converging, while the higher-frequency $H > 0$ pattern is typically associated with temperature observations fluctuating randomly, similar to a drunkard's walk. According to Lovejoy (2013) and Lovejoy et al. (2014), this converging regime is an example of "macroweather," a category of low-frequency weather. Macroweather persists for very long timescales for the GCM control simulations; in practice, it lasts for 10–30 years (industrial) and 50–100 years (pre-industrial) before a new regime with $H > 0$ appears. Climate change is frequently associated with this new regime (see Franzke et al. 2013 as well as Lovejoy et al. 2013, Fig. 5). Numerous studies have investigated macroweather scaling, such as Bunde et al. (2005), Østvand et al. (2014), Rypdal and Rypdal (2014), Eichner et al. (2003), Kantelhardt et al. (2006) and Rybski et al. (2006).

The transition from "macroweather" to climate (at scale c) seems to be due to the fact that over macroweather timescales—where fluctuations are "cancelling" each other out—other slower processes, which likely include external climate forcings and other slow (internal) land-ice or biogeochemical processes, gradually become more dominant. Eventually, at scale *c*, their variability takes over. The specific value of the global transition scale c during the pre-industrial Holocene and the possibility of large regional variations in c during the Holocene are major points of contention; Lovejoy provides additional information. These factors imply that the data from Greenland ice cores may not be universally representative.

26.1.3 The Numerical Modeling of Scaling

Several studies have investigated the responses of GCMs to low-frequency control runs (Vyushin et al. 2004; Zhu et al. 2006; Fraedrich et al. 2009; Lovejoy et al. 2013; Fredriksen and Rypdal 2016). These investigations have demonstrated that the lack of a typical timescale for simulating long-time convergence causes the responses to scale down to their lowest frequencies. The scaling exponents are empirically found to be very small, leading to an "ultra-slow" convergence of general circulation models (GCMs) (Lovejoy et al. 2013). The wider ramifications of this scaling have not been generally accepted, despite past research concentrating on the statistical connections inherent in the scaling statistics (Varotsos et al. 2009, 2015a, b).

This behavior has more recently been positioned within the conceptual structure of GCM climate algorithm thanks to scaling fluctuation analysis (Lovejoy et al. 2013). According to some theories, the climate processes of slow internal variability that interact with external forcings e.g., SF and VF, as well as more recent human influences, are what cause climate to emerge. The low-frequency variability in GCMs, spanning multiple centuries, is primarily driven by external forcings. Nonetheless, incorporating slow processes such as land ice or biogeochemical processes may introduce additional sources of low-frequency internal variability. The main concern for GCMs in the context of the past millennium is whether they can model accurately the change from a climate phase with growing $H > 0$ to one with falling "macroweather" fluctuations ($H < 0$). In a publication by Lovejoy et al. (2013), an analysis was conducted on four GCMs that simulated the climate of the last millennium. The study revealed that the variability in the low-frequency spectrum (particularly below a 100-year timescale) that was deduced from these models, was relatively weak. This weakness was attributed to the SF, which was found to be weak when employing solar reconstructions that are sunspot-based with positive values of H. Additionally, for strong VF, the kind of statistical forcing (with negative values of H) also played a role in this variability. Previous studies by Lovejoy and Schertzer (2012a), Bothe et al. (2013a, b), and Zanchettin et al. (2013) have also highlighted similar findings. For further insights into the dynamics of centennial timescales, refer to Zanchettin et al. (2010).

26.1.4 The Missing Understanding

Examining the responses to SF and VF at multi-centennial periods raises the topic of linearity. When reducing climate forcings to their corresponding radiative forcings, it is common to assume that the combined response of SF and VF is approximately the total of each of their distinct effects. However, Mann et al. (2005) have already pointed out that this additivity assumption does not hold in the Zebiak Cane (ZC) model. To better understand this issue and measure the degree of sub-additivity as it changes with time, Lovejoy and Varotsos conducted an analysis.

Clement et al. (1996) also emphasized the high sensitivity to minimal forcings and poor sensitivity to big forcings resulting from the nonlinear model response as another linear/nonlinear issue. Systems that exhibit different statistical behaviors for strong and weak events are often referred to as "intermittent" and display varying degrees of clustering. Lovejoy and Varotsos quantitatively investigated this phenomenon and confirmed that it is very noticeable for VF. Additionally, they found that the ZC model, with a GCM, exhibits significantly less intermittency, indicating that the model successfully attenuates the forcing in a strong and nonlinear manner. Lovejoy and Varotsos established analysis methodologies to resolve these problems and used them to analyze model results spanning the past millennium. Regrettably, despite our examination of the NASAGISS E2-R simulations from the last millennium, there is a lack of comprehensive last-millennium GCM simulations combining forcings and responses that are either volcanic, solely solar, or both. As a result, Lovejoy and Varotsos (2016) have resorted to utilizing the simplified ZC model results presented by Mann et al. (2005), despite the fact that even this model lacks the control runs required to measure internal variability directly.

The ZC model, while deficient in certain crucial mechanisms such as deep ocean dynamics, does exhibit sources of low-frequency variability. For instance, Goswami and Shukla (1991) identified multidecadal and multicentennial nonlinear variability in their 360-year control runs, ascribed to the interplay of low-level convergence, atmospheric heating, and anomalies in Sea Surface Temperature (SST). Furthermore, Mann et al. (2005) highlighted the variability on a centennial scale in their millennium ZC simulations as a key factor (Varotsos 2013).

26.2 Climate Simulation Models for Solar and Volcanic Forcings

In the previous millennium's pre-industrial era, the composition of the atmosphere remained relatively stable, while the Earth's orbital parameters experienced minor variations. Climate models in GCM simulations primarily utilized SF and VF during this time period. Although some models also incorporated reconstructed land use changes, the impact of these forcings was relatively weak. Research by Minnis et al. (1993) highlighted the significant cooling effect of volcanic aerosols following Mount Pinatubo's eruption in 1991.

Further research by Shindell et al. (2003) highlighted the impact of solar irradiance variability and volcanic aerosols on pre-industrial climate change. The best consistency with historical and proxy data was discovered when SF and VF were combined.

However, these forcings elicit different responses due to the contrasting effects they have on the stratosphere and surface. SF tends to reinforce each other, while VF exhibits opposing effects. Furthermore, the temporal variabilities of volcanic and solar activities, alongwith the seasonality, have been statistically associated with the onset of El Niño-Southern Oscillation (ENSO) events. Decreased solar irradiance results in cooling of the surface and stratosphere, whereas surface cooling is induced by volcanic eruptions but warms the sunlit lower stratosphere thanks to aerosol heating. This results in a lower stratosphere meridional gradient that is higher but a tropopause area gradient that is lower (e.g., Varotsos et al. 1994).

According to Vyushin et al. (2004), VF improves the efficiency of atmosphere–ocean models in terms of low-frequency variability scaling when compared to other forcings. However, Blender and Fraedrich (2004) provided a comment on this suggestion, referencing earlier studies in the field. In 2005, Weber used simulations with a climate model powered by reconstructed forcings to study how the temperature in the Northern Hemisphere responded to SF and VF between 1000 and 1850. According to the study's findings, the response to VF never stabilizes because of how frequently eruptions occur in comparison to the climate system's dissipation period. In contrast, the response to SF stabilizes at interdecadal timescales.

Mann et al. (2005) used the ZC model for the combined atmosphere–ocean system in the tropical Pacific to examine how variations in natural radiative forcing affected El Niño from 1000 and 1999. They found that the variability in past El Niño records is accurately reproduced by the combined feedback of radiative SF and VF. In their analysis of the timescale dependency of different solar and volcanic reconstructions, Lovejoy and Schertzer (2012a) pointed out discrepancies in SF based on ice core ^{10}Be isotopes or sunspots, with isotope-based reconstructions declining and sunspot-based reconstructions rising with timescale. The discrepancy between features based on quality and quantity prompts questions about how accurate the solar reconstructions are. The two volcanic reconstructions, on the other hand, showed comparable statistical features, with considerable strength at the annual and occasionally multiannual levels and rapidly reducing with rising periods ($H < 0$).

Mann et al. (2005) created a 100-realization ensemble for SF and VF, and combined forcings during the previous thousand years using the Zebiak and Cane (1987) ZC model of the tropical Pacific combined atmosphere–ocean system. The following source provided by Lovejoy and Varotsos (2016) with the model's forcings and mean responses: ftp://ftp.ncdc.noaa.gov/pub/data/paleo/climate_forcing/mann2005/mann2005.txt. It should be noted that no anthropogenic effects were considered in their analysis.

A geometric factor of 1.57 was applied by Mann et al. (2005) to the Crowley VF reconstruction to account for the narrow range of latitudes (between 30° N and 30° S). The forcings and average responses derived from the ZC algorithm are shown in Fig. 26.1a. Reconstructed solar (brown), SF amplified by five (orange), and VF (black) make up the forcings. It is noteworthy that the SF series, which is based on sunspots rather than ^{10}Be, shows a higher resolution and a wandering pattern throughout the last few centuries. The responses for the SF alone (top), the VF alone (middle), and the combined forcings (bottom) are shown in the bottom graph. These response series have been vertically offset by 2.5K, 1.5K, and 0.5K, respectively, to improve clarity. Furthermore, the equivalent simulated responses from the GISS-E2-R model are shown in Fig. 26.1b, as documented by Lovejoy et al. (2013) and Schmidt et al. The industrial era that began after 1900 has been disregarded because of the predominance of manmade forces. Based on sunspots from 1610, the SF utilized in this analysis is the same as the one used in the ZC model.

The response to SF alone is shown in the top row of the graph, while the response to both SF and the Crowley-reconstructed VF series (which is the same as utilized in the ZC algorithm) is shown in the middle series. The SF and reconstructed VF series from Gao et al. are used in the bottom series.

For clarity, we have offset each of these response series vertically by 1K. It is important to note that the land area of the Northern Hemisphere, which is slightly different in geography from the simulations made by ZC algorithm, was averaged over in these simulations. The high frequencies show some fluctuation, but the low frequencies seem similar despite this difference (Fig. 26.1).

26.3 Scaling Fluctuation Analysis—Haar Analysis

Lovejoy and Varotsos (2016) computed the fluctuations ΔF(t) (forcings, Wm^{-2}) and $\Delta T(\Delta t)$ (responses, K) in order to separate the variability based on timescale t. Although it is customary to characterize desrepancies in terms of absolute differences, as in $\Delta T\ (\Delta t) = T(t + \Delta t) - T(t)$, this method is insufficient for the purposes of their investigation. Rather, they ought to take into account the absolute difference in the averages between t and $t + \Delta t/2$ and $t + \Delta t/2$ and $t + \Delta t$. When defining fluctuations, this approach is equivalent to employing Haar wavelets as opposed to "poor man's" wavelets (differences). The variations in a scaling domain show a power law association with the time lag:

$$\Delta T = \varphi \Delta t^H, \tag{26.1}$$

where H is a dynamical variable that controls the process (e.g., a dynamical flux), and whose average is independent of the timeframe (i.e., lag Δt). This suggests that, on average, fluctuations tend to increase with scale when $H > 0$ and decline when $H < 0$, showing that the average fluctuation behavior is $\langle \Delta T \rangle \approx \Delta t^H$. Harold Edwin Hurst is honored by the symbol "H" (Hurst 1951). While H is equal to his eponymous exponent in the context of quasi-Gaussian statistics, the H used here is valid in the more general multifractal scenario and usually differs from.

Variances, which are known as fluctuations, are appropriate for fluctuations that grow larger with size ($H > 0$). The rate at which mean variances rise with time lag when $H > 0$ is a direct indicator of how important low frequencies are becoming in relation to high frequencies. Still, variations tend to rise in physical systems even in the case where $H < 0$. This is due to the fact that as the time lag Δt rises, the correlations between $\langle T(t + \Delta t)T(t)\rangle$ tend to weaken, increasing the mean square variances

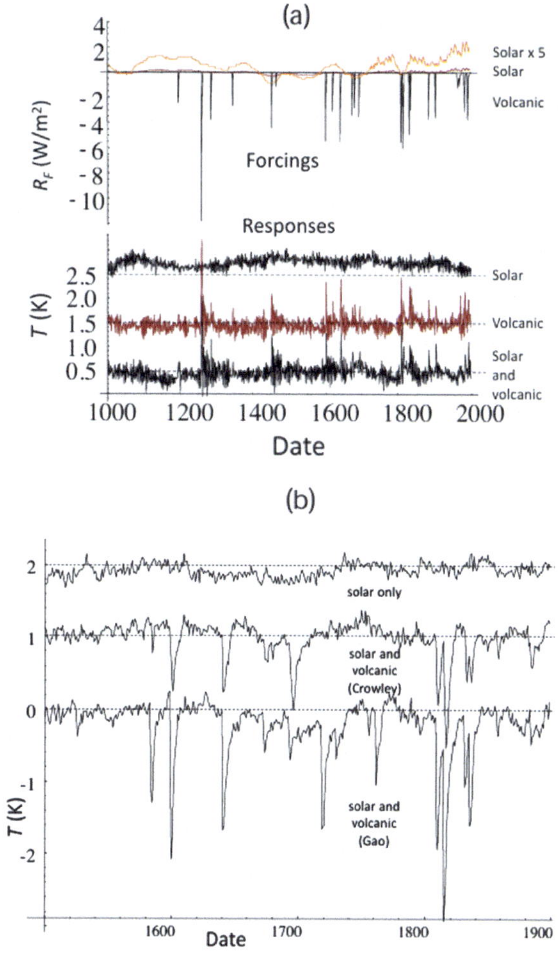

Fig. 26.1 **a** For the Zebiak–Cane model, the first graph shows the radiative forcings (RF) and responses (T) from AD 1000 to 2000. The whole simulation region is covered by the integration of these data. **b** This section presents the GISS-ER-2 responses at a yearly resolution, averaged over the northern hemisphere's land that the absolute temperature readings are not very important because these values are anomalies (Lovejoy and Varotsos 2016)

$(\langle \Delta T(\Delta t)2\rangle)$. In terms of mathematics, a stationary process has $\langle \Delta T(\Delta t)^2 = \langle (T(t+\Delta t) - T(t))^2 = 2(\langle T^2\rangle - \langle T(t+\Delta t)T(t)\rangle)$.

This suggests that variances are unable to adequately describe fluctuations when $H < 0$. The variations in this instance are dominated by the high-frequency features, which keep them from falling as scale t increases.

Since its value might closely approximate the difference fluctuation in regions where $H > 0$ with adequate calibration, the Haar fluctuation—which is helpful for $-1 < H < 1$—is very simple to understand. Conversely, it can be made to mimic another readily interpreted "anomaly fluctuation" in places where $H < 0$. The latter is the mean series over a time interval Δt, with the general mean removed [Lovejoy and Schertzer (2012b) refer to this as a "tendency" fluctuation, which may not make as much sense].

The way in which averaging a (mean zero) process (the anomaly) over longer time scales lessens its variability is demonstrated by the decrease in Haar fluctuations as lag Δt grows. In order to precisely align the raw Haar fluctuation with the associated difference and anomaly fluctuations (for time scales with $H > 0$ and $H < 0$, respectively), the calibration includes increasing the raw fluctuation by a factor of 2. This suggests that the Haar fluctuations can be interpreted as anomalies in regions where $H < 0$ and as differences with good precision in places where $H > 0$.

Other techniques, such as detrended fluctuation analysis, are also useful for estimating exponents, but because their fluctuations are the standard deviations of the residues from polynomial regressions on the running sum of the original series, they are more difficult to interpret. Actually, there's a common presentation of the DFA fluctuation function with no units.

After the estimation is finished, the statistical characteristics of the fluctuations can be examined to measure the fluctuations' variations over time. For quantifying these fluctuations, $Sq(\Delta t)$, the qth order structure function, is especially helpful.

$$S_q(\Delta t) = \langle \Delta T(\Delta t)^q \rangle \tag{26.2}$$

In Eq. (26.2), $Sq(\Delta t)$ is defined as the product of the fluctuations $\Delta T(t)$ raised to the power of q, with ensemble averaging denoted by the symbol " ⟨ ⟩ ". It is worth noting that while q can theoretically take any value, in this context, it is restricted to values greater than 0 to avoid potential divergences. Divergences are expected to occur for any q less than 0 in the case of multifractals.

In a scaling regime, the behavior of $Sq(\Delta t)$ follows a power law as shown in Eq. (26.3).

$$S_q(\Delta t) = \langle \Delta T(\Delta t)^q \rangle \propto \Delta t^{\xi(q)}; \quad \xi(q) = qH - K(q), \tag{26.3}$$

The exponent $\xi(q)$ is composed of two components: qH, which is linear, and $K(q)$, which is convex and generally nonlinear. $K(1)$ is equal to 0. Strong non-Gaussian and multifractal variability, or "intermittency," is characterized by the function $K(q)$. With Gaussian processes, however, $K(q)$ is always equal to 0.

The exponent of the root-mean-square (RMS) fluctuation, $S_2(\Delta t)^{1/2}$, or $S(\Delta t)$, is $\xi(2)/2 = H - K(2)/2$. It is noteworthy to mention that Lovejoy and Varotsos (2016) found that $\xi(2)/2 \approx H = \xi(1)$ when the intermittency is modest ($K(q) \approx 0$). Moreover, the power law spectra's exponent (β) and exponent H can be connected exactly as follows: $\beta = 1 + \xi(2) = 1 + 2H - K(2)$. The Wiener-Khintchin theorem's corollary, this relationship is especially helpful because the spectrum is a second-order statistic. In the unique scenario where $K(2)$ is small, the widely accepted relation $\beta \approx 1 + 2H$ is valid. Here, $\beta > 1$ is correlated with $H > 0$, and $\beta < 0$ is correlated with $H < 0$.

It is crucial to remember that a greater value of q indicates the scaling of strong events, whilst a lower value of q indicates the scaling of weak events in order to comprehend the consequences of the nonlinear function $K(q)$. It is important to note that q is not restricted to whole numbers. When the clustering of strong and weak events differs, the scaling features also change. The codimension, an additional exponent that determines this grouping precisely, is uniquely determined by $K(q)$.

The estimates of the Haar fluctuations for the SF and VF are shown in Fig. 26.2a. Two distinct ^{10}Be reconstructions are represented by the two curves on the right (Shapiro et al. 2011; Steinhilber et al. 2009). Their exponents are roughly the same and the amplitudes decrease quickly with scale, despite the fact that at any given scale, their disparate assumptions result in amplitudes that differ by almost a factor of 10 (Fig. 26.2).

The solar reconstruction utilized is a hybrid that combines a lower resolution ^{10}Be-based reconstruction (Fig. 26.2b, bottom) covering a longer time period with an annual resolution sunspot-based reconstruction (Fig. 26.2b, top). The two curves on the right in Fig. 26.2a correspond to two distinct reconstructions of ^{10}Be. It is evident that the statistical features are completely different, with the ^{10}Be reconstruction "canceling" ($H < 0$) and the sunspot variations "wandering" ($H > 0$). The sunspot data were from Shapiro et al. (2011), and the sunspot data were for the "background" (that is, without an 11-year cycle; see Wang et al. 2005, for details).

In both reconstructions, the amplitudes differ by over a factor of 10 at any given timescale, yet the Haar fluctuations become smaller as the scale increases ($H <$ 0.3). The qualitative difference between the "wandering" ($H > 0$, sunspot-based) and canceling ($H < 0$, ^{10}Be-based) solar reconstructions is seen in Fig. 26.2b (top). The early, low-resolution portion of the ^{10}Be data (1000–1610) was interpolated to yearly resolution using a close-to-linear interpolation technique for the "spliced" reconstruction utilized here. Consequently, we find that $H > 1$ is observed over the scale range of 1–50 years, whereas the $H < 0$ part is hardly discernible over the range of 100–600 years (about corresponding to the length of the reconstruction's ^{10}Be part).

On the other hand, it has been proposed since Hasselmann's (1976) work that variables may respond linearly to space–time-averaged forcings if they are sufficiently space–time-averaged. The nonlinear deterministic dynamics of these low-frequency phenomenological models serve as a source of random perturbations, leading to a stochastic model which is usually taken to be linear. However such models are only meaningful if the high-frequency and low-frequency processes are separated on a physical scale. Since Panofsky and Van der Hoven's work in 1955, there has been evidence of a significant break at scales of two to ten days. Van der Hoven first proposed this break as the "scale of migratory pressure systems of synoptic weather map scale" in 1957, and Kolesnikov and Monin proposed it as the "synoptic maximum" in 1965.

According to Penland and Sardeshmuhk, Newman et al. (2003), and Sardeshmukh and Sura (2009), the system is viewed as a multivariate Ornstein–Uhlenbeck (OU) process from the standpoint of linear stochastic modeling, sometimes known as "linear inverse modeling" (LIM). An OU process behaves like white noise itself at low frequencies, but at high frequencies, it can be thought of as the integral of white noise. These regimes correlate to weather and macroweather, respectively, in the setting of LIMs. Newman (2013) recently showed that, over 1–2-year timeframes, global temperature hindcasts have a somewhat better predictive ability than GCMs.

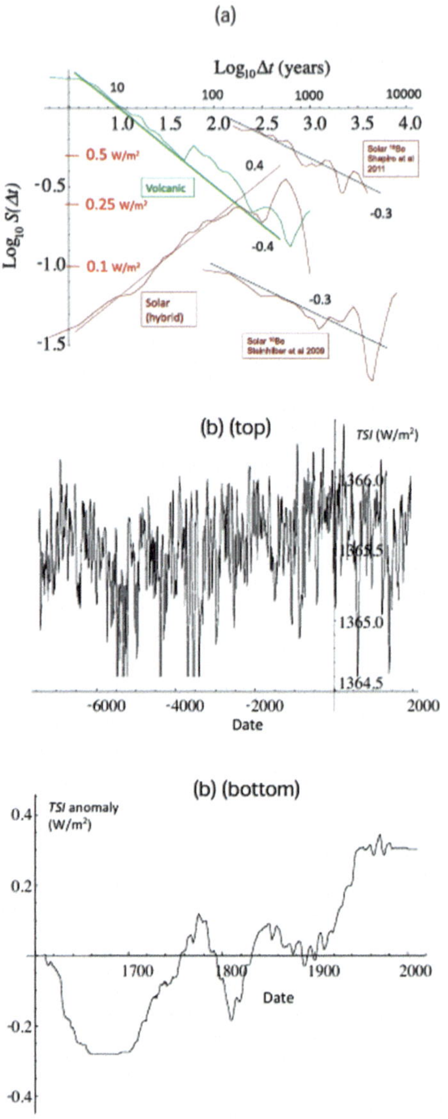

Fig. 26.2 **a** For lags t ranging from 2 to 1000 years, the RMS Haar fluctuation $S(t)$ for the solar and volcanic reconstructions employed in the ZC simulation (left). The sunspot-based reconstruction (**b**, top) and the ^{10}Be-based reconstruction (**b**, bottom) are "spliced" together to create the hybrid solar. **b** A comparison between a recent ^{10}Be reconstruction (bottom, total TSI mean plus anomaly–since 7362BC; see (**a**) for a fluctuation analysis, H 0.3) similar to that "spliced" onto the sunspot reconstruction for the period 1000–1610 and the sunspot-derived total solar irradiance (TSI) anomaly (top, used in the ZC and GISS simulations back to 1610, H 0.4)

Within the more general scaling hypothesis of Lovejoy and Schertzer (1986), the transition point denotes the length of time that planetary structures have existed. Lovejoy and Schertzer (2010) provided more quantitative support for this idea by analyzing the turbulent energy rate density. The low- and high-frequency regime scaling behavior showed spectra that were clearly different from Ornstein–Uhlenbeck (OU) processes, especially in the region $0.2 < \beta_1 < 0.8$. Nowadays, people refer to these two different regimes as "weather" and "macroweather" (Lovejoy and Schertzer

2013). At lower frequencies, the main departure from the traditional Local Interaction Model (LIM) is noticeable.

The LIM value of $\beta_l = 0$ (white noise) is unpredictable at low frequencies, even though the difference in the parameter l might seem negligible. On the other hand, actual values that fall between 0.2 and 0.8 (mostly depending on the location) have a significant potential for predictability (which can diverge as β_l approaches 1). In 2015, Lovejoy et al. (2015) and Lovejoy introduced the "Scaling Macroweather Model" (SLIMM), which is a set of fractional-order (yet linear) stochastic differential equations that show the ability to predict global mean temperatures up to ten years in advance. Although weather and macroweather regimes have unique statistical properties, linear stochastic models like LIM or SLIMM can be used as reliable approximations over large ranges.

Regardless of their type—SLIM or SLIMM—these linear stochastic models clearly represent the weather/macroweather transition and provide predictive power up to macroweather scales that could reach decades. A distinct class of phenomenological models, based on radiative energy balances, is frequently used, nevertheless, for longer timescales.

Energy balance models are typically characterized by nonlinearity, including tipping points and bifurcations, and are particularly focused on slower climate scale phenomena, such as sea ice–albedo feedbacks. These models can be either deterministic or stochastic, and they are usually zero- or one-dimensional in space, either averaged across the whole Earth or over particular latitude bands. Numerous research have contrasted these two methods; Dijkstra (2013) offers an overview of the more current stochastic "random dynamical systems" technique and the conventional deterministic dynamical systems method.

Despite the fact that energy balance models are often nonlinear, some academics contend that linear models can nevertheless be applied to sizes as large as millennials and multimillennials. Furthermore, there is empirical evidence of stochastic linearity between forcings and responses in the macroweather domain, as demonstrated by linear regressions' capacity to lessen the effects of SF, VF, and human-caused forcing. In the case of anthropogenic forcing, where globally averaged CO_2 radiative forcings act as a linear proxy for all anthropogenic forcings, this evidence has been statistically proven. This radiative forcing has produced residues with amplitudes that are similar to estimates of natural variability from GCMs when it is regressed on similarly averaged temperatures.

Stochastic linearity in the system is confirmed by the finding that the global temperature is the sum of residuals indicating natural variability and an anthropogenic component calculated by regression. This is because, in the absence of human interference, the residues that have been detected are probably going to change. Actually, it was discovered that only the residues' statistical characteristics closely matched those of the pre-industrial multiproxy statistics after performing fluctuation analysis on them (Varotsos et al. 2020; Cracknell and Varotsos 2012; Kondratyev and Varotsos 2013; Varotsos and Ghosh 2017).

26.4 Nonintermittent Events, Climate Responses and Power-Law

Changes in boundary conditions lead to climate change, according to GCMs. Climate forcings are the term for these boundary conditions, with human forcings being of particular importance. The responses are reasonably linear because these forcings are typically about 1% of the mean solar input. This allows the reduction of forcings to a common denominator, called "equivalent radiative forcing," on the assumption that the various forcings are "additive" or add linearly. Additionally, linearity suggests that climatic sensitivities are unaffected by how much the forcings fluctuate. Nonetheless, there are limitations to both the independence of climatic sensitivities and the linearity assumption. Energy balance models, for example, reject linearity at millennial and longer scales and favor nonlinear albedo responses to orbital fluctuations. Similarly, in the setting of large and abrupt volcanic forcings, the linearity of climate sensitivity has been questioned at monthly and annual scales.

Since the linearity assumption is used so frequently, it is imperative to quantitatively determine its bounds. Utilizing numerical climate models is the most effective way to accomplish this. Since land use changes had little effect on land use during the pre-industrial era and were predominantly driven by SF and VF in the simulations from the previous millennium, they offer an especially useful background. A suite of fully connected GCMs should ideally include control runs for internal variability in addition to solar-only, volcanic-only, and combined SF and VF.

This would enable the evaluation of each response alone as well as in combination. Regretfully, the volcanic-alone responses are absent from the set of GCM outputs that are currently accessible, which only consists of the GISS E2-R millennium simulations with solar-only and SF plus VF. The ZC model, a condensed climate model created by Mann et al. (2005), produced a complete set of external forcing responses, which Lovejoy and Varotsos (2016) also considered.

Lovejoy and Varotsos (2016) first evaluated the forcings' timescale-based variability by examining fluctuations, in line with an earlier study. The difference between the averages of the first and second halves of time intervals was used to calculate these fluctuations, or "Haar" fluctuations. The two separate regimes—one where average fluctuations rise with timescale ($H > 0$) and another where they drop with scale ($H < 0$)—were only possible thanks to this precise characterization. The SF generally increased with increasing timelines, but it had little effect at annual scales. Conversely, the volcanic forcing exhibited notable strength at yearly scales but quickly decreased, eventually reaching a position after about 200 years where both forcings were roughly equal. Lovejoy and Varotsos (2016) assessed and measured their non-additivity (nonlinearity) by looking at the response to combined forcing. The ZC model's additivity of radiative forcings was effective up to about 50-year scales, according to direct analysis (Fig. 26.2a, b). At 400-year scales, however, negative feedback interactions between SF and VF resulted in a 1.5–2 reduction in the combined effect. At these scales, this "subadditivity" led to comparatively small combined effects.

Although this result is statistically stable for the ZC millennium simulations, more research is needed to determine the cause of the nonlinearity and reproduce the findings using fully-coupled GCMs. If it were possible to measure internal variability using the output from ZC control runs, the conclusions would also be strengthened. Several more simulations with various forcings are needed to settle this issue definitively.

Lovejoy and Varotsos (2016) used the idea of linear and scaling systems to investigate the possibility of nonlinear reactions to extreme events like volcanic eruptions. In these kinds of systems, the moment order is a linear function of the difference in the structure function exponents for the forcings and responses. Higher order moments are more susceptible to unusually high values, whereas lower order moments are mostly dominated by frequently occurring low values. Lovejoy and Varotsos (2016) separated the nonlinear component $\xi(q)$ of the structural function exponents—represented by the function $K(q)$—through trace moment analysis. In relation to the "spikiness" of the signal, this component measures the variability's intermittent and multifractal nature.

Notably, the results showed that the intermittency of volcanoes was far stronger than that of solar systems. However, the model responses were almost nonintermittent and very smoothed in both scenarios, almost like a Gaussian distribution. As a result, Lovejoy and Varotsos (2016) came to the conclusion that the model's reactions to strong events weren't as sensitive as those to more frequent, less powerful forcing events.

Lovejoy and Varotsos (2016) looked at our climate system model's outputs and found evidence that, for at least 50 years, the climate system's response to forcing events is still roughly linear, especially for weak and nonintermittent events. On the other hand, abrupt and sporadic occurrences such as volcanic eruptions, which can break linearity at shorter durations, progressively lose their significance as timescales increase (with a scaling exponent $H \approx -0.3$). Practically speaking, a wide variety of macroweather timescales ($\approx$10 days to over 50 years) may be covered by linear stochastic models. However, given their possible importance, it might be worthwhile to create tailored coupled climate model experiments to look into this issue more thoroughly.

References

Blender R, Fraedrich K (2004) Comment on "Volcanic forcing improves atmosphere–ocean coupled general circulation model scaling performance" by D. Vyushin, I. Zhidkov, S. Havlin, A. Bunde, and S. Brenner. Geophys Res Lett 31:22213. https://doi.org/10.1029/2004GL020797

Bothe O, Jungclaus JH, Zanchettin D (2013a) Consistency of the multi-model CMIP5/PMIP3-past1000 ensemble. Clim past 9:2471–2487. https://doi.org/10.5194/cp-9-2471-2013

Bothe O, Jungclaus JH, Zanchettin D, Zorita E (2013b) Climate of the last millennium: ensemble consistency of simulations and reconstructions. Clim past 9:1089–1110. https://doi.org/10.5194/cp9-1089-2013

Bunde A, Eichner JF, Kantelhardt JW, Havlin S (2005) Long term memory: a natural mechanism for the clustering of extreme events and anomalous residual times in climate records. Phys Rev Lett 94:1–4. https://doi.org/10.1103/PhysRevLett.94.048701

Clement AC, Seager R, Cane MA, Zebiak SE (1996) An ocean dynamical thermostat. J Clim 9:2190–2196

Cracknell AP, Varotsos C, Cracknell AP, Varotsos CA (2012) Atmospheric ozone and climate. In: Remote Sensing and Atmospheric Ozone: Human Activities versus Natural Variability, pp 485–557

Dijkstra H (2013) Nonlinear climate dynamics. Cambridge University Press, Cambridge, p 357

Eichner JF, Koscielny-Bunde E, Bunde A, Havlin S, Schellnhuber H-J (2003) Power-law persistence and trends in the atmosphere: a detailed study of long temperature records. Phys Rev E 68:046133. https://doi.org/10.1103/PhysRevE.68.046133

Fraedrich K, Blender R, Zhu X (2009) Continuum climate variability: long-term memory, scaling, and 1f-noise. Int J Mod Phys B 23:5403–5416

Franzke J, Frank D, Raible CC, Esper J, Brönnimann S (2013) Spectral biases in tree-ring climate proxies. Nat Clim Change 3:360–364. https://doi.org/10.1038/Nclimate1816

Fredriksen H-B, Rypdal K (2016) Scaling of atmosphere and ocean temperature correlations in observations and climate models. J Clim 29:1253–1268. https://doi.org/10.1175/JCLI-D-150457.1

Goswami BN, Shukla J (1991) Aperiodic variability in the Cane Zebiak Model. J Clim 6:628–638

Hansen J, Sato MKI, Ruedy R, Nazarenko L, Lacis A, Schmidt GA, Bell N (2005) Efficacy of climate forcings. J Geophys Res 110:D18104. https://doi.org/10.1029/2005JD005776

Hasselmann K (1976) Stochastic climate models, part I: Theory. Tellus 28:473–485

Hurst HE (1951) Long-term storage capacity of reservoirs. Trans Am Soc Civ Eng 116:770–808

Huybers P, Curry W (2006) Links between annual, Milankovitch and continuum temperature variability. Nature 441:329–332. https://doi.org/10.1038/nature04745

Kantelhardt JW, Koscielny-Bunde E, Rybski D, Braun P, Bunde A, Havlin S (2006) Long-term persistence and multifractality of precipitation and river runoff record. J Geophys Res 111:D01106. https://doi.org/10.1029/2005JD005881

Kolesnikov VN, Monin AS (1965) Spectra of meteorological field f luctuations. Izv Atmos Ocean Phys 1:653–669

Kondratyev KY, Varotsos C (2013) A review on greenhouse effect and ozone dynamics over Greece. Atmos Ozone Dyn Observ Mediterranean Region 53:175

Lovejoy S (2013) What is climate? Eos 94:1–2

Lovejoy S, Schertzer D (1986) Scale invariance in climatological temperatures and the local spectral plateau. Ann Geophys 4B:401–410

Lovejoy S, Schertzer D (2010) Towards a new synthesis for atmospheric dynamics: space-time cascades. Atmos Res 96:1–52. https://doi.org/10.1016/j.atmosres.2010.01.004

Lovejoy S, Schertzer D (2012a) Stochastic and scaling climate sensitivities: solar, volcanic and orbital forcings. Geophys Res Lett 39:L11702. https://doi.org/10.1029/2012GL051871

Lovejoy S, Schertzer D (2012b) Low frequency weather and the emergence of the climate. In: Sharma AS, Bunde A, Baker DN, Dimri VP (eds) Extreme events and natural hazards: the complexity perspective. AGU Monographs, Washington, D.C., pp 231–254

Lovejoy S, Schertzer D (2013) The weather and climate: emergent laws and multifractal cascades. Cambridge University Press, Cambridge, p 496

Lovejoy S, Schertzer D, Varon D (2013) Do GCMs predict the climate … or macroweather? Earth Syst Dyn 4:439–454. https://doi.org/10.5194/esd-4-439-2013

Lovejoy S, Muller JP, Boisvert JP (2014) On Mars too, expect macroweather. Geophys Res Lett 41:7694–7700. https://doi.org/10.1002/2014GL061861

Lovejoy S, del Rio Amador L, Hébert R (2015) The ScaLIng Macroweather Model (SLIMM): using scaling to forecast global scale macroweather from months to decades. Earth Syst Dyn 6:637–658. https://doi.org/10.5194/esd-6-637-2015

Lovejoy S, Varotsos C (2016) Scaling regimes and linear/nonlinear responses of last millennium climate to volcanic and solar forcings. Earth System Dynamics 7(1):133–150

Mann ME, Cane MA, Zebiak SE, Clement A (2005) Volcanic and solar forcing of the tropical pacific over the past 1000 years. J Clim 18:447–456

Meehl GA, Washington WM, Ammann CM, Arblaster JM, Wigley TML, Tebaldi C (2004) Combinations of natural and anthropogenic forcings in Twentieth-century climate. J Clim 17:3721–3727

Minnis P, Harrison EF, Stowe LL, Gibson GG, Denn FM, Doelling DR, Smith WL Jr (1993) Radiative climate forcing by the Mount Pinatubo eruption. Science 259:1411–1415

Newman M (2013) An empirical benchmark for decadal forecasts of global surface temperature anomalies. J Clim 26(5260):5269. https://doi.org/10.1175/JCLI-D-12-00590.1

Newman MP, Sardeshmukh PD, Whitaker JS (2003) A study of subseasonal predictability. Mon Weather Rev 131:1715–1732

Østvand L, Nilsen T, Rypdal K, Divine D, Rypdal M (2014) Long-range memory in internal and forced dynamics of millennium-long climate model simulations. Earth Syst Dyn 5:295–308. https://doi.org/10.5194/esd-5-295-2014

Pelletier JD (1998) The power spectral density of atmospheric temperature from scales of 10 2 to 106 yr, Earth Planet. Sci Lett 158:157–164

Pielke R (1998) Climate prediction as an initial value problem. Bull Am Meteorol Soc 79:2743–2746

Roques L, Chekroun MD, Cristofol M, Soubeyrand S, Ghi M (2014) Parameter estimation for energy balance models with memory. Proc R Soc A 470:20140349. https://doi.org/10.1098/rspa.2014.0349

Rybski D, Bunde A, Havlin S, von Storch H (2006) Long-term persistance in climate and the detection problem. Geophys Res Lett 33:L06718. https://doi.org/10.1029/2005GL025591

Rypdal M, Rypdal K (2014) Long-memoryeffects in linear response models of Earth's temperature and implications for future global warming. J Clim 27:5240–5258. https://doi.org/10.1175/JCLI-D-1300296.1

Sardeshmukh PD, Sura P (2009) Reconciling non-gaussian climate statistics with linear dynamics. J Clim 22:1193–1207

Shapiro AI, Schmutz W, Rozanov E, Schoell M, Haberreiter M, Shapiro AV, Nyeki S (2011) A new approach to long-term reconstruction of the solar irradiance leads to large historical solar forcing. Astron Astrophys 529:A67. https://doi.org/10.1051/00046361/201016173

Shindell DT, Schmidt GA, Miller RI, Mann ME (2003) Volcanic and solar forcing of climate change during the preindustrial era. J Clim 16:4094–4107

Steinhilber F, Beer J, Frohlich C (2009) Total solar irradiance during the Holocene. Geophys Res Lett 36:L19704. https://doi.org/10.1029/2009GL040142

Varotsos CA (2013) The global signature of the ENSO and SST-like fields. Theor Appl Climatol 113:197–204

Varotsos CA, Ghosh S (2017) Impacts of climate warming on atmospheric phase transition mechanisms. Theor Appl Climatol 130:1111–1122

Varotsos C, Kalabokas P, Chronopoulos G (1994) Association of the laminated vertical ozone structure with the lower stratospheric circulation. J Appl Meteorol 33:473–476

Varotsos C, Efstathiou M, Tzanis C (2009) Scaling behaviour of the global tropopause. Atmos Chem Phys 9:677–683. https://doi.org/10.5194/acp-9-677-2009

Varotsos CA, Lovejoy S, Sarlis NV, Tzanis CG, Efstathiou MN (2015a) On the scaling of the solar incident flux. Atmos Chem Phys 15(7301–7306):2015. https://doi.org/10.5194/acp-15-7301

Varotsos CA, Tzanis C, Cracknell AP (2015b) Precursory signals of the major El Niño Southern Oscillation events. Theor Appl Climatol. https://doi.org/10.1007/s00704-015-1464-4

Varotsos C, Mazei Y, Efstathiou M (2020) Paleoecological and recent data show a steady temporal evolution of carbon dioxide and temperature. Atmos Pollut Res 11(4):714–722

Vyushin D, Zhidkov I, Havlin S, Bunde A, Brenner S (2004) Volcanic forcing improves atmosphere-ocean coupled, general circulation model scaling performance. Geophys Res Lett 31:L10206. https://doi.org/10.1029/2004GL019499

Wang Y-M, Lean JL, Sheeley NRJ (2005) Modeling the Sun's magnetic field and irradiance since 1713. Astrophys J 625:522–538

Watson AJ, Lovelock JE (1983) Biological homeostasis of the global environment: the parable of Daisyworld. Tellus B 35:284–289

Zanchettin D, Rubino A, Jungclaus JH (2010) Intermittent multidecadal-to-centennial fluctuations dominate global temperature evolution over the last millennium. Geophys Res Lett 37(L14702):2010. https://doi.org/10.1029/2010GL043717

Zanchettin D, Rubino A, Matei D, Bothe O, Jungclaus JH (2013) Multidecadal-to-centennial SST variability in the MPI-ESM simulation ensemble for the last millennium. Clim Dyn 40:1301–1318

Zebiak SE, Cane MA (1987) A Model El Niño-Southern Oscillation. Mon Weather Rev 115:2262–2278

Zhu X, Fraederich L, Blender R (2006) Variability regimes of simulated Atlantic MOC. Geophys Res Lett 33:L21603. https://doi.org/10.1029/2006GL027291

Afterword

The work must be completed. "You can't embrace the boundless," said Kozma Prutkov 150 years ago, repeatedly emphasizing this important statement, to the point that "Spit in the eyes of anyone who says, 'You can embrace the boundless.'".

Convection, which has many new and undeveloped aspects, is highlighted in a separate large Chap. 13. For example, the "forest" of convective columns in the atmosphere and the deep ocean await understanding. Such "forests we have seen in the laboratory experiments but their positions at the phase diagrams Ra-Ta had to be appeared. I heard about the former from the late Nikita Fedorovich, a corresponding member Glazovskiy, but no real fieldwork was performed to the fall of the Soviet Union.

1. The proportionality of the average square of velocity to time can be interpreted as the average energy per unit mass. Cumulative distributions—$N(\geq E)$ are used to describe random events such as earthquakes, which represent the number of events in a given period with energy greater than or equal to a specified value. This characteristic has the dimension of frequency and therefore $N(\geq E) = c_1\varepsilon/E$, where ε is the rate of energy generation that causes the studied process, c_1 is a dimensionless number appearing in comparison with experimental data. The other parameters defining the process in the study may form a dimensionless similarity criterium Π on which the dimensionless number c_1 may depend (see Chap. 3). This is known as the Gutenberg-Richter law, the distribution of asteroids, Chap. 11.4, the number of rocks depending on their mass on the surface of Mars, and the distributions for the surface relief of celestial bodies, etc.
2. The proportionality of the average square of the distance between events to the cube of time, i.e. $\langle x^2(t)\rangle = c_1\varepsilon t^3$. This quantity is related to area. This is the cumulative distribution for the areas of lithospheric plates and disk-shaped galaxies by mass. In the latter case, gravity plays a role, while in the former case, it is the geothermal flux and the limited spherical surface of the Earth.
3. Events and processes unfold not only in time but also in space, which is reflected in the existence of a mixed moment for the complete equation ANK34 (with a

G. Golitsyn and C. Varotsos, *The Stochastic Nature of Environmental Phenomena and Processes*, https://doi.org/10.1007/978-3-031-77015-9

substantial probability distribution function dp/dt, not partial $\partial p/\partial t$), specifically $\langle u_i x_i \rangle = \varepsilon t^2$ (see the fundamental solution of the FPK equation, given in MY75, §24, and the method of its solution proposed by A. A. Lushnikov in Chap. 1.3). This leads to Richardson-Obukhov turbulent diffusion law and the structural function for Kolmogorov-Obukhov 1941 small-scale turbulence.

4. A cubic dependence of the mean square of the coordinate on time has another interpretation if the process occurs only (or predominantly) in one dimension. For example, wind waves are predominantly vertical, at least in terms of measuring displacements on the water surface in that direction, and if they are proportional to t^3, then their frequency spectrum will be ω^{-4}. However, there is a numerical coefficient in this dependence that may depend on the similarity parameters present in the process. In the case of wind waves, this is their age, which slightly alters the exponent, as described in Chap. 6.
5. Almost all celestial bodies rotate depending on the size of the phenomena, and the frequency of rotation can lead to significant changes in processes. Typically, they stabilize movements to some extent (like a bicycle!). The outcome in all processes depends on the excitation and the duration of its action. Rotation limits the excitation to some extent, but there can still be elements of randomness. An example of this are hurricanes. Their area is proportional to t^3, and their speeds are proportional to $t^{1/2}$. As a result, their strength is approximately $t^4 \sim \omega^{-4}$. In fluids, the effect is manifested through the Coriolis force $2\omega\sin\theta$—that's why hurricanes do not occur in the $\pm$ 5° latitude band, where there are no spiral vortices on the surface of the ocean (Chap. 9). In laboratory experiments on the influence of rotation on turbulence, this is manifested in the fact that the spatial spectrum decreases faster with the increase of the wave number of Kolmogorov-Obukhov turbulence (Chap. 5), i.e., k^{-2} instead of $k^{-5/3}$ for the usual cumulative distribution of discrete volumes by areas. This is the distribution of lithospheric plates by areas and the same distribution of spiral galaxies by masses.

   Cumulative distribution of discrete volumes by areas will be $N(\geq S) \sim (\varepsilon/S)^{1/3}$. This is the distribution of lithospheric plates by areas and the same distribution of spiral galaxies by masses assuming that masses are proportional to areas, Chap. 12.

   Power-law cumulative distributions have also been obtained in social sciences. There, they result from the analysis of empirical data, but without attempts to construct any models or reasons for their appearance. An attempt at such analysis was made by Maslov (2006) for economics and for text analysis (Newman 2005; Jeon and McCoy 2005). Let's just give a simple example—Zipf's law:

$$\ln r + \ln \omega_n = \text{const}, \tag{1}$$

   where r is the rank of a word, i.e. its number in the frequency list in decreasing order of frequency, ω_n—the frequency, i.e. the number of times this word is used in the text (Maslov and Maslova 2006). This means that the product of these two quantities is approximately constant. Thus, formula (1) is a direct consequence in logarithmic form of the cumulative relationship $\langle ui^2 \rangle = \varepsilon t \equiv E$. But this is true within certain limits of r and ω_n. In other cases, nonlinear averaging must be used, which goes beyond the scope of this text. A brief description is necessary.

6. The work of ANK34 is interesting for the practice of scientific research because it generalized the use of the Fokker–Planck equation for systems with many variables different from what we are used to. Instead of the second derivative concerning the coordinate of the probability distribution function, its second derivative concerning velocities is written, making the diffusion coefficient the rate of kinetic energy generation ε. Here, this is used for two-dimensional turbulence, describing the process of vortex merging, which most simply and vividly depicts the reverse cascade of energy transfer from small to large scales.
7. The probabilistic laws of A.N. Kolmogorov, developed by his school into practical methods of their application, serve as a reliable physical–mathematical basis for the theory of similarity and dimensionality. Although the latter itself became a full-fledged theory by the last quarter of the twentieth century, it still did not fit into the minds of many in the middle of the last century.

Now, new concepts of intermediate asymptotics, self-similarity of the second kind, and a sequence of rules for analyzing reactions have formed in this theory.

It is appropriate to recall the first epigraph belonging to Gibbs in the book: "The aim of science is to find a point of view from which the problem can be solved most simply and naturally." Here it should be added—given our current knowledge of the world around us. It turns out that a similar thought was expressed two thousand years ago by the famous ancient astronomer Ptolemy about the geocentric system of the world, which was then the simplest and most natural for him. And this view lasted for about one and a half thousand years! Our understanding of the world is developing tens of times faster, and we can hope for even more progress. A short and concise presentation of the content and ideas of the first part of this book with the latest results on tornadoes can be found in Golitsyn (2024).

The last epigraph belongs to Kolmogorov, who expressed this thought about half a century ago. Since then, our knowledge and understanding of the surrounding world have deepened to a tremendous extent, which greatly strengthens and specifies this statement. Therefore, it should be repeated and used to conclude the book: "Chance is a necessary, if not the most important, element of the universe, but it has a certain order that leads to specific, often stable structures. Stability is limited in time and space, which are unique in specific situations."

Chapters 15 through 26 present an analysis of the hidden properties in the "noise" obtained after the removal of trends and conventional cycles in natural spatio-temporal series. This noise provides information about the long-range correlations in several environmental timeseries. These are investigated using Haar Analysis and Detrended Fluctuation Analysis and their modified versions. The long and short memory effect in natural spatio-temporal fluctuations is investigated to show that most of the future extreme events in nature are the responses of past disturbances occurred due to natural or anthropogenic activities. In this context, Greek Poet G. Seferis (Nobel 1963) said: "Erasing a piece from the past is like erasing a corresponding piece from the future".

Based on this "memory effect" and applying the well-established "Natural time Analysis" (that is based on the new concept of "natural time") new tools for forecasting and nowcasting of future disastrous phenomena are presented. Special attention is paid in the atmospheric greenhouse gases, global warming, land and sea temperature, mean sea level, air-pollution, ozone hole dynamics, spectral solar radiation, El Nino/La Nina phenomenon, albedo and reflectance, cosmic rays, tropospheric and stratospheric wind field (e.g., Varotsos 2025, Varotsos et al., 2024a, b).

The aforementioned discussion in the Chaps. 15–26 is made in terms of the theories of the complex, multiphase and complicated systems, fractals, chaos, critical phenomena, self-organised criticality, crossover, tipping and bifurcation points, etc. (Bak 1997; Schroeder 2001; Sornette 2003).

Prof. GS Golitsyn

Prof. CA Varotsos

References

Bak PE (1997) How the nature works. The science of self-organized criticality. Oxford University Press, Oxford, p 212

Golitsyn GS (2024) ANKolmogorov's 1934 paper is the basis for explaining the statistics of natural phenomena of the macrocosm. Phys Usp 67(1):80–90

Jeon Y-P, McCoy BJ (2005) Evolution of power law distributions in science and society. Phys Rev E 72:037104

Maslov VP, Maslova TV (2006) On Zipf's law and rank distributions in linguistics and semiotics. Matem Notes 80(5):718–732

Maslov VP (2006) Quantum economics. M.: Fizmatlit, 92 p

Newman MEJ (2005) Power laws, Pareto distributions and Zipf law. Contemp Phys 46(5):323–351. https://doi.org/10.1080/00107505000252444

Schroeder M (2001) Fractals, chaos, power laws. RHD, M. Izhevsk, 328 p

Sornette D (2003) Critical phenomena in natural sciences. In: Chaos, fractals, self- organization and disorder: concepts and tools, 2nd ed. Springer, Singapore, 434 p

Varotsos C, Sarlis NV, Mazei Y, Saldaev D, Efstathiou M (2024a) A Composite Tool for Forecasting El Niño: the case of the 2023–2024 event. Forecasting 6(1):187–203

Varotsos CA, Golitsyn GS, Mazei Y, Sarlis NV, Xue Y, Mavromichalaki H, Efstathiou MN (2024b) On the observed time evolution of cosmic rays in a new time domain. Acta Astronaut 225:436–43

Varotsos CA (2025) Remote sensing and extreme environmental phenomena. Remote Sens Lett 16(4):434–48

G. Golitsyn and C. Varotsos, *The Stochastic Nature of Environmental Phenomena and Processes*, https://doi.org/10.1007/978-3-031-77015-9

The manufacturer's authorised representative in the EU is Springer Nature Customer Service Centre GmbH, Europaplatz 3, 69115 Heidelberg, Germany. If you have any concerns regarding our products, please contact ProductSafety@springernature.com

Printed and bound by CPI Group (UK) Ltd, Croydon, CR0 4YY
01/10/2026
02225965-0007